Evolutionary Ecology

Evolutionary Ecology

Second Edition

Eric R. Pianka
The University of Texas at Austin

Harper & Row, Publishers
NEW YORK, HAGERSTOWN, SAN FRANCISCO, LONDON

Project Editor: Holly Detgen
Designer: Andrea C. Goodman
Production Supervisor: Kewal K. Sharma
Compositor: Bi-Comp, Incorporated
Printer and Binder: Halliday Lithograph Corporation
Art Studio: Vantage Art, Inc.

Evolutionary Ecology, Second Edition

Copyright © 1978 by Eric R. Pianka

Library of Congress Cataloging in Publication Data

Pianka, Eric R
    Evolutionary ecology.

    Bibliography: p.
    Includes index.
    1. Ecology. 2. Evolution. I. Title.
QH541.P5   1978   574.5'2   77-17841
ISBN 0-06-045224-2

To Ronald Fisher and Robert MacArthur
whose penetrating insight and inductive genius
laid so much of the foundation of modern ecology.

Contents

9 Biogeography   305

Preface

Language forces us to express ourselves in a one-dimensional stream of words. But nature is seldom so simple and tractable; instead, she has many different dimensions. This is particularly true of ecology because its subject matter includes many complexly interrelated concepts and phenomena involving several levels of organization. Thus there is no such thing as an "ideal" outline or a perfect sequential order for presentation of the subject treated here. To obtain an overview of modern ecology, a student needs to assimilate a great many ideas. Ideally, a reader would know everything in this book even before beginning to read it! Perhaps the only solution is to read it twice. To help a reader follow a train of thought different from that I have chosen to use here, various chapters and sections are cross-referenced.

Even as the first edition went into production, my own thinking on reproductive tactics and niche theory had advanced enough so that these sections were outdated. In addition to complete revisions of these sections, major changes in this second edition are the deletion of the chapter on man and his environment and the consolidation of physiological ecology, previously scattered throughout the book, into a new chapter (I thank Richard Dapson for suggesting this). In Chapter 4, numerous physiological processes—including tolerance curves, sensory capacities, and metabolic pathways—are viewed in an evolutionary context; the notion of integrated suites of adaptations is also developed. This new chapter helps to bridge the gap between the material on the physical environment, the interface between climate and vegetation, and hard-core population ecology.

The entire text and references have been thoroughly updated, with extensive revisions and addition of new material throughout: diffuse competition, the niche overlap hypothesis, fugitive species, resource matrices, guild structure, curvilinear competition isoclines, alternatives to competition coefficients, future prospects for competition studies, transition matrices, an improved presentation of prey–predator interactions (neutral stability, limit cycles, and types of functional responses are clarified), simple symbiosis equations and graph, an improved discussion of community stability, and a short new section on applied biogeography and the design of nature preserves. Plants and the plant–animal interface are given greater emphasis than in the first edition, with new or improved sections on leaf tactics, secondary succession, tree species diversity in tropical rain forests, and defense tactics of cryptic versus apparent plants.

Perhaps the strongest criticism of the first edition—aside from a comment made in private by a prominent young ecologist that the periodic table of niches is nonsense (which may be true)—was that it reported the literature uncritically and threatened "to transmute today's speculation into tomorrow's dogma." I certainly hope not. But if an elementary text is to provide the reader with principles and an organized overview, it must gloss over some controversial areas and untested ideas. Otherwise a student would be left confused and bewildered. Moreover, a really critical reporting of "the literature" would require a massive and unwieldy volume, one quite unsuitable for a short course. The best way to appreciate and evaluate the literature is to read it. Here I try to provide a selected, but hopefully solid, foundation from which one can dip (or jump!) into the vast sea of original work in ecology. This book is my own peculiar blend and distillation of what I think are significant facts, ideas, and principles; these represent the residue remaining after considerable sifting and sorting of many other facts and ideas. My approach is abstract and conceptual, and I strive to provide an overview of the subject matter. More than anything else, the book is an image of a part of my mind, a part I trust mirrors some sort of reality external to it and common to all of us.

I am very grateful to my students for helping me, indeed, sometimes *forcing* me, to clarify ideas and improve my presentation of them. My own thinking has been indelibly molded by my students, teachers, and colleagues, particularly by Thomas Frazzetta, John Gillespie, Henry Horn, Raymond Huey, Larry Lawlor, Robert MacArthur, William Neill, Gordon Orians, Daniel Otte, Robert Paine, William Parker, Joseph Schall, Christopher Smith, and Mary Willson. I am especially indebted to Gordon Orians for introducing me to "selective thinking" (that is, using the theory of natural selection the way that Ronald Fisher used it). I thank Gordon Orians and my wife, Helen, for carefully reading an early draft of the first edition and for offering many valuable suggestions for its improvement. John Avis, Eric Charnov, LaMont Cole, Richard Dapson, Paul Ehrlich, Raymond Huey, Larry Lawlor, Thomas Schoener, Eliot Tramer, Robert Whittaker, Henry Wilbur, and Edward Wilson also made many useful comments. Amy Kramer provided me with much early encouragement, without which I probably would never have completed this project.

ERIC R. PIANKA

Evolutionary Ecology

Introduction

1

Domain of Ecology: Definitions and Groundwork

Ecologists, along with all scientists, assume that an organized reality exists in nature and that principles can be formulated which will adequately reflect that natural order. A fundamental and important way in which biological phenomena can be ordered is by simple and direct enumeration, as in the classification of organisms or biotic communities. Thus we recognize different ecological systems such as tundra, desert, prairie, savanna, deciduous forest, coniferous forest, and rain forest. Early ecology was primarily descriptive; the forefathers of the science spent most of their time describing, itemizing, and classifying various ecological elements. This process was absolutely necessary before modern ecology could develop. Founded and firmly based upon this older body of descriptive information, modern ecology seeks to develop general theories with predictive powers that can be compared against the real world. Modern ecologists want to understand and to explain, in general terms, the origin and mechanisms of interactions of organisms with one another and with the non-living world. In order to build such general theories of nature, ecologists construct "models" of reality which generate testable predictions. If the predictive powers of a model fail, it is either discarded or revised. Models and theories that do not conform adequately to reality are gradually replaced by those that better reflect the real world. The great complexity of ecological systems necessitates the use of graphical and mathematical models, so much so that modern ecologists often employ nearly as much mathematics as biology. However, the development of sound ecological principles depends equally as much upon what might be called "biological intuition," and there is certainly no substitute for a firm foundation in natural history. Models based on erroneous biological assumptions, no matter how elegant and elaborate, can hardly be expected to reflect nature accurately! Hence a good background for comprehension of the field of ecology includes some biology and mathematics as well as a solid basis in general science.

Ecology and environment are words frequently encountered in the news and popular media, almost invariably in conjunction with humanity and its environment. As often as not they are misused, especially by politicians and other advertisers. Many people use "ecology" to refer primarily to human ecology. Every year I begin my courses by telling students that the basic science of ecology is not synonymous with a study of the effects of people on their own surroundings and on other organisms, but in fact represents a much broader class of subject matter. Yet some students always complain that there is not enough "ecology" in my courses, presumably because they feel that our particular devastating problems are given short shrift. Some problems facing us today illustrate what can happen when ecological systems are not used wisely in accordance with sound ecological principles; as such, the content of this book is pertinent to human ecology. However, throughout the book, emphasis is given to principles of *basic* ecology, particularly as these principles apply to and can be interpreted in terms of the theory of natural selection. Major concepts and principles are stressed more than detail, but references are given at the end of each chapter for those desiring to delve more deeply into particular subjects.

Ecology has been variously defined as "scientific natural history," "the study of the structure and function of nature," "the sociology and economics of animals," "bionomics," "the study of the distribution and abundance of organisms," and "the study of the interrelationships between organisms and their environments." The last of these definitions is probably the best, with "environment" being defined as *the sum total of all physical and biological factors impinging on a particular organismic unit*. For "organismic unit" one can substitute either "individual," "family group," "population," "species," or "community." Thus we may speak of the environment of an individual or the environment of a population, but to be precise, a particular organismic unit should be understood or specified. The environment of an individual contains fewer elements than the environment of a population, which in turn is a subset of the environment of the species or community.

To avoid the apparent circularity in the above definition, ecology might be better defined as *the study of the relations between organisms and the totality of the physical and biological factors affecting them or influenced by them*. Thus ecologists begin with the organism and seek to understand how the organism affects its surroundings and how these surroundings in turn affect the organism.

Environment includes everything from sunlight and rain to soils and other organisms. An organism's environment consists not only of other plants and animals encountered directly (such as foods, trees used for nesting sites, predators, and competitors), but also of purely physical processes and inorganic substances such as daily temperature fluctuations and oxygen and carbon dioxide concentrations. Of course the latter may be affected by other organisms which are then indirectly a part of the environment of the first organism. Indeed, any remote connection or interaction between two organismic units means that each is part of the other's environment.

Because there are direct or indirect interactions between almost all organisms in a given area, the biotic component of the environment of most organisms is extremely complex. Coupling this great complexity with a multifaceted physical environment makes ecology an exceedingly broad subject. No other discipline seeks to explain such a variety of phenomena at so many different levels. As a consequence, ecology takes in aspects of many other fields, including physics, chemistry, mathematics, computer science, geography, climatology, geology, oceanography, economics, sociology, psychology, and anthropology. Ecology is properly classified as a branch of biology; students of ecology attempt to interweave and correlate the subdisciplines of biology such as evolution, genetics, systematics, morphology, physiology, ethology (behavior), as well as various taxonomic subdivisions of biology like algology, entomology, ichthyology, herpetology, mammalogy, and ornithology. Sometimes "plant ecology" is distinguished from "animal ecology." As basic as this distinction may be, however, it is most unfortunate; plants and animals inevitably constitute part of one another's environments, and their ecologies should always be considered together. Interactions between plants and animals have recently been the subject of considerable thought and field study. Another promising frontier in modern ecology is the interface between population genetics and population ecology; as the principles and theories of each of these disciplines gradually fuse into one another, many potent new insights into population biology are emerging.

Obviously no one could master all of such an enormous field, and as a result there are many different kinds of ecologists with a wide variety of perspectives on the subject matter of the science. The breadth of ecology, combined with its youth and great relevance to human problems, makes it a fascinating and exciting field, with real potential for growth and refinement. Young sciences, and especially complex biological ones like ecology, can be characterized as "soft" sciences, in that they are not as precise as older and better established "hard" sciences such as chemistry and physics. As every science matures, it becomes more and more abstract and its hypotheses are refined and improved until they eventually attain the status of "laws," as in the familiar laws of chemistry and physics. Ecology at present has few firm laws but many hypotheses, and much work and testing of these hypotheses remains. The one concept closest to deserving the status of "law" in ecology, and one which is shared with all of biology, is *natural selection* (see pp. 9–11).

There is a natural sequence to the subject matter of ecology, proceeding from the inorganic to the organic world, which is diagrammed in Figure 1.1. Here components of ecological systems are considered in

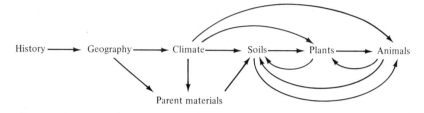

Figure 1.1   Diagrammatic representation of the subject matter of ecology showing the natural sequence proceeding from the inorganic to the organic world. Many other arrows and feedback loops could be added, but those depicted are of major importance. Chapter 2 deals with history, geography, and climate; Chapter 3 briefly treats climate, soils, and plants. Chapter 4 examines how animals cope with their physical environments. Remaining chapters examine the interactions between and among plants and animals, especially at the population and community levels.

order of generally increasing complexity (from left to right in the figure). In this book the organismic world is treated in much greater detail than the nonorganismic world.

The climate, soils, bacteria, fungi, plants, and animals at any particular place together constitute an *ecosystem.* Thus each ecosystem has both abiotic (nonliving) and biotic (living) components. The biotic components of an ecosystem, or all the organisms living in it, taken together, comprise an ecological *community.* The abiotic components can be separated into inorganic and organic, while the biotic components are usually classified as producers, consumers, and decomposers. Producers, sometimes called *autotrophs,* are the green plants that trap solar energy and convert it into chemical energy. Consumers, or *heterotrophs,* are all the animals that either eat the plants or one another; all heterotrophs are thus directly or indirectly dependent on plants for energy. Several levels of consumers are recognized (primary, secondary, and tertiary) depending on whether they eat plants directly or other herbivorous or carnivorous animals. Decomposers, also heterotrophs, are often bacteria and fungi; they function in the ecosystem by breaking down plant and animal material into simpler components and thereby returning nutrients to the autotrophs. Decomposers are therefore essential in recycling matter within an ecosystem.

As indicated above, plants and animals in ecosystems can be considered at several different levels: individuals, family groups, populations, species, and communities (subsequent chapters deal with each organismic level of organization, but most emphasis is given to populations). None of these levels of organization can be adequately understood in isolation because each exerts strong influences upon the others. Every individual is simultaneously a member of a population, a species, and a community, and it must therefore be adapted to cope with each and must be considered in that context. An individual's fitness—its ability to perpetuate itself as measured by its reproductive success—is determined not only by its status within its own population but by the various interspecific associations of its species, and especially by the particular community in which it finds itself. Similarly, every community is composed of many populations and numerous individuals, which determine many, but by no means all, of its properties. At each level of organization, important new properties emerge that are not properties of the preceding level. Thus individuals have a fixed genetic makeup and live or die, whereas populations have gene frequencies, birth rates, and death rates. All these population parameters (and others) can change in time as the composition of a population changes in response to a changing environment.

Basic Mendelian Genetics

Although a background in genetics is certainly not essential for appreciation of many populational and ecological phenomena, it is a useful aid for application to some such phenomena and is required for a full understanding of others. Precise rules of inheritance were unknown when Darwin (1859) developed the theory of natural selection, but they were formulated a short time afterward (Mendel, 1865). Darwin accepted the mechanism of inheritance in vogue at the time: *blending inheritance.* Under the blending inheritance hypothesis the genetic makeups of both parents are blended in their progeny, and all offspring produced by sexual reproduction should be genetically intermediate between their parents; genetic variability is thus lost rapidly unless new variation is continually being produced. (Under blending inheritance and random mating, genetic variability is *halved* each generation.) Darwin was forced to postulate extremely high mutation rates to maintain the genetic variability observed in most organisms (Fisher, 1930), and he was well aware of the inadequacy of knowledge on inheritance. Mendel's discovery of *particulate inheritance* was one of the major breakthroughs in biology.

Mendel performed breeding experiments with different varieties of peas, paying particular attention to a single trait at a time. He had two types which bred "true" for yellow and green peas, respectively. When a purebred green pea plant was crossed with a purebred yellow pea plant, all progeny, or individuals of the first filial generation (F_1), had yellow peas. However, when these F_1 plants were crossed with each other or self-fertilized, about one out of every four offspring in the second filial generation (F_2) had green peas. Furthermore, only about one-third of the yellow F_2 pea plants bred true; the other two-thirds, when self-fertilized, produced some offspring with green peas. All green pea plants bred true. Mendel proposed a very simple hypothesis to explain his results and performed many other breeding experiments on a variety of other traits which corroborated and confirmed his interpretations. Subsequent work has strengthened his hypothesis, although it has also led to certain modifications and improvements.

Mendel postulated that each pea plant had a double dose of the "character" controlling pea color, but that only a single dose was transmitted into each of its sexual cells, or *gametes* (pollen and ovules or sperm and eggs). Purebred plants, with identical doses, produced genetically identical single-dosed gametes; F_1 plants, on the other hand, with two different doses, produced equal numbers of the two kinds of gametes, half bearing the character for green and half that for yellow. In addition, Mendel proposed that yellow masked green whenever the two occurred together in double dose; thus all F_1 plants had yellow peas, but when self-fertilized produced some F_2 progeny with green peas. All green pea plants, which had a double dose of green, bred true.

Modern terminology for various aspects of Mendelian inheritance is as follows: (1) the "character" or "dose" controlling a particular trait is termed an *allele;* (2) its position on a chromosome (below) is termed its *locus;* (3) a single dose is the *haploid* condition, designated by n, while the double-dosed condition, designated by $2n$, is *diploid* (*polyploids*, such as triploids and tetraploids, are designated by still higher numbers); (4) the set of alternative alleles which may occur at a given locus (there can be only two alleles in a diploid individual, but there may be more than two in any given population) is termed a *gene;* (5) purebred diploid individuals with identical alleles are *homozygotes*, homozygous for the trait concerned; (6) individuals with two different alleles, such as the F_1 plants above, are *heterozygotes*, heterozygous at that locus; (7) an allele which masks the expression of another allele is said to be *dominant* whereas the one that is masked is *recessive;* (8) unlinked alleles separate, or *segregate*, from each other in the formation of gametes; (9) whenever heterozygotes or two individuals homozygous for different alleles mate, new combinations of alleles arise in the following generation by *reassortment* of the genetic material; (10) observable traits of an individual (i.e., yellow or green in the above example) are aspects of its *phenotype*, which includes all observable characteristics of an organism; and (11) whether or not an organism breeds true is determined by its *genotype*, which is the sum total of all its genes.

Occasionally, in some organisms, pairs of alleles with *incomplete dominance* occur. In such cases the phenotype of the heterozygote is intermediate between that of the two homozygotes; that is, phenotype accurately reflects genotype and vice versa. Presumably, alleles conferring advantages upon their bearers usually evolve dominance over time because such dominance ensures that a maximal number of the organism's progeny and descendants will benefit from possession of that allele. The apparent rarity of incomplete dominance is further evidence that dominance has evolved. Moreover, so-called wild-type alleles, that is, those most prevalent in natural wild populations, are nearly always dominant over other alleles occurring at the same locus. Geneticists have developed numerous theories of "the evolution of dominance," but the exact details of the process have not been completely resolved.

Cytological observations of appropriately prepared cell nuclei confirm Mendel's hypothesis beautifully (Figure 1.2). Microscopic examination of such cells reveals elongated dense bodies in cell nuclei;

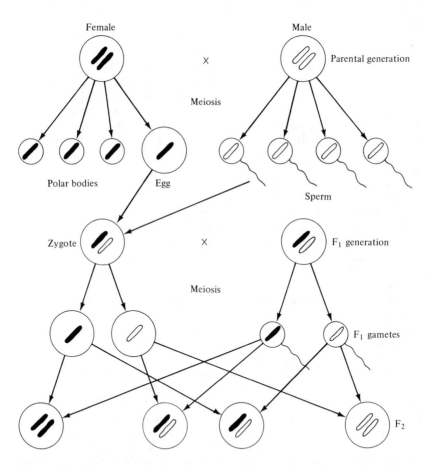

Figure 1.2 Diagrammatic representation of the cytological events in cell nuclei showing how the two parental genomes are sorted and recombined in the next generation, or the F_2. For simplicity, only one pair of chromosomes is shown and the complex events of the reduction division (meiosis) are omitted.

these are the *chromosomes*, which contain the actual genetic material, deoxyribonucleic acid, or DNA. Nuclei of diploid cells, including the zygote (the fertilized ovule or egg) and the somatic (body) cells of most organisms always contain an even number of chromosomes. (The exact number varies widely from species to species, with as few as two in certain arthropods to hundreds in some plants.) Moreover, pairs of distinctly similar *homologous chromosomes* are always present and often easily detected. Gametes, however, contain only half the number of chromosomes found in diploid cells, and, except in polyploids, none of them are homologous. Thus haploid cells contain only one full set of different chromosomes and alleles, or one *genome*, whereas diploid cells contain two. During the reduction division *(meiosis)* in which diploid gonadal cells give rise to haploid gametes, homologous chromosomes separate (Figure 1.2). Later, when male and female gametes fuse to form a diploid *zygote* which will develop into a new diploid organism, homologous chromosomes come together again. Hence one genome in every diploid organism is of paternal origin while the other is of maternal ancestry. Because each member of a pair of homologous chromosomes separates from its homologue independently of other chromosome pairs, the previous generation's chromosomes are reassorted with each reduction division. Thus the genetic material is regularly rearranged and mixed up by the dual processes of meiosis and the actual fusion of gametes.

Many different loci and allelic systems occur on each chromosome. Two different traits controlled by different alleles located on the same chromosome do not segregate truly independently but are statistically associated with or dissociated from one another. This is the phenomenon of *linkage*. During meiosis homologous chromosomes can effectively exchange portions by means of *crossovers;* this process is referred to as *recombination*. Because the frequency of occurrence of crossovers between two loci is a function of the distance between them on the chromosome, geneticists can use crossover frequencies to "map" the effective distance between loci, as well as their positions relative to one another on the chromosome. By means of close linkage whole blocks of statistically associated alleles can be passed on to progeny as a functionally integrated unit of "coadapted" alleles.

In many organisms a single pair of chromosomes, termed *sex chromosomes*, determine the sex of their bearer (the remaining chromosomes, which are not involved in sex determination, are *autosomes*). Typically one homologue of the sex chromosome pair is smaller. In the diploid state, an individual heterozygous for the sex chromosomes is *heterogametic*. In mammals, males are the heterogametic sex with an XY pair of sex chromosomes, and females are the homogametic sex with an XX pair. Since male–male matings are impossible, the homozygous genotype YY can never occur. In birds and some other organisms, the female is the heterogametic sex.

Natural Selection

The theory of natural selection is a truly fundamental unifying theory of life. A thorough appreciation of it is essential background to understanding modern ecology. Natural selection comes as close to being a "fact" as anything in biology, including Mendelian genetics. While there is no such thing as "proof" in science (except in mathematics where all postulates are taken as given), an enormous body of data has been amassed in support of the theory of natural selection over the last century.

Although natural selection is not a difficult concept, it is frequently misunderstood. A common misconception is that natural selection is synonymous with evolution. It is not. Evolution refers to temporal changes of any kind, whereas natural selection specifies one particular way in which these changes are brought about. There are other possible mechanisms of evolution besides natural selection, such as the inheritance of acquired characteristics, gene flow, meiotic drive, and genetic drift. Another frequent misconception is that natural selection occurs mainly through differences between organisms in death rates, or *differential mortality.*

Selection may proceed in a much more subtle and inconspicuous way. Whenever one organism leaves more successful offspring than another, in time its genes will come to dominate the population gene pool. Eventually the genotype leaving fewer offspring must become extinct in a stable population, unless there are concomitant changes conferring an advantage on it as it becomes rarer. Thus, ultimately, *natural selection operates **only by differential reproductive success.*** Differential mortality can be selective *only* to the degree that it creates differences between individuals in the number of reproductive progeny they produce.

Hence Darwin's choice of words, such as "struggle for existence" and "survival of the fittest," have had a most unfortunate consequence. They have tended to make people think in terms of a dog-eat-dog world and to consider such things as predation and fighting over food as the prevalent means of selection. All too often natural selection is couched in terms of differential death rates, and the strongest and fastest individuals are considered to have a selective advantage over weaker and slower individuals. But, if this were the case, every species would continually gain in strength and speed. Because this is not happening, selection against increased strength and speed (counterselection) must be occurring and must limit the process.

Animals are sometimes too aggressive for their own good; an extremely aggressive individual may spend so much time and energy chasing other animals that it spends less than average time and energy on mating and reproduction, and as a result leaves fewer offspring than average. Likewise an individual can be too submissive and spend too much time and energy running away from other animals. Under *stable* conditions, *intermediates* in a population leave more descendants, on the average, than do extreme phenotypes. We say that they are more "fit." An individual's "fitness" is measured by the proportion of its genes left in the population gene pool. Selection of this sort, which continually crops the extremes and tends to hold constant the intermediate or average phenotype, is termed *stabilizing selection* (Figure 1.3*a*). In a stable environment, genetic recombination increases populational variance each generation, while stabilizing selection reduces it to approximately what it was in the previous generation.

However, in a *changing* environment, average individuals (modal phenotypes) may not be the most fit members of the population. Under such a situation, *directional selection* occurs and the population mean shifts toward a new phenotype (Figure 1.3*b*) better adapted to the altered environment. Eventually, of course, unless the environment continues to change, an equilibrium is reached in which the population is readjusted to the new environment, whereupon stabilizing selection resumes.

A third type of selection, *disruptive selection*, takes place when two or more phenotypes with high fitnesses are separated by intermediate phenotypes of lower fitness (Figure 1.3*c*). This usually occurs in distinctly heterogeneous environments with a discrete number of

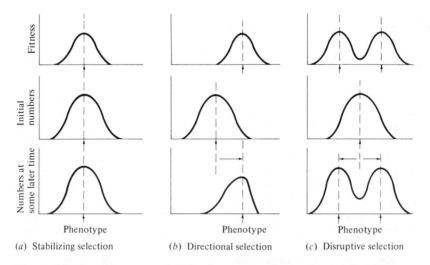

Figure 1.3 Graphic portrayal of the three types of selection. (*a*) Stabilizing selection, which occurs in stable environments, holds the modal phenotype constant. (*b*) Directional selection takes place in a changed environment and causes a shift in the modal phenotype. (*c*) Disruptive selection, with two or more modal phenotypes, occurs in patchy environments with more than one discrete phase.

different "patches." Disruptive selection is one mechanism that produces and maintains polymorphisms, such as the green–brown polymorphisms of some insects (Chapter 6). For instance, some butterflies (commonly called "leaf butterflies") mimic leaves; one population may contain both green and brown animals, with the former matching living leaves and the latter dead ones. Through appropriate behavior and selection of resting sites, each phenotype has a relatively high fitness; in contrast a butterfly with a phenotype intermediate between green and brown would match its surroundings less well and have a considerably lower fitness.

Although natural selection actually operates on phenotypes of individuals (that is, an organism's immediate fitness is determined by its total phenotype), the effectiveness of selection in changing the composition of a population depends upon the *heritability* of phenotypic characteristics, or the percentage of variability attributable to genotype. Thus, because nongenetic traits are not inherited, differential reproduction by different phenotypes stemming from such nontransmittable traits cannot alter a population's gene pool. Different genotypes may often have fairly similar phenotypes, and thus similar fitnesses. Selection may even favor alleles that are "good mixers" and work well with a wide variety of other genes to increase their bearer's fitness (Mayr, 1959). Conversely, of course, identical genotypes can develop into rather different phenotypes under different environmental conditions (see also pp. 215–216).

A word of warning is now appropriate. Overenthusiastic proponents of natural selection have sometimes used it to explain observed biological phenomena in a somewhat after-the-fact manner. Thus one might say that an animal "does what it does because that particular behavior increases its fitness." Used in this way, natural selection can be misleading; it is so pervasive and powerful that nearly any observable phenomenon can be interpreted as a result of selection, even though some may not be. Hence there is a real danger of circularity in such arguments. One should always consider alternative explanations for various biological phenomena.

Some examples of natural selection are given on pages 216–217.

Self-replicating Molecular Assemblages

Life began with the first self-replicating molecular assemblage; moreover, natural selection begins to operate as soon as any complex of molecules begins to replicate itself. No copying device is perfect, and some variants of the molecular assemblage produced are bound to be better than others in their abilities to survive and replicate themselves under particular environmental conditions. As resources become depleted, competition can occur among various self-replicating units.

Furthermore, given enough time, some inferior variants presumably become extinct. Thus each molecular unit maximizes its own numbers at the expense of all other such units. Even as originally simple self-replicating units become more and more elaborate, eventually attaining the complex form of present-day organisms, the same principles of natural selection are in effect throughout. Thus we can make certain statements about life that are entirely independent of the precise mechanism of replication. For example, should there be self-replicating molecular assemblages on Mars or any other distant planet, the fact that they probably will not obey the laws of Mendelian genetics would not drastically alter their basic attributes as living material. *Natural selection and competition are inevitable outgrowths of heritable reproduction in a finite environment.* Hence natural selection exists independently of life on Earth and principles developed here will persist as long as assemblages of molecules replicate themselves anywhere in the cosmos.

Once a self-replicating entity arises, qualitatively new phenomena exist that are not present in an inanimate world. To reproduce, living organisms (or replicating molecular assemblages) must actively gather other molecules and energy; that is, they must have some sort of acquisition technique. Direct and indirect disputes over resources place those units best able to acquire materials and energy (and best able to transform them into offspring) at a selective advantage over other such units that are inferior at these processes. Thus natural selection has the same effect as an efficiency expert, optimizing the use of available resources in reproduction.

Obviously the ultimate endpoint of these processes would be for the one best organismic unit to take over all matter and energy and to exclude all others. This has not occurred for a variety of reasons, as discussed earlier, but especially because of the great variability of the Earth's surface, both in space and in time (Chapter 2).

Units of Selection

Does natural selection act on individuals or on entire groups of individuals such as families, populations, species, communities, and/or ecosystems? What is the "unit" of selection? These questions are often discussed both by geneticists and by ecologists, but there is some disagreement as to correct answers.

One frequently reads statements similar to this: "reproductive capacity might become so great as to be detrimental *to a species*. The many deleterious effects of overcrowding are well known" (from Cole, 1954b, p. 104, italics mine). Stressing that natural selection operates by means of differential reproductive success among individuals, Fisher (1958a) cautioned against arguments invoking "the benefit of the species." In spite of his warning, it has become vogue to consider and interpret many behavioral and ecological attributes of species as having evolved for the benefit of the group rather than the individual.

Any individual sacrificing its own reproductive success for the benefit of a group is obviously at a selective disadvantage (within that group) to any other individual not making such a sacrifice. In order for such traits to evolve, natural selection must therefore act on entire groups of organisms, which in turn must possess differential rates of survivorship and reproduction (that is, differential fitness). Because the course of selection within groups cannot be altered by selection acting between groups (Wright, 1931), group selection requires some very special conditions. Most importantly, a population or species must be broken up into distinct interbreeding isolates. Then, for group selection to operate, selfish subgroups must go extinct faster than selfishness arises within altruistic subgroups and the majority of newly founded isolates must be altruistic.

Williams (1966a) reemphasized, restated, and expanded the argument against group selection, pointing out that classical Darwinian selection at the level of the individual is adequate to explain the majority of putatively "group-selected" attributes of populations and species, such as those suggested by Wynne-Edwards (1962) and Dunbar (1960, 1968, 1972). Williams reminds us that group selection has more conditions and is therefore a more onerous process than classical natural selection; furthermore he urges that it be invoked only after the simpler explanation has clearly failed. Although group selection is certainly possible, it is quite improbable that it could actually oppose natural selection at the individual level except under most unusual circumstances. A special form of selection at the level of the individual, *kin selection* (pp. 166–169), may frequently be the mechanism behind phenomena described as group selection. We return to this issue from time to time in later chapters (see also p. 219).

Limiting Factors and Tolerance Limits

Ecological events and their outcomes, such as growth, reproduction, photosynthesis, primary production, and population size, are often regulated by the availability of one or a few factors or requisites in short supply, while other resources and raw materials present in excess may go partially unused. This principle has become known as the "law of the minimum" (Liebig, 1840). For instance, in arid climates, primary production (the amount of solar energy trapped by green plants) is strongly correlated with precipitation (Figure 1.4); here water is a "master limiting factor." Of many different factors that can be limiting, frequently among the most important are various nutrients, water, and temperature.

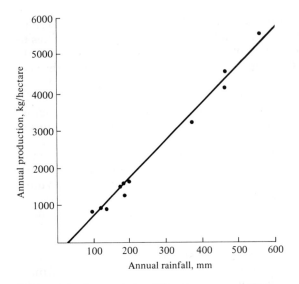

Figure 1.4   An example of the strong correlation
between annual rainfall and primary production along
a precipitation gradient in a desert region of South West
Africa. [Adapted from Odum (1959) after Walter
(1939).]

When considering populations, we often speak of those which are
food-limited, predator-limited, or climate-limited. Populations may be
limited by other factors as well; for example, density of breeding pairs
of blue tits *(Parus caeruleus)* in an English woods was doubled by
addition of many new nesting boxes (Lack, 1954, 1966), an indication
that nest sites were limiting. However, limiting factors are not always
so clear-cut, but may usually interact so that a process is limited simultaneously by several factors, with a change in any one of them resulting
in a new equilibrium. For instance, both increased food availability and
decreased predation pressures might result in a larger population size.

A related concept, developed by Shelford (1913b), is now known as
the "law of tolerance." Too much or too little of anything can be detrimental to an organism. In the early morning, a desert lizard finds itself
in an environment which is largely too cold, whereas later in the day its
environment is too hot. The lizard compensates somewhat for this by
spending most of its time during the early morning in sunny places,
while later on most of its activities take place in the shade. Each lizard
has a definite optimal range of temperature, with both upper and lower
limits of tolerance. More precisely, when measures of performance
(such as fitness, survivorship, or foraging efficiency), are plotted
against important environmental variables, bell-shaped curves usually
result (see also pp. 74–77).

Levels of Approach to Science

Why do migratory birds fly south in the autumn? A physiologist might tell us that decreasing daylength (photoperiod) stimulates hormonal changes which in turn alter bird behavior so that they become restless. Eventually their wanderlust gets the upper hand and they head south. In contrast, an evolutionist would most likely explain that, by virtue of reduced winter mortality, those birds which flew south lived longer and therefore left more offspring than their nonmigratory relatives. Over a long period of time, natural selection resulted in intricate patterns of migratory behavior by means of differential reproductive success.

The physiologist's answer concerns the *mechanism* by which avian migratory behavior is influenced by *immediate* environmental factors, whereas the evolutionist's response is couched in terms of what might be called the *"strategy"* by which individual birds have left the most offspring in response to *long-term* consistent patterns of environmental change (i.e., high winter mortality). The difference between them is in outlook, between thinking in an "ecological" time scale (now time) or in an "evolutionary" time scale (geological time). At the physiologist's level of approach to science the physiologist's answer is complete, as is the evolutionist's answer at the evolutionist's own level. Mayr (1961) has termed these the "how?" and "why?" approaches to biology. They have also been called the "functional" and "evolutionary" explanations and the "proximate" and "ultimate" factors influencing an event (Baker, 1938). Neither is more correct; a really thorough answer to any question must include both, although often only the first can be tested directly. Nor are these two ways of looking at biological phenomena mutually exclusive; ecological events can always be profitably considered within an evolutionary framework and vice versa. Moreover, physiological mechanisms surely must evolve in response to environmental demands (see Chapter 4).

The evolutionary approach to biological questions is relatively new, and has resulted in a major revolution in biology during the last 50 years. Before then, most biologists merely accepted as immutable a broad range of phenomena, such as the fact that sex ratios are often near equality (50 : 50), without considering why such facts might be so or how they could have evolved. Although we may not fully understand the causes and consequences of many populational phenomena, we can be confident that all have an evolutionary explanation. This is true of a broad spectrum of observations and facts, such as: (1) some genes are dominant, others are recessive; (2) some organisms live longer than others; (3) some organisms produce many more offspring than others; (4) some organisms are common, others are rare; (5) some organisms are generalists, others are specialists; (6) some species are promiscuous, some polygamous, and some monogamous; (7) some species migrate, others do not; and (8) more species coexist in some areas than in others. All these variables are subject to the effects of natural selection.

Population biologists are now thinking in an evolutionary time scale, and we have made substantial progress toward a theoretical understanding of why many of the above differences occur. Ronald Fisher was one of the first to recognize the power of rigorous application of the genetical theory of natural selection to population biology (Fisher, 1930), and his book has become a classic. Numerous other biologists have expanded, experimented with, and built upon Fisher's groundwork. Lack (1954, 1966, 1968) showed that reproductive rates are subject to natural selection. Others have worked on the evolution of dominance, mating systems, sex ratio, old age, life history phenomena, reproductive tactics, foraging tactics, community structure, and so on. Many of these topics, the subject matter of evolutionary ecology, are taken up in Chapters 5, 6, and 7.

Selected References

Domain of Ecology: Definitions and Groundwork

Allee (1951); Allee *et al.* (1949); Andrewartha (1961); Andrewartha and Birch (1954); Billings (1964); Clarke (1954); Collier *et al.* (1973); Connell, Mertz, and Murdoch (1970); Daubenmire (1947, 1968); Dawson and King (1971); Elton (1927, 1958); Greig-Smith (1964); Harper (1967); Hazen (1964, 1970); Kendeigh (1961); Kershaw (1964); Knight (1965); Krebs (1972); Levins (1966); MacArthur (1972); MacArthur and Connell (1966); MacFadyen (1963); May (1973, 1976); Maynard Smith (1968); Odum (1959, 1963, 1971); Oosting (1958); Pielou (1969); Platt (1964); Ricklefs (1973); Shelford (1963); Smith (1966); Watt (1973); Whittaker (1970); Wilson and Bossert (1971).

Basic Mendelian Genetics

Darlington and Mather (1949); Ehrlich and Holm (1963); Ford (1931, 1964); Maynard Smith (1958); Mendel (1865); Mettler and Gregg (1969).

Natural Selection

Birch and Ehrlich (1967); Darwin (1859); Dobzhansky (1970); Ehrlich and Holm (1963); Emlen (1973); Fisher (1930, 1958a, 1958b); Ford (1964); Haldane (1932); Kettlewell (1956, 1958); Lewontin (1974); MacArthur (1962); Maynard Smith (1958); Mayr (1959); Mettler and Gregg (1969); Orians (1962); Pianka (1976b); Salthe (1972); Williams (1966a); Wilson and Bossert (1971); Wright (1931).

Self-replicating Molecular Assemblages

Bernal (1967); Blum (1968); Calvin (1969); Ehrlich and Holm (1963); Fox and Dose (1972); Jukes (1966); Oparin (1957); Ponnamperuma (1972); Salthe (1972); Wald (1964).

Units of Selection

Boorman and Levitt (1972, 1973); Brown (1966); Cole (1954b); Darlington (1971); Darnell (1970); Dawkins (1976); Dunbar (1960, 1968, 1972); Emerson (1960); Emlen (1973); Eshel (1972); Fisher (1958a); Gilpin (1975); Levins (1970, 1975); Lewontin (1970); Maynard Smith (1964); Van Valen (1971); Wade (1976, 1977); Wiens (1966); Williams (1966a, 1971); D. S. Wilson (1975); E. O. Wilson (1973, 1976); Wright (1931); Wynne-Edwards (1962, 1964, 1965a, 1965b).

Limiting Factors and Tolerance Limits

Ehrlich and Birch (1967); Errington (1956); Hairston, Smith, and Slobodkin (1960); Lack (1954, 1966); Liebig (1840); Murdoch (1966); Odum (1959, 1963, 1971); Shelford (1913b); Terborgh (1971); Walter (1939).

Levels of Approach to Science

Baker (1938); Fisher (1930); Lack (1954, 1966, 1968); MacArthur (1959, 1961); MacArthur and Connell (1966); Mayr (1961); Orians (1962).

The Physical Environment

2

Earth supports an enormous variety of organisms. Plants range from microscopic short-lived aquatic phytoplankton to small annual flowering plants to larger perennials to gigantic ancient Sequoia trees. Animals, while they never attain quite the massive size of a redwood tree, include forms as diverse as marine zooplankton, jellyfish, sea stars, barnacles, clams, snails, fish, whales, beetles, butterflies, worms, frogs, lizards, sparrows, hawks, bats, elephants, and lions. Different species have evolved and live under different environmental conditions. Some organisms are relatively specialized either in the variety of foods they eat or in the microhabitats they exploit, whereas others are more generalized; some are widespread, occurring in many different habitats, whereas still others have more restricted habitat requirements and geographic ranges. Temporal and spatial variation in the physical conditions for life often make possible or even actually necessitate variety among organisms, both directly and indirectly. Of course, interactions among organisms also contribute to the maintenance of this great diversity of life. Before considering such biological interactions, we examine briefly the nonliving world, which sets the background for all life and which often strongly influences the ecology of any particular organismic unit. A major factor in the physical environment is climate, which in turn is the ultimate determinant of water availability and the thermal environment; moreover, as we shall see in Chapter 3, the latter two interact to determine the actual amount of solar energy that can be captured by plants (primary productivity) at any given time and place. Finally, because climate is a major determinant of both soils and vegetation, there is a close correspondence between particular climates and the types of natural biological communities that exist under those climatic conditions. The interface between climate and vegetation is considered in Chapter 3.

Major global and local patterns of climate are described briefly in this chapter. Entire books have been devoted to some of these subjects, and the reader interested in greater detail is referred to the references at the end of the chapter.

Major Determinants of Climate

The elements of climate (sun, wind, and water) are complexly interrelated. Incident solar energy produces thermal patterns that, coupled with Earth's rotation and movements around its sun, generate the prevailing winds and ocean currents. These currents of air and water in turn strongly influence the distribution of precipitation, both in time and in space.

The amount of solar energy intercepting a unit area of Earth's surface varies markedly with latitude for two reasons. First, at high latitudes a beam of light hits the surface at an angle, and its light energy is spread out over a large surface area. Second, a beam that intercepts the atmosphere at an angle must penetrate a deeper blanket of air, and hence more solar energy is reflected by particles in the atmosphere and radiated back into space. (Local cloud cover also affects the amount of the sun's energy that reaches the ground.) A familiar result of both these effects is that average annual temperatures tend to decrease with increasing latitude (Table 2.1). The poles are cold and the tropics are generally warm (seasons are discussed later).

Table 2.1   Average Annual Temperature (°C) at Different Latitudes

Latitude	Year	January	July	Range
90°N	−22.7	−41.1	−1.1	40.0
80°N	−18.3	−32.2	2.0	34.2
70°N	−10.7	−26.3	7.3	33.6
60°N	−1.1	−16.1	14.1	30.2
50°N	5.8	−7.1	18.1	25.2
40°N	14.1	5.0	24.0	19.0
30°N	20.4	14.5	27.3	12.8
20°N	25.3	21.8	28.0	6.2
10°N	26.7	25.8	27.2	1.4
Equator	26.2	26.4	25.6	0.8
10°S	25.3	26.3	23.9	2.4
20°S	22.9	25.4	20.0	5.4
30°S	16.6	21.9	14.7	7.2
40°S	11.9	15.6	9.0	6.6
50°S	5.8	8.1	3.4	4.7
60°S	−3.4	2.1	−9.1	11.2
70°S	−13.6	−3.5	−23.0	19.5
80°S	−27.0	−10.8	−39.5	28.7
90°S	−33.1	−13.5	−47.8	34.3

Source: Adapted from Haurwitz and Austin (1944).

Water in the atmosphere is warmed by heat radiating from Earth's surface; much of this heat is radiated back to Earth again. The result is the so-called greenhouse effect, which leads to the retention of heat, keeping Earth relatively warm even at night when there is temporarily no influx of solar energy. Without this effect, Earth's surface would cool to many degrees below zero—like the dark side of the moon's surface. Thus the atmosphere buffers day–night thermal change.

Hot air rises. The ground and air masses above it receive more solar energy at low latitudes than at higher ones (Figure 2.1). Thus tropical air masses, especially those near the equator, are warmed relatively more than temperate air masses, and an equatorial zone of rising air is created. These equatorial air masses cool as they rise and eventually move northward and southward high in the atmosphere above Earth's surface (Figure 2.2a). As this cold air moves toward higher latitudes, it sinks slowly at first and then descends rapidly to the surface at the so-called horse latitudes of about 30°N and 30°S. At ground level at these latitudes, some air moves toward the equator again and some of it moves toward the poles. (The amount of air in the atmosphere is finite, so air masses leaving one place must always be replaced by air coming from somewhere else; thus a closed system of circulating air masses is set up.) An idealized diagram of the typical vertical and horizontal movements of atmospheric currents is shown in Figure 2.2. At the surface, the equator is a zone of convergence of air

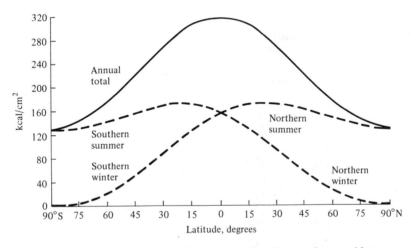

Figure 2.1   Estimated amount of incoming solar radiation that would intercept Earth's surface in the absence of an atmosphere as a function of latitude. The six-month period from the spring equinox to the fall equinox (see Figure 2.7) is labeled "northern summer" and "southern winter" while the six months from the fall equinox to the spring equinox represent the "northern winter" and the "southern summer." [After Haurwitz and Austin (1944).]

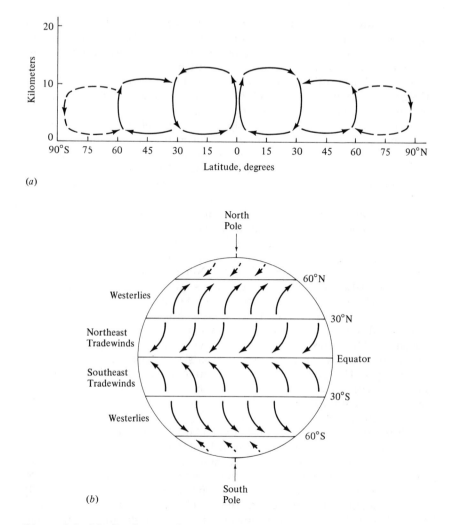

Figure 2.2   Idealized atmospheric circulation patterns. *(a)* Vertical profile against latitude. *(b)* Prevailing wind currents on earth's surface. These belts of moving air move north and south with the seasons. [After MacArthur and Connell (1966), MacArthur (1972), and others.]

masses, while they are diverging at the horse latitudes. Between latitudes 0° and 30°, surface air generally moves toward the equator; between latitudes 30° and 60°, it generally moves away from the equator. As air masses move along the surface, they are slowly warmed and eventually rise again.

Movements of air masses are not strictly north–south as suggested by Figure 2.2a; instead they acquire an east–west component due to the rotation of Earth about its axis (Figure 2.2b). Earth rotates from west to east. A person standing on either pole would rotate slowly around and do a full "about face" each 24 hours. (Near the North Pole the ground moves counterclockwise under one's feet, whereas near the South Pole it moves clockwise; other related important differences between the hemispheres are considered in the following paragraphs.) Someone located near the equator, however, travels much farther during a 24-hour period; indeed, such a person would traverse a distance equal to Earth's circumference, or about 25,000 miles, during each rotation of the globe. Hence the velocity of a body near the equator is approximately 1000 miles/hour (relative to Earth's axis), while a body at either pole is, relatively speaking, at a standstill.

Above considerations, plus the law of conservation of momentum, dictate that objects moving north in the Northern Hemisphere must speed up, relative to Earth's surface, and thus veer toward the right. Similarly, objects moving south in the Northern Hemisphere are going slower than Earth's surface; thus they slow down relative to the surface, which means that they also veer to the right. In contrast, moving objects in the Southern Hemisphere always veer to the left: northward-moving objects slow down and southward-moving ones speed up. These forces, known collectively as the "Coriolis force," act on north–south wind and water currents to give them an east–west component. The Coriolis force is maximal at the poles, where a slight latitudinal displacement is accompanied by a large change in velocity, and minimal at the equator, where a slight latitudinal change has little effect upon the velocity of an object.

Equator-bound surface air between latitudes 0° and 30° slows relative to the surface and veers toward the west in both hemispheres, producing winds from the east ("easterlies"); these constitute the familiar trade winds, known as the northeast trades between 0° and 30°N and the southeast trades between 0° and 30°S. Between latitudes 30° and 60°, surface air moving toward the poles speeds up (again, relative to earth's surface) and veers toward the east, producing the familiar prevailing "westerly" winds at these latitudes in both hemispheres (Figure 2.2b).

These wind patterns, coupled with the action of the Coriolis force on water masses moving north to south, drive the world's ocean currents; in the Northern Hemisphere ocean waters rotate generally clockwise, while they rotate counterclockwise in the Southern Hemisphere (Figure 2.3). During their movement westward along the equator, oceanic waters are warmed by solar irradiation. (These waters also "pile up" on the western sides of oceanic basins; in Central America the sea level of the Atlantic is several feet higher than that of the Pacific.) As this warm equatorial water approaches the eastern sides of land masses, it is diverted northward and/or southward to

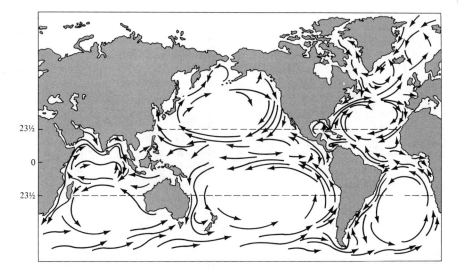

Figure 2.3   Major surface currents of the world's oceans.

higher latitudes, carrying equatorial heat toward the poles along the eastern coasts of continents. Cold polar waters flow toward the equator on west coasts (this is the main reason the Pacific Ocean off southern California is cold but at the same latitude the Atlantic Ocean off Georgia is quite warm).

Heat is molecular movement. Compressing a volume of air results in more collisions between molecules and increased molecular movement; hence compression causes an air mass to heat up. Exactly analogous considerations apply in the reverse case: allowing compressed air to expand decreases the number of molecular collisions and the air mass cools off. As warm air rises, atmospheric pressure decreases, and the air expands and is cooled "adiabatically," or without change in total heat content. Some of the air's own heat is used in its expansion. As descending cold air is compressed it is adiabatically warmed in a similar manner.

Warm air can carry more water vapor than cold air. As warm, water-laden equatorial air rises and cools adiabatically, it first becomes saturated with water at its dew point, and then its water vapors condense to form precipitation. As a result, regions of heavy rainfall tend to occur near the equator (Figure 2.4). In contrast, at the horse latitudes, as cold dry air masses descend and warm adiabatically, they take up water; such desiccating effects help to produce Earth's major deserts at these latitudes (see Figures 2.4 and 2.5). Deserts occur mainly on the western sides of continents, where cold offshore waters are associated with a blanket of cold, and therefore dry, air; westerly winds coming in

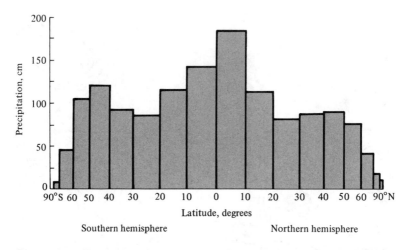

Figure 2.4   Histogram of average annual precipitation versus latitudinal zones. Note the effect of earth's major deserts in the horse latitudes. [From Haurwitz and Austin (1944).]

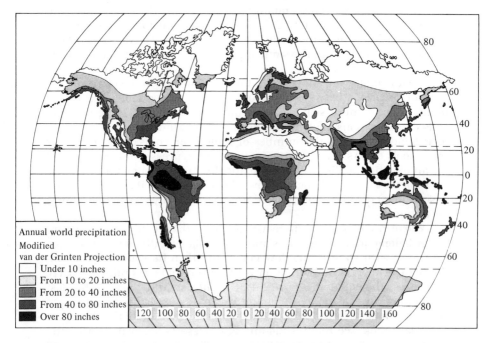

Figure 2.5   Geographic distribution of average annual precipitation. [After MacArthur and Connell (1966) after Koeppe.]

off these oceans do not contain much water to give up to the descending dry air, and precipitation must therefore be scanty. The global pattern of annual precipitation shown in Figure 2.5 conforms in general outline to the expected, but there are many local anomalies and exceptions.

Local Perturbations

Major climatic trends are modified locally by a variety of factors, most notably by the size(s) and position(s) of nearby water bodies and land masses and by topography (especially mountains). Ascending a mountain is, in many ways, comparable to moving toward a higher latitude: mountains are usually cooler and windier than adjacent valleys and generally support communities of plants and animals characteristic of lower elevations at higher latitudes (1000 feet of elevation corresponds roughly to 100 miles of latitude). In addition to this thermal effect, mountains markedly modify water availability and precipitation patterns. Water rapidly runs off a slope but sits around longer on a flatter place and soaks into the ground. For this reason, precipitation falling on a mountainside is generally less effective than an equivalent amount falling on a relatively flat valley floor. Precipitation patterns themselves are also directly affected by the presence of mountains. Consider a north–south mountain range receiving westerly winds, such as the Sierra Nevada of the western United States. Air is forced upward as it approaches the mountains, and, as it ascends, it cools adiabatically, becomes saturated with water, and releases some of its water content as precipitation on the windward side of the sierra. After going over the ridge, this same air, now cold and dry, descends, and as it warms adiabatically, it takes up much of the moisture available on the leeward (downwind) side of the mountains. The so-called rain-shadow effect (Figure 2.6) is produced, with windward slopes being relatively wet and leeward slopes being much drier. Desiccating effects of these warm dry air masses extend for many miles beyond the sierra and help to produce the Mojave Desert in southern California and Nevada.

Water has a high specific heat; that is, a considerable amount of heat energy is needed to change its temperature. Conversely, a body of water can give up a relatively large amount of heat without cooling very much. A result of these heat "sink" attributes is that large bodies of water, particularly oceans, effectively reduce temperature changes of nearby land masses. Thus coastal "maritime" climates are distinguished from inland "continental" climates, with the former being much milder and less variable. Large lakes, such as the Great Lakes, also decrease thermal changes on adjacent land masses, and produce a more constant local temperature.

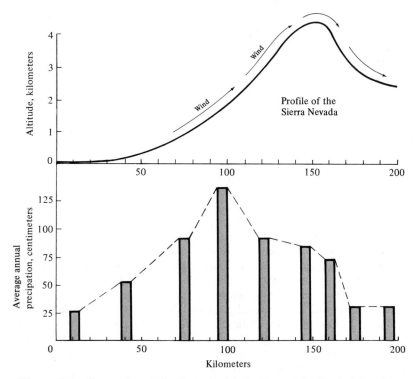

Figure 2.6   Illustration of the "rainshadow" effect of the Sierra Nevada in central California.

Variations in Time and Space

The seasons are produced by the annual elliptical orbit of Earth around its sun and by the inclination of the planet's axis relative to this orbital plane (Figure 2.7). For historical reasons, these movements and patterns have been described from the point of view of the Northern Hemisphere, although by symmetry the same events occur some six months out of phase in the Southern Hemisphere. Twice each year, at the vernal equinox (March 21) and the autumnal equinox (September 23), solar light beams intercept the surface of Earth perpendicularly on the equator (that is, the sun is "directly overhead" at its zenith and equatorial shadows point exactly east to west). At two other times of year, the summer solstice (June 22) and the winter solstice (December 22), the axis of Earth is tilted maximally with respect to the sun's rays. Viewed from the Northern Hemisphere, the axis inclines approximately $23\frac{1}{2}°$ toward the sun during the summer solstice and $23\frac{1}{2}°$ away from it during the winter solstice (see Figure 2.7). At each of the solstices, rays of light hit the surface perpendicularly (the sun is at its zenith) near the Tropic of Cancer ($23\frac{1}{2}°$N) and the Tropic of Capricorn ($23\frac{1}{2}°$S), respectively. At summer solstice, the North Pole is in the middle of its six-month period of sunlight (the "polar summer"). The excess of daylight in summer is exactly balanced by the winter deficit, so the total annual period of daylight is precisely six months at every latitude; the equator has invariate days of exactly 12 hours duration, whereas the poles receive their sunshine all at once over a six-month interval and then face six months of twilight and total darkness. Daylength is one of the most dependable indicators of seasonality and many temperate zone plants and animals rely on photoperiod as an environmental cue (Figure 2.8).

Of course, prevailing winds and ocean currents are not static as suggested by Figures 2.2 and 2.3; in fact they vary seasonally with Earth's movement about its sun. The latitudinal belt receiving the most solar radiation (the "thermal equator") gradually shifts northward and southward between latitudes $23\frac{1}{2}°$S and $23\frac{1}{2}°$N; moreover, latitudinal belts of easterlies and westerlies also move northward and southward with the seasons, producing seasonal weather changes at higher latitudes. Because of the spherical shape of Earth, seasonal changes in insolation increase markedly with increasing latitude.

Although temperatures are modified by prevailing winds, topography, altitude, proximity to bodies of water, cloud cover (Figure 2.9), and other factors, annual marches of average daily temperatures at any given place nevertheless closely reflect earth's movement around the sun. Thus average daily temperatures on the equator change very little seasonally, while those at higher latitudes usually fluctuate consid-

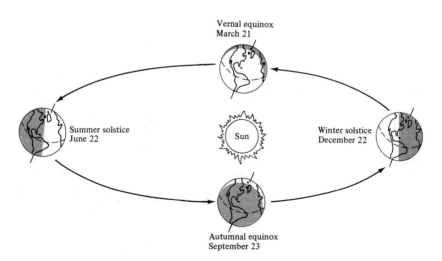

Figure 2.7   Diagram of Earth's annual elliptical orbit around the sun, which produces the seasons. [After MacArthur and Connell (1966).]

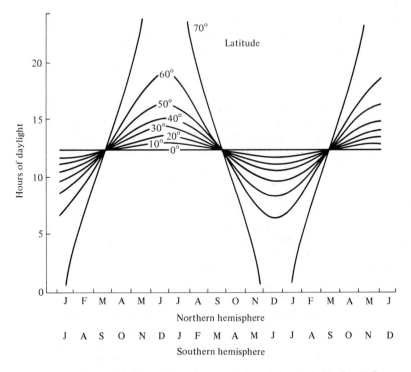

Figure 2.8   Seasonal changes in photoperiod at different latitudes [After Sadleir (1969).]

erably more (Figure 2.10); moreover, the annual range in temperature is also much greater in the temperate zones (Table 2.1).

Annual patterns of precipitation also reflect Earth's orbital movements (Figure 2.11), although precipitation patterns are perhaps modified locally more easily than thermal patterns. At very low latitudes, say 10°S to 10°N, there are often two rainfall maxima each year (Figure 2.11a), one following each equinox as the region of rising air (the thermal equator) passes over the area. Thus the sun's passing over these equatorial areas twice each year produces a bimodal annual pattern of precipitation. Bimodal annual precipitation patterns also occur in other regions at higher latitudes, such as the Sonoran Desert in Arizona (Figure 2.11b), but for different reasons. However, not all equatorial areas have such a rainfall pattern, and some have only one rainy season each year (Figure 2.11c).

During summer, air masses in the central parts of continents at high latitudes tend to warm faster than those around the periphery, which are cooled somewhat by nearby oceans. As hot air rises from the center of continents, a "low" pressure area is formed and cooler, but

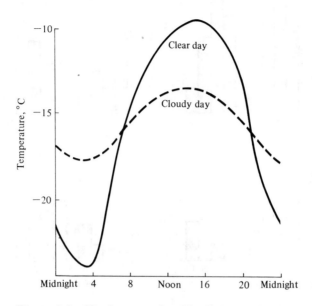

Figure 2.9   Cloud cover reduces both the amplitude and the extremes of the daily march of temperature. [After Haurwitz and Austin (1944).]

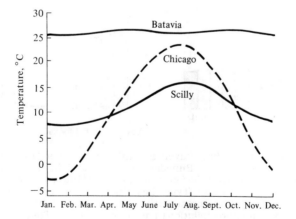

Figure 2.10   Annual marches of average daily temperature at an equatorial locality (Batavia) and two temperate areas, one coastal (Scilly) and one inland (Chicago). [After Haurwitz and Austin (1944).]

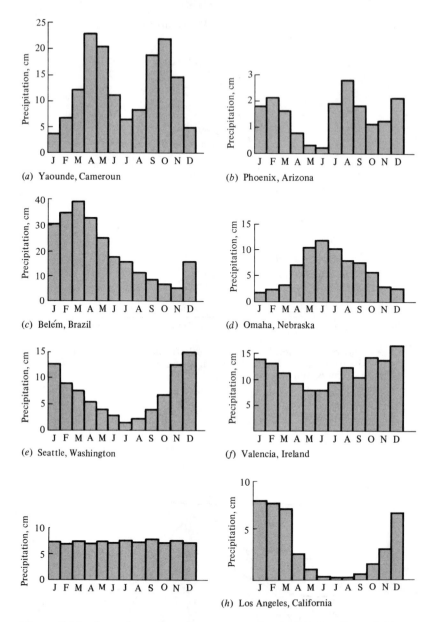

Figure 2.11 Annual marches of average yearly precipitation from eight selected localities (see text). (*a*) Bimodal equatorial rainfall. (*b*) The bimodal annual precipitation pattern of the Sonoran Desert. (*c*) Unimodal annual rainfall pattern of an equatorial area. (*d*) Typical continental summer rainfall regime. (*e*) A coastal area with winter precipitation. (*f*) A typical maritime rainfall regime with more winter precipitation than summer rain. (*g*) The annual march of precipitation in eastern North America is distributed evenly over the year. (*h*) An area with a pronounced summer drought that would support a chaparral vegetation. [After Haurwitz and Austin (1944).]

still warm and water-laden, coastal air is drawn in off the oceans and coastlines. When this moist air is warmed over the land, it rises, cools adiabatically, and releases much of its water content on the interiors of continents. Thus continental, in contrast to coastal, climates are often characterized by summer thunderstorms (Figure 2.11d). In winter, as the central regions of continents cool relative to their water-warmed

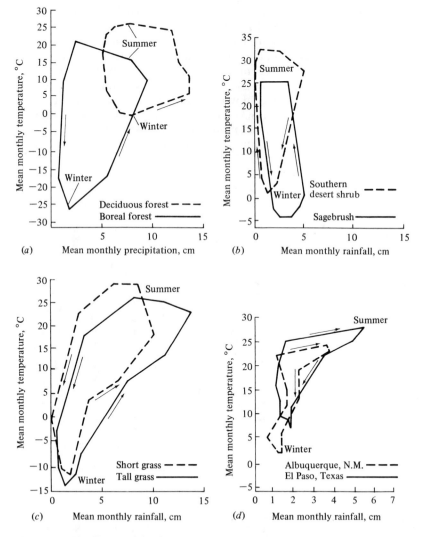

Figure 2.12 Climographs for eight different areas with different vegetation types. [After Smith (1940).]

edges, a "high" pressure area develops and winds reverse, with cold dry air pouring outward toward the coasts.

A weather "front" is produced when a cold (usually polar) air mass and a warm air mass collide. The warmer lighter air is displaced upward by the heavy cold air; as this warm air rises, it is cooled adiabatically, and, provided it contains enough water vapor, clouds form and eventually precipitation falls. Such fronts typify the boundaries between the polar easterlies and the mid-latitude westerlies, which move north to south with the seasons.

The seasonality of coastal rainfall is affected by the differential heating of land and water in another, almost opposite, way. At high latitudes, say 40° to 60°, landmasses cool in winter and become much colder than offshore oceans; along the west coasts of continents, water-rich westerly winds coming in off such a relatively warm ocean are rapidly cooled when they meet a cold landmass and hence deposit much of their moisture as cold winter rain and/or snow along the coast (Figure 2.11e). Maritime climates typically have precipitation through the entire year, with somewhat more during the winter months (Figure 2.11f). As a result of prevailing westerly winds at latitudes 30° to 60°,

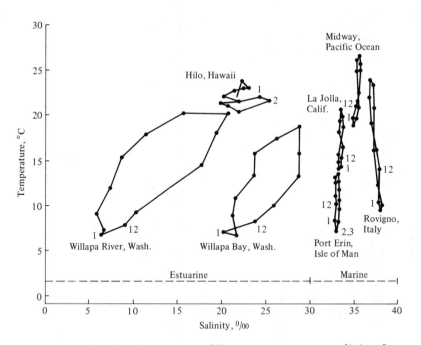

Figure 2.13   Graphs of average monthly temperature versus salinity of some estuarine and marine waters; seasonal variation is great in the brackish waters, whereas salinity varies little in true marine waters. [After Odum (1971) after Hedgepeth.]

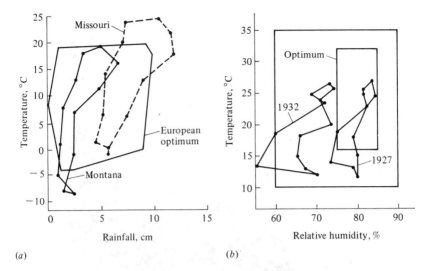

Figure 2.14 Two plots of temperature against moisture. (*a*) Climographs for an area in Montana where the Hungarian partridge was introduced successfully and a Missouri locality where its introduction failed, compared to the average climatic conditions of its European geographic range. Apparently Missouri summers are too hot and/or too wet for these birds. (*b*) Plots of temperature versus relative humidity in 1927 and 1932 in Israel superimposed upon optimal (inner rectangle) and favorable (outer rectangle) conditions for the Mediterranean fruit fly. [Note that, as drawn, these rectangles assume no interaction between temperature and humidity; in actuality the edges would presumably be rounded due to the principle of allocation (see also Chapter 4).] Damage to fruit crops by these flies was much greater in 1927. [After Odum (1959).]

east coasts at these latitudes have continental climates, whereas west coasts have maritime climates. In eastern North America, precipitation is spread fairly evenly over the year (Figure 2.11*g*). Areas with a long summer drought and winter rains (Figure 2.11*h*) typically support a vegetation of chaparral (see also next chapter).

A convenient means of graphically depicting seasonal climates is the "climograph," a plot of average monthly temperature against average monthly precipitation (Figure 2.12). Although such graphs do not reflect year to year variability in climate, they do show at a glance the changes in both temperature and precipitation within an average year, as well as the season(s) during which the precipitation usually falls. (Without actually identifying points by months, however, spring cannot be distinguished from fall since both are seasons of moderate temperatures—hence the arrows in Figure 2.12.) Exactly analogous plots are often made for any two physical variables, such as temperature versus humidity, or temperature versus salinity in aquatic systems (Figure 2.13). Many other variables of biological importance,

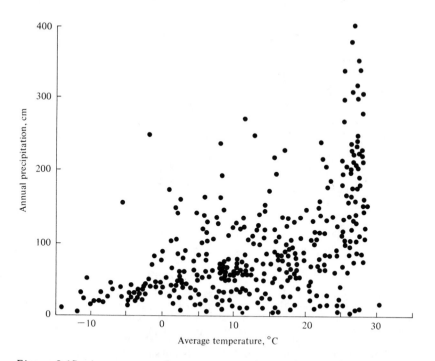

Figure 2.15 Average annual temperature and annual precipitation for many localities scattered more or less evenly over the land area of Earth. [Adapted from Ricklefs (1973) based on data of Clayton (1944).]

such as pH and dissolved phosphorus or nitrogen, can be treated similarly. When coupled with information on the tolerance limits of organisms, climographs and their analogues can be useful in predicting responses of organisms to changes in their physical environments (Figure 2.14).

Very cold climates are usually dry, whereas warmer regions show a wide range of average annual precipitation (Figure 2.15). Although there are an infinite number of different types of climates, attempts have been made to classify them. One scheme (Köppen's) recognizes five major climatic types (as well as many minor ones): tropical rainy, dry, warm temperate rainy, cool snow forest, and polar. Another classification is shown in Figure 2.16.

The Geological Past

Climatic changes over geological time, called paleoclimatology, are of considerable ecological interest because organisms have had to evolve

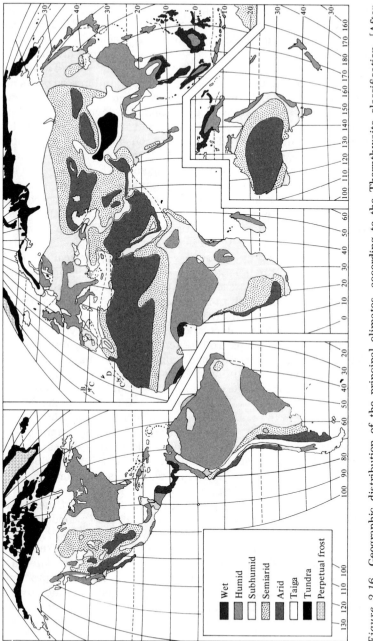

Figure 2.16 Geographic distribution of the principal climates, according to the Thornthwaite classification. [After Blumenstock and Thornthwaite (1941).]

along with such changes. A really thorough ecological study usually includes consideration of the past history of the area under study. Earth has changed in innumerable ways during the geological past; its poles have shifted and wandered, periods of orogeny (mountain building) by tectonic upheaval of Earth's crust have waxed and waned at different places and times, the continents have "drifted" and moved on its mantle, and the planet itself has alternately warmed and cooled, the latter resulting in periods of extensive glaciation. Sea levels dropped during glacial periods as water accumulated on land as snow and ice (such sea level changes controlled by glacier alterations are called "eustatic").

Although it is often extremely difficult to trace past history, a variety of techniques, some of them quite ingenious, have been developed that allow us to deduce many of the changes Earth has undergone. Perhaps the simplest and most direct way to look into the past is to examine the fossil record. Lake sediments are an ideal source of layered fossils and have often been used to follow the history of an area. Fossilized pollens in a lake's sediments are relatively easily identified; pollens of plants adapted to particular types of climates can be used as indicators of past climates, as well as the types and composition of forests that prevailed near the lake at different times (Figure 2.17). Of course, such palynological analyses are fraught with difficulties due both to variations between taxa in rates of pollen production and to differential transport and deposition. Moreover, many deposits may contain mixtures of pollens from several different communities.

A technique known as carbon dating allows estimation of the age of fossil plant remains, including pollens and charcoal. Solar radiation converts some atmospheric nitrogen into a radioactive isotope of carbon called carbon 14 (^{14}C). This ^{14}C is oxidized to carbon dioxide and is taken up by plants in photosynthesis in proportion to its abundance in the air around the plant. All radioactive isotopes emit neutrons and electrons, eventually decaying into nonradioactive isotopes. Half of a quantity of ^{14}C becomes nonradioactive carbon 12 (^{12}C) each 5600 years (this is the "half-life" of ^{14}C). When a plant dies it contains a certain maximal amount of ^{14}C. Comparison of the relative amounts of ^{14}C and ^{12}C in modern-day and fossil plants allows estimation of the age of a fossil. Thus a fossil plant with half the ^{14}C content of a modern plant is about 5600 years old, one with one-quarter as much ^{14}C is 11,200 years old, and so on. The carbon dating method has been checked against ancient Egyptian relics of known age made from plant materials; it accurately estimates their ages, confirming that the rate of production of ^{14}C and the proportion of ^{14}C to ^{12}C have not changed much over the last 5000 years. The technique allows the assignment of very accurate ages to all sorts of recently fossilized plant materials. Other methods of dating older materials have also been developed.

A similar technique makes use of the fact that the uptake of two oxygen isotopes, ^{16}O and ^{18}O, into carbonates is temperature dependent. Thus the proportion of these two isotopes in a fossil seashell presum-

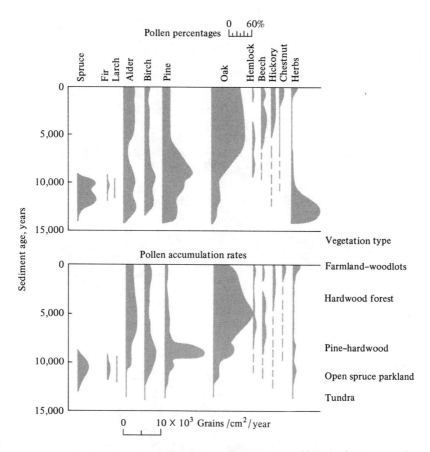

Figure 2.17 Fossil pollen profiles from dated layers of lake sediments in the northeastern United States for the period following the last ice age. Upper plot shows number of pollen grains of each species group as a percentage of the total sample. Lower plot gives estimated rates of deposition of each type of pollen and, at the right, the type of vegetation that probably prevailed in the area. [Adapted from Odum (1971) after M. Davis.]

ably reflects the temperature of an ancient ocean in which that particular mollusk lived.

There is now a massive body of evidence that the continents have actually moved or "drifted" away from one another during geological time (Figure 2.18; see also pp. 306–307). Certain types of rocks, particularly basalts, retain a magnetic "memory" of the latitude in which they were solidified. Such paleomagnetic evidence strongly indicates that the continents have moved with respect to one another. The continents are formed of light "plates" of siliceous, largely granitic rocks about 30 km thick which in turn float on denser mantlelike basaltic blocks. The

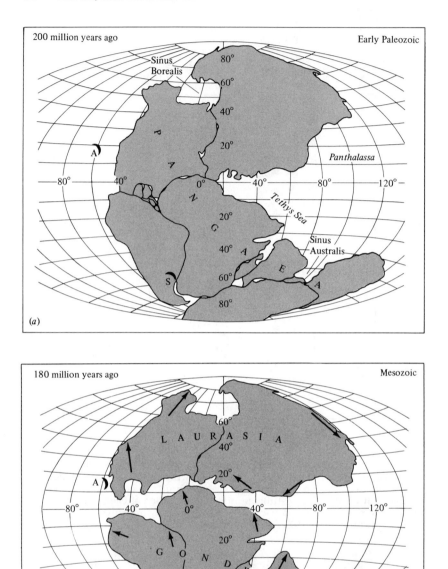

(a)

(b)

Figure 2.18 Approximate positions of major landmasses at different times in the geological past, showing their probable movements. [Adapted from Dietz and Holden (1970). The Breakup of Pangaea. Copyright © 1970 by Scientific American, Inc. All rights reserved.]

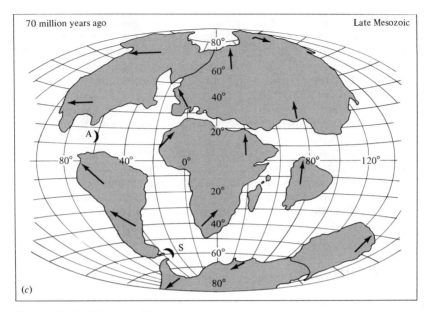

Figure 2.18   (Continued)

ocean floors are composed of a relatively thin altered top of Earth's mantle. A mountain range on the sea floor in the mid-Atlantic has recently been mapped, and this ridge is now thought to represent a region of upwelling of the mantle. Under this interpretation, as the upwelling proceeds, sea floors spread and continents move apart (Figure 2.19). Thus modern theory holds that, except for the Pacific, which

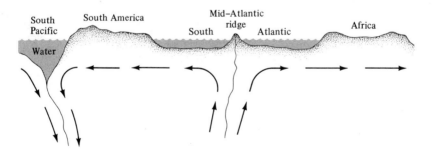

Figure 2.19   Diagrammatic cross-sectional view of the probable movements of Earth's mantle and crust that lead to sea-floor spreading and continental drift. Upwelling of deep mantle materials in the mid-Atlantic is accompanied by a surface movement away from the mid-oceanic ridge. The continents, which float on top of these moving denser materials, are carried along. At the oceanic trench on the far left these materials sink back down into the mantle, forming a closed system of circulating materials.

is shrinking, the oceans are growing with very young ocean floors in mid-ocean and progressively older floors toward the continents. Other evidence, such as the apparent ages of islands and the depths of sediments, nicely corroborates this conclusion. Imagine the effects of changing climates on plants and animals as continents drift through different latitudes! Australia, for example, became arid as it drifted into the horse latitudes; moreover, as this landmass continues to move toward the equator, its climate will gradually become wetter and more tropical.

Reasons for many of these past changes, such as polar movements and the alternate warming and cooling of the planet, are little known and may well involve solar changes. Piecing together all this diverse and sometimes conflicting evidence on Earth's past history is a most difficult and challenging task, one which occupies many of our finest scientists.

Selected References

Major Determinants of Climate; Local Perturbations;
Variations in Time and Space

Blair and Fite (1965); Blumenstock and Thornthwaithe (1941); Byers (1954); Chorley and Kennedy (1971); Collier *et al.* (1973); Finch and Trewartha (1949); Flohn (1969); Gates (1962); Haurwitz and Austin (1944); Lowry (1969); MacArthur (1972); MacArthur and Connell (1966); Taylor (1920); Thornthwaithe (1948); Trewartha (1943); U.S. Department of Agriculture (1941).

The Geological Past

Birch and Ehrlich (1967); Dansereau (1957); Darlington (1957, 1965); Dietz and Holden (1970); Hesse, Allee, and Schmidt (1951); Jelgersma (1966); Martin and Mehringer (1965); Sawyer (1966); Udvardy (1969); Wilson (1971, 1973); Wiseman (1966); Wright and Frey (1965).

The Interface Between Climate and Vegetation 3

Climate is the major determinant of vegetation. Plants in turn exert some degree of influence on climate. Both climate and vegetation profoundly affect soil development and the animals that live in an area. Here we examine some ways in which climate and vegetation interact. More emphasis is given to terrestrial ecosystems than to aquatic ones, although some aquatic analogues are briefly noted. Topics presented rather succinctly here are treated in greater detail elsewhere (see references at end of chapter).

Plant Life Forms and Biomes

Terrestrial plants adapted to a particular climatic regime often have similar morphologies, or plant growth forms. Thus climbing vines, epiphytes, and broad-leafed species characterize tropical rain forests. Evergreen conifers dominate very cold areas at high latitudes and/or altitudes, while small frost-resistant tundra species occupy still higher latitudes and altitudes. Seasonal temperate zone areas with moderate precipitation usually support broad-leafed, deciduous trees, while tough-leafed (sclerophyllous) evergreen shrubs, or so-called chaparral-type vegetation, occur in regions with winter rains and a pronounced long water deficit during spring, summer, and fall. Chaparral vegetation is found wherever this type of climate prevails, including southern California, Chile, Spain, Italy, southwestern Australia, and the northern and southern tips of Africa (see Figure 3.1), although the actual plant species comprising the flora usually differ. Areas with very predictable and stable climates tend to support fewer different plant life forms than regions with more erratic climates (Figure 3.2). In general, there is a close correspondence between climate

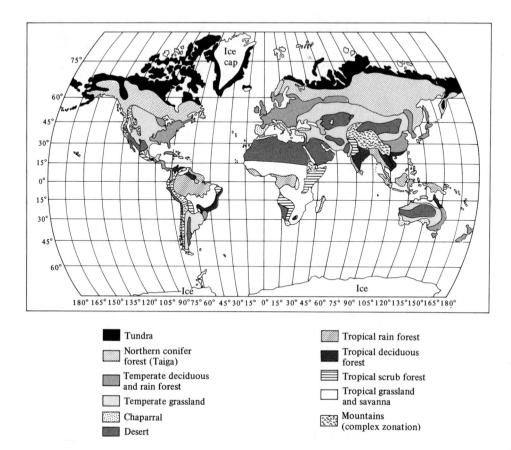

Tundra

Northern conifer forest (Taiga)

Temperate deciduous and rain forest

Temperate grassland

Chaparral

Desert

Tropical rain forest

Tropical deciduous forest

Tropical scrub forest

Tropical grassland and savanna

Mountains (complex zonation)

Figure 3.1   Geographic distribution of major vegetation types. [After MacArthur and Connell (1966) after Odum.]

and vegetation (compare Figures 2.5 and 2.16 with Figure 3.1); indeed, climatologists have sometimes used vegetation as the best indicator of climate! Thus rain forests occur in rainy tropical and rainy warm-temperature climates, forests exist under more moderate mesic climates, savannas and grasslands prevail in semiarid climates, and deserts characterize still drier climates. Topography and soils, of course, also play a part in the determination of vegetation types, which are sometimes termed "plant formations." Such major communities of characteristic plants and animals are known as biomes. Classification of natural communities is discussed later in this chapter.

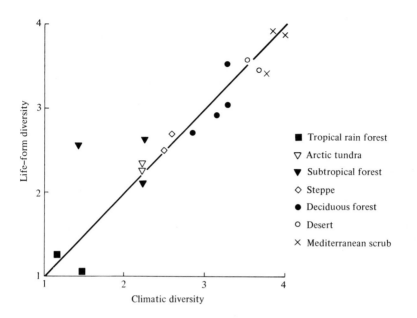

Figure 3.2   Rainfall unpredictability or "climatic diversity" is positively correlated with plant life form diversity. [From May (1975) after T. Givnish.]

Microclimate

Even in the complete absence of vegetation, major climatic forces, or macroclimates, are expressed differently at a very local spatial level, which has resulted in the recognition of so-called microclimates. Thus the surface of the ground undergoes the greatest daily variation in temperature, and daily thermal flux is progressively reduced with both increasing distance above and below ground level (Figure 3.3). During daylight hours the surface intercepts most of the incident solar energy and rapidly heats up, whereas at night this same surface cools more than its surroundings. Such plots of temperature versus height above and below ground are called thermal profiles. An analogous type of graph, called a bathythermograph, is often made for aquatic ecosystems by plotting temperature against depth (see Figure 3.23).

Daily temperature patterns are also modified by topography even in the absence of vegetation. A slope facing the sun intercepts light beams more perpendicularly than does a slope facing away from the sun; as a result a south-facing slope in the Northern Hemisphere receives more solar energy than a north-facing slope, and the former heats up faster and gets hotter during the day (Figure 3.4). Moreover, such a south-

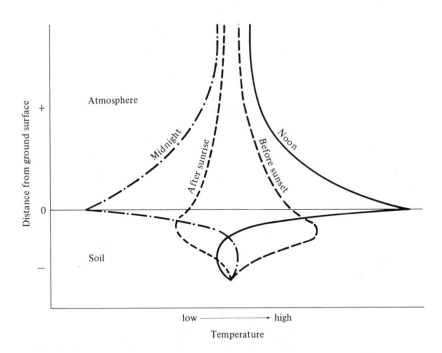

Figure 3.3 Idealized thermal profile showing temperatures at various distances above and below ground at four different times of day. [After Gates (1962).]

facing slope is typically drier than a north-facing one because it receives more solar energy and therefore more water is evaporated.

By orienting themselves either parallel to or at right angles to the sun's rays, organisms (and parts of organisms such as leaves) may decrease or increase the total amount of solar energy they actually intercept. Leaves in the brightly illuminated canopy often droop during midday, whereas those in the shaded understory typically present their full surface to incoming beams of solar radiation. Similarly, many desert lizards position themselves on the ground perpendicular to the sun's rays in the early morning when environmental temperatures are low, but during the high temperatures of midday these same animals may climb up off the ground into cooler air temperatures and orient themselves parallel to the sun's rays by facing into the sun (see also pp. 86–89).

The major effect of a blanket of vegetation is to moderate most daily climatic changes, such as changes in temperature, humidity, and wind. (Plants, however, generate daily variations in concentrations of oxygen and carbon dioxide through their photosynthetic and respiratory activities.) Thermal profiles at midday in cornfields at various stages of growth are shown in Figure 3.5, demonstrating the marked

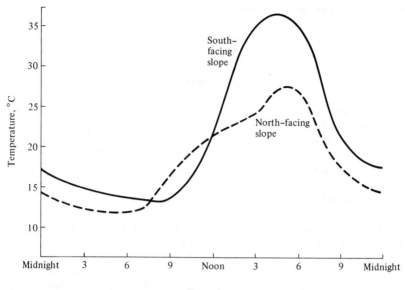

Figure 3.4   Daily marches of temperature on an exposed south-facing slope (solid line) and a north-facing slope (dashed line) during late summer in the northern hemisphere. [After Smith (1966) after van Eck.]

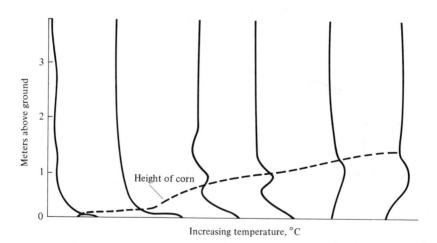

Figure 3.5   Temperature profiles in a growing cornfield at midday, showing the effect of vegetation on thermal microclimate. [After Smith (1966).]

reduction in ground temperature due especially to shading. In the mature field, air is warmest at about a meter above ground. Similar vegetational effects on microclimates occur in natural communities. For example, a patch of open sand in a desert might have a daily thermal profile somewhat like that shown in Figure 3.3, whereas temperatures in the litter underneath a nearby dense shrub would vary much less with the daily march of temperature.

Humidities are similarly modified by vegetation, with relative humidities within a dense plant being somewhat greater than those of the air in the open adjacent to the plant. An aphid may spend its entire lifetime in the very thin zone (only about a millimeter thick) of high humidity that surrounds the surface of a leaf. Moisture content is more stable, and therefore more dependable, deeper in the soil than it is at the surface, where high temperatures periodically evaporate water to produce a desiccating effect.

Wind velocities are also reduced sharply by vegetation and are usually lowest near the ground (Figures 3.6 and 3.7). Moving currents of air promote rapid exchange of heat and water; hence an organism cools or warms more rapidly in a wind than it does in a stationary air mass at the same temperature. Likewise, winds often carry away moist air and replace it with drier air, thereby promoting evaporation and water loss; the desiccating effects of such dry winds can be extremely important to an organism's water balance (see also pp. 89–90).

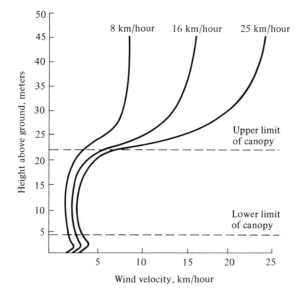

Figure 3.6   Wind velocities within a forest vary relatively little with changes in the wind velocity above the canopy. [After Fons (1940).]

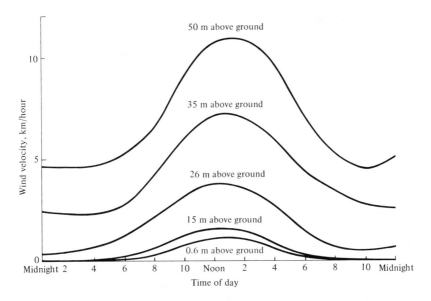

Figure 3.7   Daily march of average wind velocities during June at various heights inside a coniferous forest in Idaho. [After Smith (1966) after Gisborne.]

In aquatic systems, water turbulence parallels wind in many ways, and rooted vegetation around the edges of a pond or stream reduces water turbulence. At a more microscopic level, algae and other organisms that attach themselves to underwater surfaces (so-called periphyton) create a thin film of distinctly modified microenvironment in which water turbulence, among other things, is reduced. Localized spatial patches with particular concentrations of hydrogen ions (pH), salinities, dissolved nitrogen and phosphorus, etc., form similar aquatic microhabitats.

By actively or passively selecting such microhabitats, organisms can effectively reduce the overall environmental variation they encounter and enjoy more optimal conditions than they could without microhabitat selection. Innumerable other microclimatic effects could be cited, but these should serve to illustrate their existence and their significance to plants and animals.

Primary Production and Evapotranspiration

In terrestrial ecosystems, climate is by far the most important determinant of the amount of solar energy plants are able to capture as chemical energy, or the gross primary productivity (Table 3.1). In

Table 3.1   Net Primary Productivity and World Net Primary Production for the Major Ecosystems

	Area[a] (10⁶km²)	Net Primary Productivity, Per Unit Area[b] (dry g/m²/yr) normal range	mean	World Net Primary Production[c] (10⁹ dry tons/yr)
Lake and stream	2	100–1500	500	1.0
Swamp and marsh	2	800–4000	2000	4.0
Tropical forest	20	1000–5000	2000	40.0
Temperate forest	18	600–2500	1300	23.4
Boreal forest	12	400–2000	800	9.6
Woodland and shrubland	7	200–1200	600	4.2
Savanna	15	200–2000	700	10.5
Temperate grassland	9	150–1500	500	4.5
Tundra and alpine	8	10–400	140	1.1
Desert scrub	18	10–250	70	1.3
Extreme desert, rock, and ice	24	0–10	3	0.07
Agricultural land	14	100–4000	650	9.1
Total land	149		730	109.0
Open ocean	332	2–400	125	41.5
Continental shelf	27	200–600	350	9.5
Attached algae and estuaries	2	500–4000	2000	4.0
Total ocean	361		155	55.0
Total for Earth	510		320	164.0

Source: Adapted from Whittaker (1970), *Communities and Ecosystems.* Reprinted with permission of Macmillan Publishing Co., Inc. © Copyright Robert H. Whittaker, 1970.

[a] Square kilometers + 0.3861 = square miles.
[b] Grams per square meter × 0.01 = t/ha, × 0.1 = dz/ha or m centn/ha (metric centers, 100 kg, per hectare, 10⁴ square meters), × 10 = kg/ha, × 8.92 = lbs/acre.
[c] Metric tons (10⁶ g) × 1.1023 = English short tons.

warm arid regions, water is a master limiting factor, and, in the absence of runoff, primary production is strongly positively correlated with rainfall in a linear fashion (see pp. 13–14 and Figure 1.4). Above about 80 cm of precipitation/year, primary production slowly decreases with increasing precipitation and then levels off (asymptotes) (Figure 3.8). Notice that some points fall below the line in this figure, presumably because there is some water loss by runoff and seepage into groundwater supplies.

Evapotranspiration refers to the release of water into the atmosphere as water vapor, both by the physical process of evaporation and by the biological processes of transpiration and respiration. The amount of water vapor thus returned to the atmosphere is strongly dependent on temperature, with greater evapotranspiration at higher temperatures. The *theoretical* temperature-dependent amount of water that *could* be "cooked out" of an ecological system, given its input of solar energy and provided that much water fell on the area, is

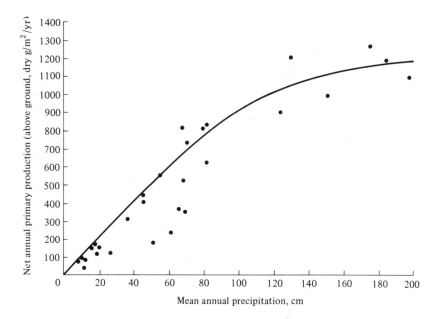

Figure 3.8   Net primary productivity (above ground) plotted against average annual precipitation. [From Whittaker (1970). Reprinted with permission of Macmillan Publishing Co., Inc., from *Communities and Ecosystems* by Robert H. Whittaker. © Copyright Robert H. Whittaker, 1970.]

called its *potential evapotranspiration* (PET). In many ecosystems water is frequently in short supply, so *actual evapotranspiration* (AET) is somewhat less than potential (clearly AET can never exceed PET and is equal to PET only in water-saturated habitats). Actual evapotranspiration can be thought of as the reverse of rain, for it is the amount of water that actually goes back into the atmosphere at a given spot.

The potential evapotranspiration for any spot on earth is determined by the same factors that regulate temperature, most notably latitude, altitude, cloud cover, and slope (topography). There is a nearly one-to-one correspondence between PET and temperature, and an annual march of PET can be plotted in centimeters of water. By superimposing the annual march of precipitation on these plots (Figure 3.9), seasonal changes in water availability are depicted graphically. A water deficit occurs when PET exceeds precipitation; a water surplus exists when the situation is reversed. During a period of water surplus, some water may be stored by plants and some may accumulate in the soil as soil moisture, depending upon runoff and the capacity of soils to hold water; during a later water deficit, such stored water can be used by plants and released back into the atmosphere. Winter rain is generally

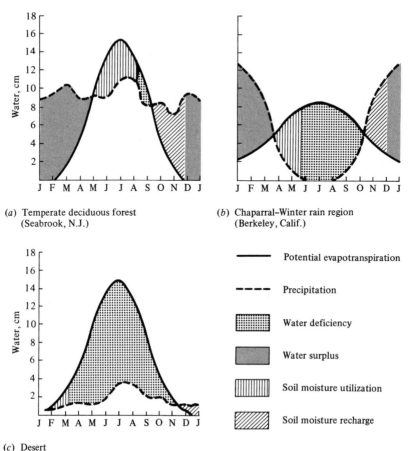

(a) Temperate deciduous forest
(Seabrook, N.J.)

(b) Chaparral–Winter rain region
(Berkeley, Calif.)

(c) Desert
(Albuquerque, N.M.)

Potential evapotranspiration

Precipitation

Water deficiency

Water surplus

Soil moisture utilization

Soil moisture recharge

Figure 3.9   Plots of annual march of potential evapotranspiration superimposed upon the annual march of precipitation for three ecologically distinct regions showing the water relations of each. (*a*) Temperate deciduous forest, (*b*) chaparral with winter rain, (*c*) desert. [From Odum (1959) after Thornthwaite.]

much less effective than summer rain because of the reduced activity (or complete inactivity) of plants in winter; indeed two areas with the same annual march of temperature and total annual precipitation may differ greatly in the types of plants they support and in their productivity, depending upon their seasonal patterns of precipitation. For example, an area receiving about 50 cm of precipitation annually supports either a grassland vegetation or chaparral, depending upon whether the precipitation falls in summer or winter, respectively.

Rosenzweig (1968) found that net annual primary production above ground is strongly correlated with actual evapotranspiration, or AET (Figure 3.10). This correlation is remarkable in that AET was crudely estimated using only monthly macroclimatic statistics with no allowance for either runoff and water-holding capacities of soils or for groundwater usage. Rosenzweig suggested that the reason for the observed correlation is that AET measured simultaneously two of the most important factors limiting primary production on land: water and solar energy. Photosynthesis, that fundamental process upon which nearly all life depends for energy, is represented by the chemical equation

$$6CO_2 + 12H_2O \xrightarrow[\text{energy}]{\text{solar}} C_6H_{12}O_6 + 6O_2 + 6H_2O$$

where $C_6H_{12}O_6$ is the energy-rich glucose molecule. Carbon dioxide concentration in the atmosphere is fairly constant at about 0.03 percent

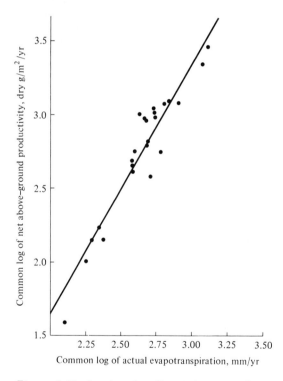

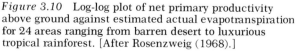

Figure 3.10   Log-log plot of net primary productivity above ground against estimated actual evapotranspiration for 24 areas ranging from barren desert to luxurious tropical rainforest. [After Rosenzweig (1968).]

and does not strongly influence the rate of photosynthesis, except under unusual conditions of high water availability and full or nearly full sunlight when CO_2 is limiting (Meyer, Anderson, and Bohning, 1960). Rosenzweig notes that, on a geographical scale, each of the other two requisites for photosynthesis, water and solar energy, are much more variable in their availabilities and more often limiting; moreover, AET measures the availability of both. Temperature is often a rate-limiting factor and markedly affects photosynthesis; it, too, is presumably incorporated into an estimated AET value. Primary production may be influenced by nutrient availability as well, but in many terrestrial ecosystems these effects are often relatively minor. In aquatic ecosystems, however, nutrient availability is often a major determinant of the rate of photosynthesis (see pp. 68–71). Primary production in aquatic systems of fairly constant temperature (such as the oceans) is often strongly affected by light. Because light intensity decreases rapidly with depth in lakes and oceans, most primary productivity is concentrated near the surface (Figure 3.11*a*). However, short wavelengths (blue light) penetrate deeper into water than longer ones, and some benthic (bottom-dwelling) marine "red" algae have evolved unique photosynthetic adaptations to utilize these wavelengths.

Similarly, within forests, light intensity varies markedly with height above ground (Figure 3.11*b*). Tall trees in the canopy receive the full incident solar radiation, while shorter trees and shrubs receive

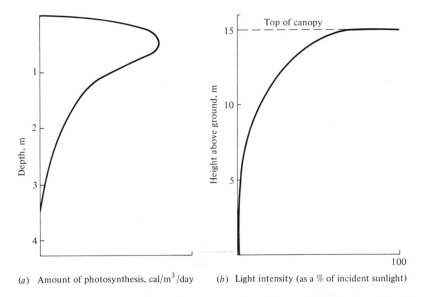

(a)   Amount of photosynthesis, cal/m^3/day   (b)   Light intensity (as a % of incident sunlight)

Figure 3.11   Vertical profiles of (*a*) the amount of photosynthesis versus depth in a lake and (*b*) light intensity versus height above ground in a forest.

progressively less light. In really dense forests, less than 1 percent of the incident solar energy impinging on the canopy actually penetrates to the forest floor. Although a tree in the canopy therefore has more solar energy available to it than a fern on the forest floor, the canopy tree must also expend much more energy on vegetative supporting tissues (wood) than the fern. (Understory plants are usually very shade tolerant and able to photosynthesize at very low light intensities. Also, herbs of the forest floor often grow and flower early in the spring before deciduous trees leaf out.) Hence each plant life form and each growth strategy has its own associated costs and profits.

Leaf Tactics

Leaves come in an almost bewildering array of sizes and shapes; some leaves are deciduous, others evergreen; their actual spatial arrangement on a given plant differs considerably both within and between species. Presumably this great diversity of leaf tactics is a result of natural selection maximizing the lifetime reproductive success of individual plants under diverse environmental conditions. Leaf tactics are influenced by a multitude of factors that include light, water availability, prevailing winds, and herbivores. When grown in the shade, individuals of many species grow larger, less dissected leaves than when grown in the sun. Similarly, shade-tolerant plants of the understory usually have larger and less lobed leaves than canopy species. Similar types of leaves often evolve independently in different plant lineages subjected to comparable climatic conditions at different geographic localities, especially among trees (Bailey and Sinnot, 1916; Ryder, 1954). Lowland wet tropical rain forest trees have large evergreen leaves with nonlobed or continuous margins, chaparral plants tend to have small sclerophyllous evergreen leaves, arid regions tend to support leafless stem succulents such as cacti or plants with entire leaf margins (especially among evergreens), plants from cold wet climates often have notched or lobed leaf margins, and so on. These repeated patterns of leaf size and shape suggest that a general theory of leaf tactics is possible.

Several models for optimal leaf size under differing environmental conditions have been developed. Efficiency of water use (grams of carbon dioxide assimilated per gram of water lost) was the measure of plant performance Parkhurst and Loucks (1971) maximized. Givnish and Vermeij (1976) developed a similar model for size and shape of vine leaves. Even these relatively simple models predict several observed patterns in leaf size, such as large leaves in warm shady wet places and small leaves in colder areas or warmer and sunnier locales (Figure 3.12).

The evergreen versus deciduous dichotomy can be approached similarly using cost-benefit arguments (Orians and Solbrig, 1977; Mil-

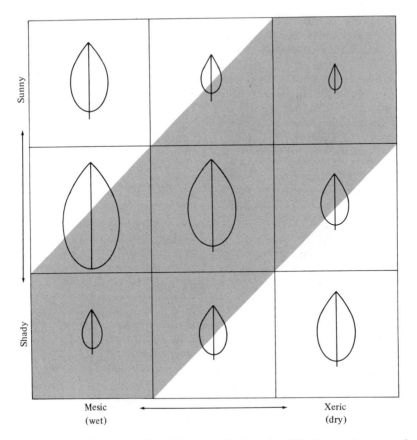

Figure 3.12 Patterns of leaf sizes predicted under differing environmental conditions from wet to dry (mesic to xeric) and shady to sunny. The shaded region represents conditions likely to prevail in nature. [From Givnish and Vermeij (1976). Copyright © 1976 by The University of Chicago Press.]

ler, 1977). In considering leaf tactics of desert plants, Orians and Solbrig contrast leaf types along a continuum ranging from the relatively inexpensive deciduous "mesophytic" leaf to the more costly evergreen "xerophytic" leaf. Mesophytic leaves photosynthesize and transpire at a rapid rate, and hence require high water availability (low "soil water potential"). In deserts such plants grow primarily along washes. In contrast, xerophytic leaves cannot photosynthesize as rapidly when abundant water is available, but they are able to extract water from relatively dry soil. Each plant leaf tactic has an advantage either at different times or in different places, thereby promoting plant life form diversity. During wet periods plants with mesophytic leaves photosynthesize rapidly, but under drought conditions they must drop their leaves and become dormant. During such dry periods, however, the slower photosynthesizers with xerophytic leaves are still able to function by virtue of their ability to extract water from dry soils. Of course, all degrees of intermediate leaf tactics exist, each of which may enjoy a competitive advantage under particular conditions of water availability (Figure 3.13). In a predictable environment, net annual profit per unit of leaf surface area determines the winning phenotype. Even a relatively brief wet season could suffice to give mesophytic leaves a higher annual profit (which accounts for the occurrence of these plant life forms in deserts).

In a discussion of leaf arrangement and forest structure, Horn (1971, 1975a, 1976) distinguishes "monolayers" from "multilayers."

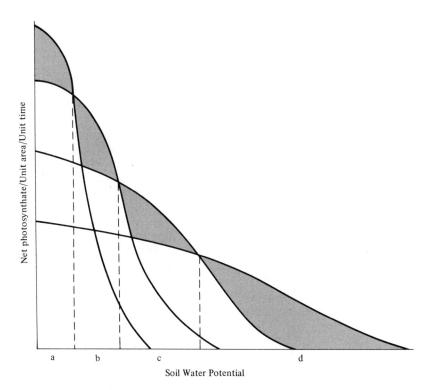

Figure 3.13 Probable relationship between efficiency of photosynthesis and water availability among different types of leaves. The most xerophytic leaf performs best under conditions of low water availability (zone d), whereas the most mesophytic leaf does best when soils are wet (zone a). Intermediate types are superior under intermediate conditions of water availability. Shaded areas indicate superiority of various leaf types under different conditions of soil water availability. [Adapted from Orians and Solbrig (1977). Copyright © 1977 by The University of Chicago Press.]

Each plant in the multilayer of a forest (usually sunnier places such as the canopy) has leaves scattered throughout its volume at several different levels, whereas monolayer plants have essentially a single blanket or shell of leaves. Plants in the multilayer gain from a geometry that allows some light to pass through to their own leaves at lower levels. Horn points out that lobing facilitates passage of light and that such plants do well in the sun (in the shade, inner leaves may respire more than they photosynthesize). In contrast, the optimal tree design in the shade is a monolayer in which each leaf typically intercepts as much light as possible (leaves are large and seldom lobed). Moreover, slow-growing monolayer plants eventually outcompete fast-growing multilayer plants that persist by regular colonization of newly vacated areas created by continual disturbance (Horn, 1976).

Soil Formation and Primary Succession

Soils are a key part of the terrestrial ecosystem because many processes critical to the functioning of ecosystems occur in the soil. This is where dead organisms are decomposed and where their nutrients are retained until used by plants and, indirectly, returned to the remainder of the community. Soils are essentially a meeting ground of the inorganic and organic worlds. Many organisms live in the soil, and the vast majority of insects spend at least a part of their life cycle there. Certain soil-dwelling organisms, such as earthworms, often play a major role in breaking down organic particles into smaller pieces which present a larger surface area for microbial action, thereby facilitating decomposition. Indeed, activities of such soil organisms sometimes constitute a "bottleneck" for the rate of nutrient cycling, and as such they can regulate nutrient availability and turnover rates in the entire community. In aquatic ecosystems, bottom sediments like mud and ooze are closely analogous to the soils of terrestrial systems.

Much of modern soil science, or pedology, was anticipated in the late 1800s by a prominent Russian pedologist named Dokuchaev. He devised a theory of soil formation, or pedogenesis, based largely on climate, although he also recognized the importance of time, topography, organisms (especially vegetation), and parent materials (the underlying rocks from which the soil is derived). The relative importance of each of these five major soil-forming factors varies from situation to situation. Figure 3.14 shows how markedly the soil changes along the transition from prairie to forest in the midwestern United States, where the only other conspicuous major variable is vegetation type. Jenny (1941) gives many other examples of how each of the five major soil-forming factors influence particular soils.

The marked effects some soil types can have on plants are well illustrated by so-called serpentine soils (Whittaker, Walker, and Kruc-

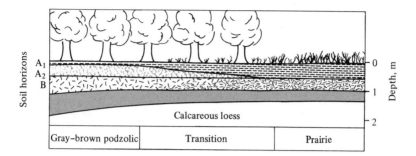

Figure 3.14 Diagram of typical soil changes along the transition from prairie (deep black topsoil) to forest (shallow topsoil) at the edge of the North American great plains. [After Crocker (1952).]

keberg, 1954), which are formed over a parent material of serpentine rock. These soils often occur in localized patches surrounded by other soil types; typically the vegetation changes abruptly from nonserpentine to serpentine soils. Serpentine soils are rich in magnesium, chromium, and nickel, but they contain very little calcium, molybdenum, nitrogen, and phosphorus. They usually support a stunted vegetation and are relatively less productive than adjacent areas with different, richer soils; indeed, entire floras of specialized plant species have evolved that are tolerant of the conditions of serpentine soils (particularly their low calcium levels). Introduced Mediterranean "weeds" have replaced native Californian coastal grasses and forbs almost everywhere except on serpentine soils, where the native flora still persists.

Soil development from bare rock, or primary succession, is a slow process that often requires centuries. Rock is fragmented by temperature changes, by the action of windblown particles, and in colder regions by the alternate freezing and thawing of water. Chemical reactions, such as the formation of carbonic acid (H_2CO_3) from water and carbon dioxide, may also help to dissolve and break down certain rock types like limestones. Such weathering of rocks releases inorganic nutrients that can be used by plants. Eventually lichens establish themselves, and as other plants root and grow, the rock is further broken up into still smaller fragments. As these plants photosynthesize, they convert inorganic materials into organic matter. Such organic material, mixed with inorganic rock fragments, accumulates, and soil is slowly formed. Early in primary succession, production of new organic material exceeds its consumption and organic matter accumulates; as soil "maturity" is approached, soil eventually ceases to accumulate (see also Figure 3.20).

There are fairly close parallels between concepts of soil development and those for the development of ecological communities;

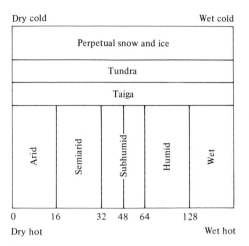

(a) Climate

(b) Vegetation

(c) Soils

Figure 3.15   Relationships between temperature and precipitation and (a) climatic types, (b) vegetational formations and (c) major zonal soil groups. [From Blumenstock and Thornthwaite (1941).]

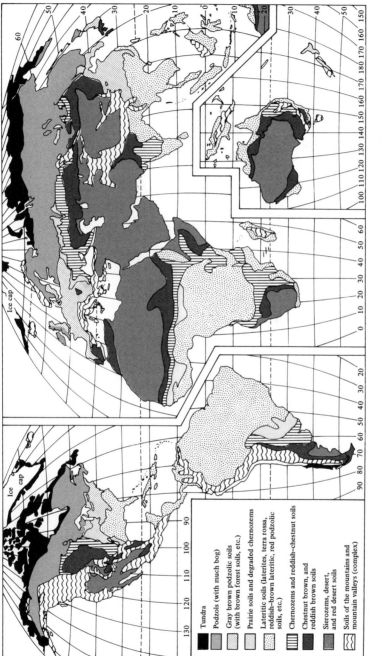

Figure 3.16 Geographic distribution of the primary soil types. Compare with Figure 3.1. [After Blumenstock and Thornthwaite (1941).]

Tundra

Podzols (with much bog)

Gray brown podzolic soils (with brown forest soils, etc.)

Prairie soils and degraded chernozems

Lateritic soils (laterites, terra rossa, reddish-brown lateritic, red podzolic soils, etc.)

Chernozems and reddish-chestnut soils

Chestnut brown, and reddish brown soils

Sierozems, desert, and red desert soils

Soils of the mountains and mountain valleys (complex)

pedologists speak of "mature" soils at the steady state, while ecologists recognize the "climax" communities that grow on and live in these same soils. These two components of the ecosystem (soils and vegetation) are intricately interrelated and interdependent; each strongly influences the other. Except in forests and rain forests, there is usually a one-to-one correspondence between them (Figure 3.15). (Compare also the geographic distribution of soil types shown in Figure 3.16 with the distribution of vegetation types shown in Figure 3.1.)

Once a mature soil has been formed, a disturbance such as the removal of vegetation by fire or human activities often results in gradual sequential changes in the organisms comprising the community. Such a temporal sequence of communities is termed a secondary succession.

Ecotones, Vegetational Continua, and Secondary Succession

Communities are seldom discrete entities; in fact they usually grade into one another in both space and time. A localized "edge community" between two other reasonably distinct communities is termed an *ecotone*. Typically such ecotonal communities are rich in species because they contain representatives from both parent communities and may also contain species distinctive of the ecotone itself. Often a series of communities grade into one another in an almost continuous fashion (Figure 3.17); such a gradient community is called an *ecocline*. Ecoclines may occur either in space or in time.

Spatial ecoclines on a more local scale have led to so-called gradient analysis (Whittaker, 1967). The abundance and actual distribution of organisms along many environmental gradients have shown that the importances of various species along any given gradient typically form bell-shaped curves reminiscent of tolerance curves (to be discussed in Chapter 4). These curves tend to vary independently of one another and often overlap broadly (Figure 3.18), indicating that each species' population has its own particular habitat requirements and width of habitat tolerance—and hence its own zone of maximal importance. Such a continuous replacement of plant species by one another along a habitat gradient is termed a *vegetational continuum* (Figure 3.18).

A temporal ecocline, or a change in community composition in time, both by changes in the relative importance of component populations and by extinction of old species and invasion of new ones, is termed a *succession*. Primary succession, as we have just seen, is the development of communities from bare rock; secondary successions are changes that take place after destruction of the natural vegetation of an area with soil. In the southeastern United States, for instance, an abandoned field allowed to change naturally is in turn invaded by an-

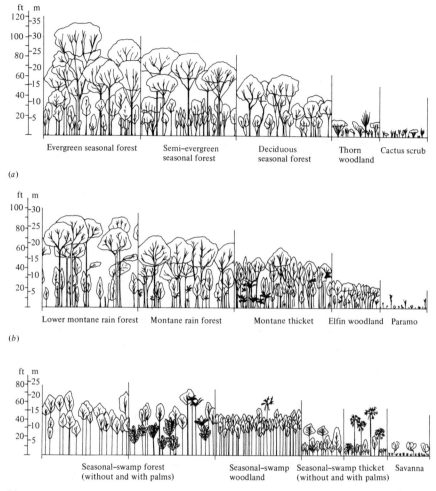

Figure 3.17   Vegetation profiles along three ecoclines. (*a*) A gradient of increasing aridity from seasonal rainforest to desert. (*b*) An elevation gradient up a tropical mountainside from tropical rainforest to alpine meadow (paramo). (*c*) A humidity gradient from swamp forest to savanna. [From Beard (1955).]

nual weeds such as crabgrass and asters, by broomsedge, by small perennials, by various shrubby species of perennials, by pine trees, and finally by oak and hickory trees (Figure 3.19). The composition of the avifauna and bird species diversity also change drastically with these successional changes in the vegetation and are part of the community succession. Plants in each stage modify the environment, presumably

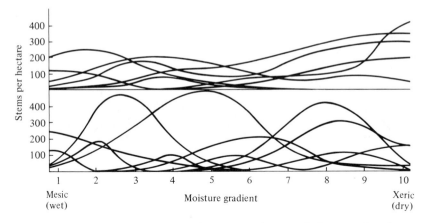

Figure 3.18 Actual distributions of some populations of plant species along moisture gradients from relatively wet ravines to dry southwest-facing slopes in the Siskiyou Mountains of northern California (above) and the Santa Catalina Mountains of Arizona (below). [After Whittaker (1967).]

making it more suitable for other species in following stages. Typically shade tolerance increases as succession proceeds. The entire process of secondary succession may take many years (over a hundred years in the above example). Only the oak–hickory forest is a stable community in a dynamic equilibrium that replaces itself; such a final stage in succession is termed its *climax*. In deserts, where the open vegetation alters microclimates very little and soil formation is virtually nonexistent, the first plants to invade are usually the climax species, and the succession, if one calls it such, is short. Earth's biomes represent the climax communities that prevail at different localities. Disturbances, both man made and natural (lightning, fires, droughts, landslides, hurricanes, floods), are often frequent enough that extensive areas have not had time enough to reach their own climax state. An equilibrium is reached whereby the proportion of a habitat supporting early successional stages is determined by the frequency of disturbance. Largely undisturbed areas may be primarily in the climax state. During the course of succession, annual production exceeds annual respiration, and organic materials accumulate to form soils and, generally, an increasingly larger biomass of plants and animals. At climax, production equals respiration and organic materials cease to accumulate (Figure 3.20).

An illuminating analysis of secondary succession was carried out by Horn (1975) for forests in the northeastern United States. Horn viewed these forests as honeycombs of independent "cells," each occupied by a particular tree. The percentages of saplings of different species observed under various species of canopy trees were used to construct a "transition matrix" (Table 3.2). Horn assumed that the proportional representation of various species of saplings under a particular canopy species reflected the probability of its replacement by that sapling species. Beginning with an observed census of the canopy species in a stand known to be 25 years old, Horn generated the composition of a predicted "climax" forest which compared reasonably closely to that actually observed in a 350-year-old forest.

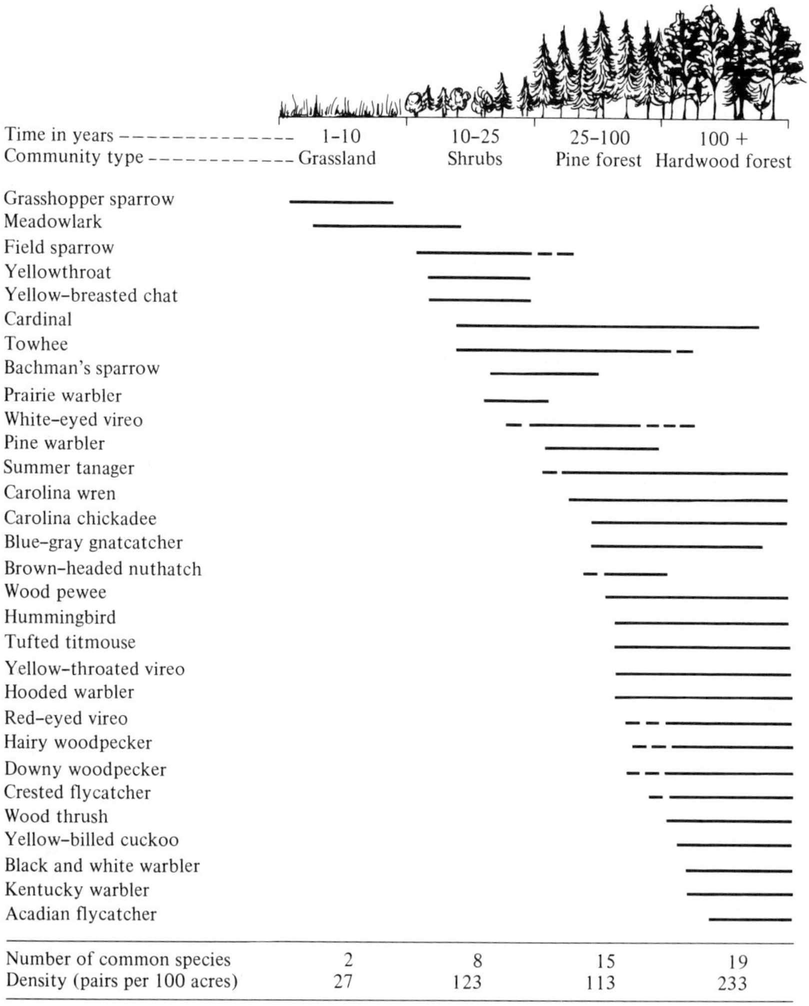

Time in years – – – – – – – – – – – – –	1–10	10–25	25–100	100 +
Community type – – – – – – – – – – – Grassland		Shrubs	Pine forest	Hardwood forest

Grasshopper sparrow				
Meadowlark				
Field sparrow				
Yellowthroat				
Yellow–breasted chat				
Cardinal				
Towhee				
Bachman's sparrow				
Prairie warbler				
White–eyed vireo				
Pine warbler				
Summer tanager				
Carolina wren				
Carolina chickadee				
Blue–gray gnatcatcher				
Brown–headed nuthatch				
Wood pewee				
Hummingbird				
Tufted titmouse				
Yellow–throated vireo				
Hooded warbler				
Red–eyed vireo				
Hairy woodpecker				
Downy woodpecker				
Crested flycatcher				
Wood thrush				
Yellow–billed cuckoo				
Black and white warbler				
Kentucky warbler				
Acadian flycatcher				

Number of common species	2	8	15	19
Density (pairs per 100 acres)	27	123	113	233

Figure 3.19 The typical pattern of secondary succession of vegetation and the avifauna on abandoned farmland in the southeastern United States. Number of bird species increases markedly with increased vertical structural complexity of the vegetation. [From MacArthur and Connell (1966) after Odum.]

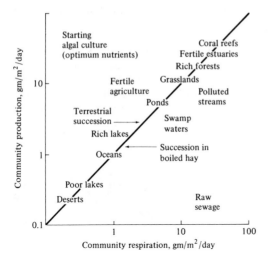

Figure 3.20 Total primary production of various communities plotted against the community's total respiration, in grams per square meter per day (proportional to cal/m²/day). Communities along the diagonal line are in equilibrium, with production equal to respiration. Production exceeds respiration (autotrophy) in those above the line, while respiration exceeds production (heterotrophy) in those below the line. [From Odum (1959) after H. T. Odum.]

Table 3.2 Transition Matrix for Institute Woods in Princeton: Percent Saplings Under Various Species of Trees

Canopy species	Sapling Species (%)											Total
	BTA	GB	SF	BG	SG	WO	OK	HI	TU	RM	BE	
Big-toothed aspen	**3**	5	9	6	6	—	2	4	2	60	3	104
Gray birch	—	—	47	12	8	2	8	0	3	17	3	837
Sassafras	3	1	**10**	3	6	3	10	12	—	37	15	68
Blackgum	1	1	3	**20**	9	1	7	6	10	25	17	80
Sweetgum	—	—	16	0	**31**	0	7	7	5	27	7	662
White oak	—	—	6	7	4	**10**	7	3	14	32	17	71
Red oak	—	—	2	11	7	6	**8**	8	8	33	17	266
Hickory	—	—	1	3	1	3	13	**4**	9	49	17	223
Tuliptree	—	—	2	4	4	—	11	7	**9**	29	34	81
Red maple	—	—	13	10	9	2	8	19	3	**13**	23	489
Beech	—	—	—	2	1	1	1	1	8	6	**80**	405

The number of saplings of each species listed in the row at the top, where the abbreviations are self-explanatory, is expressed as a percentage of the total number of saplings (last column) found under individuals of the species listed in the first column. The entries are interpreted as the percentages of individuals of species listed on the left that will be replaced one generation hence by species listed at the top. The percentage of "self-replacements" is shown in boldface.
The total number of recorded saplings is 3,286. A dash implies that no saplings of that species were found beneath that canopy; a zero, that the percentage was less than 0.5%.
The species are: BTA = *Populus grandidentata.* GB = *Betula populifolia*, SF = *Sassafras albidum,* BG = *Nyssa sylvatica*, SG = *Liquidambar styraciflua*, WO = *Quercus alba*, OK = Section *Erythrobalanus* of *Quercus,* HI = *Carya* spp., TU = *Liriodendron tulipifera,* RM = *Acer rubrum*, BE = *Fagus grandifolia.* *Source:* From Horn (1975b). Copyright © 1975 by the President and Fellows of Harvard College.

Classification of Natural Communities

Biotic communities have been classified in various ways. An early attempt to classify communities was that of Merriam (1890), who recognized a number of different "life zones" defined solely in terms of temperature (ignoring precipitation). His somewhat simplistic scheme is no longer used, but his approach did link climate with vegetation in a more or less predictive manner.

Shelford (1913a, 1963) and his students have taken a somewhat different approach to the classification of natural communities that does not attempt to correlate climate with the plants and animals occurring in an area. Rather, they classify different natural communities into a large number of so-called biomes and associations, relying largely upon the characteristic plant and animal species that compose a particular community. As such, this scheme is descriptive rather than predictive. Such massive descriptions of different communities (see, for example, Dice, 1952, and Shelford, 1963) can often be quite useful in that they allow one to become familiar with a particular community with relative ease.

Workers involved in such attempts at classification typically envision communities as discrete entities with relatively little or no intergradation between them; thus the Shelford school considers biomes to be distinct and real entities in nature rather than artificial and arbitrary human constructs. Another school of ecologists, represented by McIntosh (1967) and Whittaker (1970), takes an opposing view, emphasizing that communities grade gradually into one another and form so-called continua or ecoclines (Figures 3.17 and 3.18).

Two recent attempts to relate climate and vegetation are shown in Figures 3.21 and 3.22 (compare these with Figure 2.15). Whittaker (Figure 3.21) simply plots average annual precipitation against average annual temperature and fills in the vegetational formation typically occurring under such a climatic regime; thus he demonstrates how macroclimate determines the vegetation of an area—although he also emphasizes that these correlations are not hard and fast but that local vegetation type depends upon other factors such as soil types, seasonality of rainfall regime, and frequency of disturbance by fires and floods.

At first glance, Holdridge's scheme (Figure 3.22) appears to be much more complex and more sophisticated. But closer inspection reveals that there is really very little difference between the two. Figure 3.22 plots mean annual "biotemperature" (defined as simply mean annual temperature excluding temperatures below 0°C and above 30°C) against average annual precipitation (these two axes are labeled in boldface letters in the figure). There is a direct positive relationship between mean annual biotemperature and mean annual potential evapotranspiration (Holdridge, 1959), as indicated on the far right of Figure 3.22. Holdridge's "latitudinal regions" and "altitudinal belts"

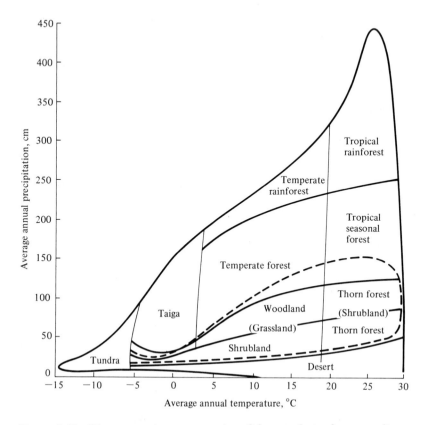

Figure 3.21 Diagrammatic representation of the correlation between climate, as reflected by average annual temperature and precipitation, and vegetational formation types. Boundaries between types are approximate and are influenced locally by soil type, seasonality of rainfall, and disturbances such as fire. The dashed line encloses a range of climates in which either grasslands or woody plants may constitute the prevailing vegetation of an area depending upon the seasonality of rainfall. Compare this figure with Figures 2.15 and 3.22. [After Whittaker (1970). Reprinted with permission of Macmillan Publishing Co., Inc., from *Communities and Ecosystems* by Robert H. Whittaker. © Copyright Robert H. Whittaker, 1970.]

are also entirely dependent upon mean annual biotemperature. The interaction of mean annual biotemperature and mean annual precipitation, in turn, determines the position on the other two axes, "potential evapotranspiration ratio" and "humidity provinces." Hence this elaborate figure actually plots only two truly *independent* variables.

These two figures neatly underscore the strong correlation between climate and vegetation mentioned at the beginning of this chapter.

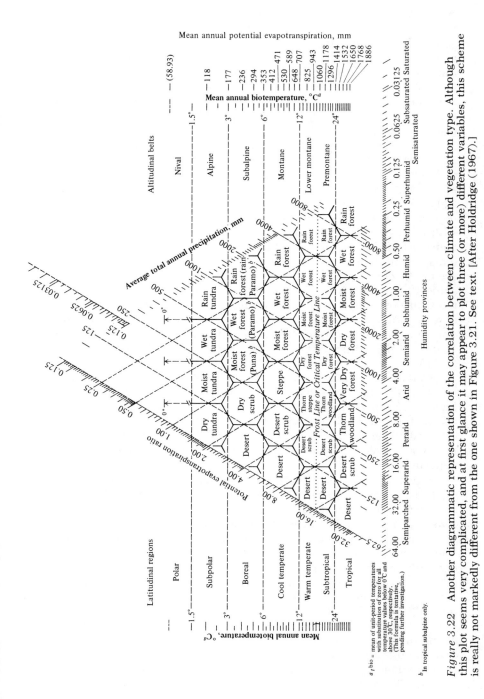

Figure 3.22 Another diagrammatic representation of the correlation between climate and vegetation type. Although this plot seems very complicated, and at first glance it may appear to plot three (or more) different variables, this scheme is really not markedly different from the one shown in Figure 3.21. See text. [After Holdridge (1967).]

Some Considerations of Aquatic Ecosystems

Although the same ecological principles presumably operate in both aquatic and terrestrial ecosystems, there are striking and interesting fundamental differences between these two ecological systems. For example, primary producers on land are sessile and many tend to be large and relatively long-lived (air does not provide much support and woody tissues are needed), while, except for kelp, producers in aquatic communities are typically free-floating, microscopic, and very short-lived (the buoyancy of water may make supportive plant tissues unnecessary; a large planktonic plant might be easily broken by water turbulence). Most ecologists study either aquatic or terrestrial systems, and various aquatic subdisciplines of ecology are recognized, such as aquatic ecology and marine ecology. Limnology is the study of freshwater ecosystems (ponds, lakes, and streams); oceanography is concerned with bodies of salt water. Because the preceding part of this chapter and most of the remainder of the book emphasize terrestrial ecosystems, certain salient properties of aquatic ecosystems, especially lakes, are briefly considered in this section. Lakes are particularly appealing subjects for ecological study in that they are self-contained ecosystems, discrete and largely isolated from other ecosystems. Nutrient flow into and out of a lake can often be estimated with relative ease. The study of lakes in fascinating; the interested reader is referred to Ruttner (1953), Hutchinson (1957b, 1967), Cole (1975), and/or Wetzel (1975).

Water has peculiar physical and chemical properties that strongly influence the organisms that live in it. As indicated earlier, water has a high specific heat; moreover, in the solid (frozen) state, its density is less than it is in the liquid state (that is, ice floats). Water is densest at 4°C and water at this temperature "sinks." Furthermore, water is nearly a "universal solvent" in that many important substances go into aqueous solution.

A typical, relatively deep lake in the temperate zones undergoes marked and very predictable seasonal changes in temperature. During the warm summer months its surface waters are heated up, and, because warm water is less dense than colder water, a distinct upper layer of warm water, termed the *epilimnion,* is formed (Figure 3.23a). (Movement of heat within a lake is due to water currents produced primarily by wind.) Deeper waters, termed the *hypolimnion,* remain relatively cold during summer, often at about 4°C; an intermediate layer of rapid temperature change, termed the *thermocline,* separates the epilimnion from the hypolimnion (Figure 3.23a). (A swimmer sometimes experiences these layers of different temperatures when diving into deep water or when in treading water his or her feet drop down into the cold hypolimnion.) A lake with a thermal profile, or bathythermograph, like that shown in Figure 3.23a is said to be "stratified" because of its layering of warm water over cold water.

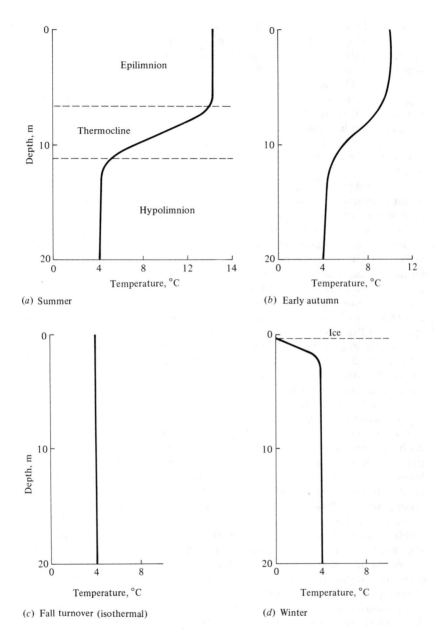

(a) Summer

(b) Early autumn

(c) Fall turnover (isothermal)

(d) Winter

Figure 3.23   Hypothetical bathythermographs showing seasonal changes typical of a deep temperate zone lake. (*a*) A stratified lake during summer. (*b*) In early autumn upper waters cool. (*c*) In late autumn or early winter the lake's waters are all at exactly the same temperature, here 4°C. (The lake is "isothermal.") (*d*) During the freezing winter months, a layer of surface ice chills the uppermost water.

With the decrease in incident solar energy in autumn, surface waters cool and give up their heat to adjacent landmasses and the atmosphere (Figure 3.23*b*). Eventually the epilimnion cools to the same temperature as the hypolimnion and the lake becomes isothermal (Figure 3.23*c*). This is the time of the "fall turnover." With winter's freezing temperatures, the lake's surface turns to ice and its temperature versus depth profile looks something like that in Figure 3.23*d*. Finally, in spring the ice melts and the lake is briefly isothermal once again (it may have a spring turnover) until its surface waters are rapidly warmed, when it again becomes stratified and the annual cycle repeats itself.

Because prevailing winds produce surface water currents, a lake's waters circulate. In stratified lakes, the epilimnion constitutes a more or less closed cell of circulating water, while the deep cold water scarcely moves or mixes with the warmer water above it. During this period, as dead organisms and particulate organic matter sink into the noncirculating hypolimnion, the lake undergoes what is known as summer stagnation. When a lake becomes isothermal, its entire water mass can be circulated and nutrient-rich bottom waters brought to the surface; limnologists thus speak of the spring and the fall "turnover." Meteorological conditions, particularly wind velocity and duration, strongly influence such turnovers; indeed, if there is little wind during the period a lake is isothermal, its waters might not be thoroughly mixed and many nutrients may remain locked in its depths. After a thorough turnover, the entire water mass of a lake is equalized and concentrations of various substances, such as oxygen and carbon dioxide, are similar throughout the lake.

Because lakes differ in their nutrient content and degree of productivity, they can be arranged along a continuum ranging from those with low nutrient levels and low productivity (*oligotrophic* lakes) to those with high nutrient content and high productivity (*eutrophic* lakes). Clear, cold, and deep lakes high in the mountains are usually relatively oligotrophic, whereas shallower, warmer, and more turbid lakes such as those in low-lying areas are generally more eutrophic. Oligotrophic lakes typically support game fish such as trout, while eutrophic lakes contain "trash" fish like carp. As they age and fill with sediments, many lakes gradually undergo a natural process of eutrophication, steadily becoming more and more productive. Mankind accelerates this process by enriching lakes with wastes, and many oligotrophic lakes have rapidly become eutrophic under our influence. A good indicator of the degree of eutrophication is the oxygen content of deep water during summer. In a relatively unproductive lake, oxygen content varies little with depth and there is ample oxygen at the bottom of the lake. In contrast, oxygen content diminishes rapidly with depth in productive lakes, and anaerobic processes sometimes characterize their depths during the summer months. With the autumn turnover, oxygen-rich waters again reach the bottom sediments and aerobic processes become possible. However, once such a lake becomes stratified, the oxygen in its deep water is quickly used up by benthic organisms (in the deep water there is little or no photosynthesis to replenish the oxygen).

These seasonal physical changes profoundly influence the community of organisms living in a lake. During the early spring and after the fall turnover, surface waters are rich in dissolved nutrients such as nitrates and phosphates and temperate lakes are very productive, whereas during the middle of summer, many nutrients are unavailable to phytoplankton in the upper waters and primary production is much reduced.

Organisms within a lake community are usually distributed quite predictably in time and space. Thus there is typically a regular seasonal progression of planktonic algae, with diatoms most abundant in the winter, changing to desmids and green algae in the spring, and gradually giving way to blue-green algae during the summer months. The composition of the zooplankton also varies seasonally. Such temporal heterogeneity may well promote a diverse plankton community by periodically altering competitive abilities of component species, hence facilitating coexistence of many species of plants and animals in the relatively homogeneous planktonic environment (Hutchinson, 1961). Although plankton are moved about by water currents, many are strong enough swimmers to select a particular depth. Such species often actually "migrate" vertically during the day and/or with the seasons, being found at characteristic depths at any given time (Figure 3.24).

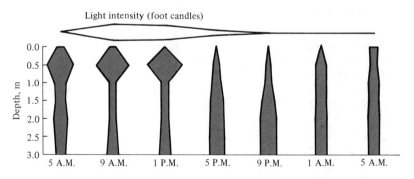

Figure 3.24 Many freshwater planktonic animals move vertically during the daily cycle of illumination somewhat as shown here (widths of bands represent the density of animals at a given depth at a particular time). [After Hutchinson (1967) after Cowles and Brambel.]

Selected References

Allee *et al.* (1949); Andrewartha and Birch (1954); Clapham (1973); Clarke (1954); Colinvaux (1973); Collier *et al.* (1973); Daubenmire (1947, 1956, 1968); Kendeigh (1961); Knight (1965); Kormondy (1969); Krebs (1972); Lowry (1969); Odum (1959, 1971); Oosting (1958); Ricklefs (1973); Smith (1966); Watt (1973); Weaver and Clements (1938); Whittaker (1970).

Plant Life Forms and Biomes

Cain (1950); Clapham (1973); Givnish and Vermeij (1976); Horn (1971); Raunkaier (1934); Whittaker (1970, 1975).

Microclimate

Collier *et al.* (1973); Gates (1962); Geiger (1966); Lowry (1969); Schmidt-Nielsen (1964); Smith (1966).

Primary Production and Evapotranspiration

Collier *et al.* (1973); Gates (1965); Horn (1971); Meyer, Anderson, and Bohning (1960); Odum (1959, 1971); Rosenzweig (1968); Whittaker (1970, 1975); Woodwell and Whittaker (1968).

Leaf Tactics

Bailey and Sinnott (1916); Clements (1904); Esser (1946a, 1946b); Gentry (1969); Givnish and Vermeij (1976); Horn (1971, 1975a, 1975b, 1976); Howland (1962); Janzen (1976); Miller (1977); Opatowski (1945, 1946); Orians and Solbrig (1977); Parkhurst and Loucks (1971); Ryder (1954); Vogel (1970).

Soil Formation and Primary Succession

Black (1968); Burges and Raw (1967); Crocker (1952); Crocker and Major (1955); Doeksen and van der Drift (1963); Eyre (1963); Fried and Broeshart (1967); Jenny (1941); Joffe (1949); Oosting (1958); Richards (1974); Schaller (1968); Waksman (1952); Whittaker, Walker, and Kruckeberg (1954).

Ecotones, Vegetational Continua, and Secondary Succession

Clements (1920, 1949); Horn (1971, 1975a, 1975b, 1976); Kershaw (1964); Loucks (1970); Margalef (1958b); McIntosh (1967); Pickett (1976); Shimwell (1971); Terborgh (1971); Whittaker (1953, 1965, 1967, 1969, 1970, 1972).

Classification of Natural Communities

Braun-Blanquet (1932); Clapham (1973); Dice (1952); Gleason and Cronquist (1964); Holdridge (1947, 1959, 1967); McIntosh (1967); Merriam (1890); Shelford (1913a, 1963); Tosi (1964); Whittaker (1962, 1967, 1970, 1975).

Aquatic Ecosystems

Clapham (1973); Cole (1975); Ford and Hazen (1972); Frank (1968); Frey (1963); Grice and Hart (1962); Henderson (1913); Hochachka and Somero (1973); Hutchinson (1951, 1957b, 1961, 1967); Mann (1969); National Academy of Sciences (1969); Perkins (1974); Russell-Hunter (1970); Ruttner (1953); Sverdrup *et al.* (1942); Watt (1973); Welch (1952); Wetzel (1975); Weyl (1970).

Physiological Ecology

4

The concern of environmental physiology, or ecophysiology, is *how* organisms function within, adapt and respond to, and exploit their physical environments. Physiological ecologists are interested primarily in the immediate functional and behavioral mechanisms by which organisms cope with their abiotic environments. Physiological mechanisms clearly must reflect ecological conditions; moreover, mutual constraints between physiology and ecology dictate that both must evolve together in a synergistic fashion.

A fundamental principle of physiology is the notion of *homeostasis*, the maintenance of a relatively stable internal state under a much wider range of external environmental conditions. Homeostasis is achieved not only by physiological means but also by appropriate behavioral responses. An example is temperature regulation in which an organism maintains a fairly constant body temperature over a considerably greater range of ambient thermal conditions (homeostasis is never perfect). Homeostatic mechanisms have also evolved that buffer environmental variation in humidity, light intensity, and concentrations of various substances such as hydrogen ions (pH), salts, and so on. By effectively moderating spatial and temporal variation in the physical environment, homeostasis allows organisms to persist and be active within a broad range of environmental conditions, thereby enhancing their fitness. The subject of environmental physiology is vast; see references at the end of this chapter.

Physiological Optima and Tolerance Curves

Physiological processes proceed at different rates under different conditions. Most, such as rate of movement and photosynthesis, are temperature dependent (Figure 4.1). Other processes vary with availability of various materials such as water, carbon dioxide, nitrogen, and hydrogen ions (pH). Curves of performance, known as *tolerance curves* (Shelford, 1913b), are typically bell shaped and unimodal, with their peaks representing optimal conditions for a particular physiological process and their tails reflecting the *limits of tolerance*. Some individuals and species have very narrow peaked tolerance curves; in others these curves are considerably broader. Broad tolerance curves are described with the prefix eury- (eurythermic, euryhaline, etc.), whereas steno- is used for narrow ones (stenophagous, etc.). An organism's use of environmental resources such as foods and microhabitats can profitably be viewed similarly, and performance can be measured in a wide variety of units such as survivorship, reproductive success, foraging efficiency, and fitness.

Performance curves can sometimes be altered during the lifetime of an individual as it becomes exposed to different ambient external conditions. Such short-term alteration of physiological optima is known as *acclimation* or acclimatization (Figure 4.1). Within certain design constraints (Brown and Feldmeth, 1971), tolerance curves clearly must change over evolutionary time as natural selection molds them to reflect changing environmental conditions. However, very little is known about the evolution of tolerance; most researchers have merely described the range(s) of conditions tolerated or exploited by particular organisms. Tolerance curves are often taken almost as given and im-

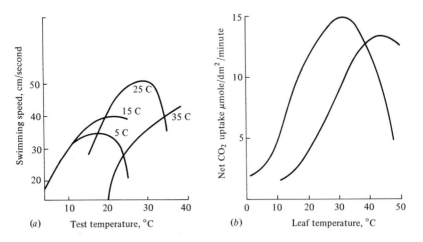

Figure 4.1   Two plots of performance against temperature. (*a*) Goldfish swimming speed versus temperature for fish acclimated to different thermal conditions; in most cases performance peaks near the temperature to which fish are accustomed. [From Ricklefs (1973) after Fry and Hart (1948).] (*b*) Photosynthetic rate versus leaf temperature in the plant *Atriplex lentiformis* at two different localities. [From Mooney, Björkman, and Berry (1975).]

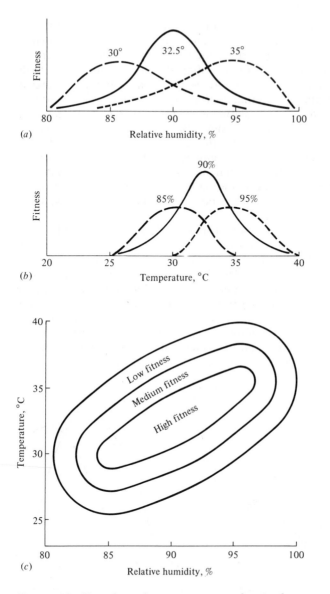

(a) Relative humidity, %

(b) Temperature, °C

(c) Relative humidity, %

Figure 4.2 Hypothetical response curves showing how two variables can interact to determine an organism's fitness. Fitness is reduced at extremes of either temperature or humidity, and the range of humidities tolerated is less at extreme temperatures than at intermediate ones.

mutable, with little or no consideration of the ecological and evolutionary forces that shape them.

Performance or tolerance is often sensitive to two or more environmental variables. For example, the fitness of a hypothetical organism in various microhabitats might be a function of relative humidity (or vapor pressure deficit), somewhat as shown in Figure 4.2a. Assume that fitness varies similarly along a temperature gradient (Figure 4.2b). Figure 4.2c combines humidity and temperature conditions to show variation in fitness with respect to both variables simultaneously (a third axis, fitness, is implicit in this figure). The range of thermal conditions tolerated is narrower at very low and very high humidities than it is at intermediate and more optimal humidities. Similarly, an organism's tolerance range for relative humidity is narrower at extreme temperatures than it is at more optimal ones. The organism's thermal optimum depends upon humidity conditions (and vice versa). Fitness reaches its maximum at intermediate temperatures and humidities. Hence temperature tolerance and tolerance of relative humidities interact in this example. The concept of a single fixed optimum is in some ways an artifact of considering only one environmental dimension at a time.

Energetics of Metabolism and Movement

The *total amount of energy* needed per unit time for maintenance increases with increasing body weight (Figure 4.3). However, because small animals have relatively high ratios of body surface to body volume, they generally have much higher metabolic rates and hence greater energy requirements *per unit of body weight* than do larger animals (Figure 4.4). Animals that maintain relatively constant internal body temperatures are known as *homeotherms;* those whose temperatures vary widely from time to time, usually approximating the temperature of their immediate environment, are called *poikilotherms.* These two terms are sometimes confused with a related pair of useful terms. An organism that obtains its heat from its external environment is an *ectotherm;* one that produces most of its own heat internally by means of oxidative metabolism is known as an *endotherm.* All plants and the vast majority of animals are ectothermic; the only continuously endothermic animals are found among birds and mammals (but even many of these become ectothermic at times). Some poikilotherms (large reptiles and certain large fast-swimming fish such as tuna) are at times at least partially endothermic. Certain ectotherms (many lizards and temperate-zone flying insects) actually regulate their body temperatures fairly precisely during periods of activity by appropriate behavioral means: thus ectotherms at times can be homeotherms! An active bumblebee or desert lizard may have a

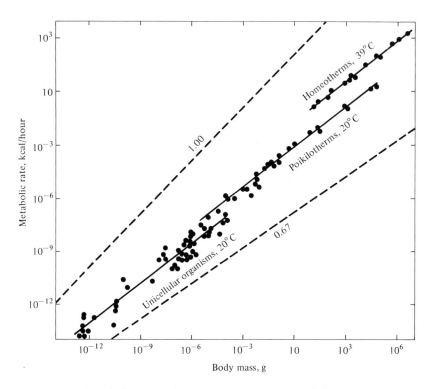

Figure 4.3 Metabolic rates of a variety of organisms of different sizes (log-log plot). Total oxygen consumption increases with increasing body size. [From Schmidt-Nielsen (1975).]

body temperature as high as that of a bird or mammal (the layman's terms "warm-blooded" and "cold-blooded" can thus be quite misleading and should be abandoned). Because energy is required to maintain a constant internal body temperature, endotherms have higher metabolic rates, as well as higher energy needs (and budgets), than ectotherms of the same body weight. There is a distinct lower limit on body size for endotherms—about the size of a small hummingbird or shrew (2–3 grams). Indeed, both hummingbirds and shrews have very high metabolic rates and hence rather precarious energetic relationships; they depend upon continual supplies of energy-rich foods. Small hummingbirds would starve to death during cold nights if they did not allow their body temperatures to drop and go into a state of torpor.

Body size, diet, and movements are complexly intertwined with the energetics of metabolism. Large animals require more matter and energy for their maintenance than small ones, and in order to obtain it they usually must range over larger geographical areas than smaller animals with otherwise similar food requirements. Food habits also influence movements and home range size. Because the foods of herbivorous animals that eat the green parts of plants (such as grazers, which eat grasses and ground-level vegetation, and browsers, which eat tree leaves) are usually quite dense, these animals usually do not have very large home ranges. In contrast, carnivores and those herbivores that must search for their foods (such as granivores and frugivores, which eat seeds and fruits, respectively) frequently spend much of their foraging time and energy in search, and they range over considerably larger areas. McNab (1963) termed the first group "croppers" and the latter "hunters." Croppers generally exploit foods which occur in relatively dense supply, while hunters typically utilize less dense foods. Hunters are often territorial, but croppers seldom defend territories. Croppers and hunters are not discrete but in fact grade into each other (Figure 4.5). A browser that eats only the leaves of a rare tree might be more of a hunter than a granivore that eats the seeds of a very common plant. However, such intermediates are uncommon enough that separation into two categories is useful for many purposes. Figure 4.6 shows the correlation between home range size and body weight for a variety of mammalian species, here separated into croppers and hunters. Analogous correlations, but with different slopes

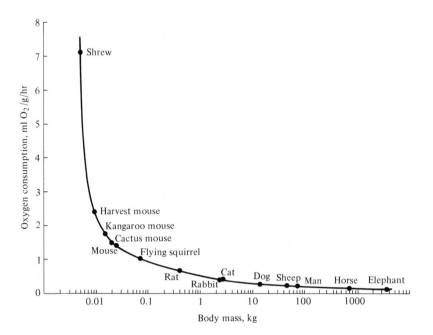

Figure 4.4 Semilogarithmic plot of rates of oxygen consumption *per unit body mass* for a wide variety of mammals plotted against body mass. [From Schmidt-Nielsen (1975).]

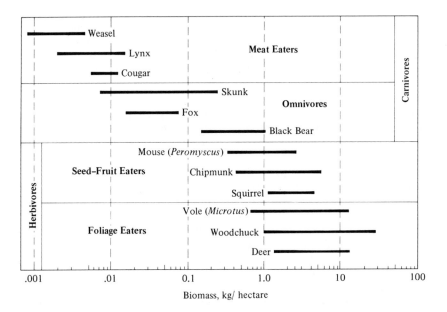

Figure 4.5   Biomass, in kilograms/hectare, of various mammals, arranged according to their food habits. Although mammals as a class vary over five orders of magnitude, the range among those that eat any given food is considerably smaller. Meat eaters and omnivores are much less dense than herbivores. [From Odum (1959) after Mohr (1940).]

and/or intercepts, have been obtained for birds and lizards (Schoener, 1968b; Turner, Jennrich, and Weintraub, 1969). Very mobile animals, like birds, frequently range over larger areas than less mobile animals such as terrestrial mammals and lizards. In areas of low productivity such as deserts, most animals may be forced to range over a greater area to find adequate food than they would be in more productive regions. Large home ranges or territories usually result in low densities, which in turn markedly limit possibilities for the evolution of sociality. Thus McNab (1963) points out that complex social behavior has usually evolved only in croppers and/or exceptionally mobile hunters.

The metabolic cost of movement varies with both an animal's body size and its mode of locomotion. The cost of moving a unit of body weight some standard distance is actually less in larger animals than in smaller ones (Figure 4.7). Terrestrial locomotion is the most expensive mode of transportation, flight is intermediate in cost, and swimming is the most economical means of moving about—provided body shape is fusiform and bouyancy is neutral (Figure 4.7).

Physiologists have documented numerous consistent size-related trends in organs and metabolic properties. For example, among mam-

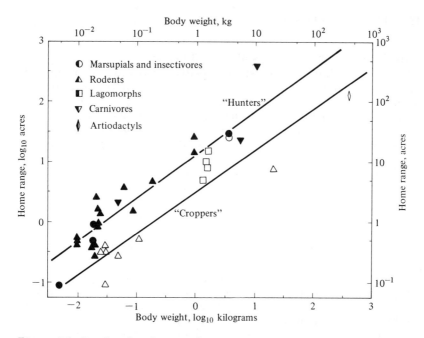

Figure 4.6 Log-log plot of average home range area against mean body weight for a variety of mammals, separated into croppers (open symbols and lower regression line) and hunters (closed symbols and upper line). [After McNab (1963).]

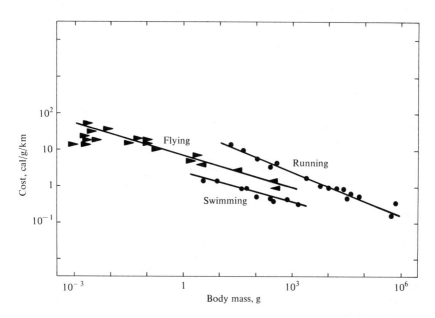

Figure 4.7 Comparison of the energetic cost of moving a unit of body weight one kilometer for different modes of locomotion. [From Schmidt-Nielsen (1975).]

mals, heart weight is always about 0.6 percent of total body weight while blood volume is almost universally about 5.5 percent of body mass, over a great range of body sizes (these organ systems are thus directly proportional to size). Other physiological attributes, such as lung surface in mammals, vary directly with metabolic rate rather than size. Some organ systems, however, such as the kidney and liver, are not scaled directly either to size or metabolic rate (Schmidt-Nielsen, 1975). Such "physiological rules" apparently dictate available avenues for physiological change, thereby constraining possible ecological adaptations.

Energy Budgets and the Principle of Allocation

Any organism has a certain limited amount of energy available to devote to growth, maintenance, and reproduction. The way in which an organism allocates this energy and other resources among various conflicting demands is of fundamental interest because such apportionments provide insight into how the organism copes with its environment. Moreover, since any individual has a finite energy budget, its capacity for regulation and homeostasis is necessarily limited. Organisms stressed along any one environmental variable are thus able to tolerate a lesser range of conditions along other environmental variables (see Figure 4.2). Various tolerance and performance curves are presumably subject to certain constraints (Levins, 1968); for example, their breadth (variance) usually cannot be increased without a simultaneous reduction in their height (and vice versa). This useful notion of tradeoffs, known as the *principle of allocation,* is helpful in interpreting and understanding numerous ecological phenomena.

As an example of allocation, imagine an animal of a given size and mouthpart anatomy. A certain size of prey item is optimal while other prey are suboptimal because they are either too large or too small for efficient capture and swallowing. Any given animal has its own "utilization curve" which indicates the actual numbers of prey of different sizes taken per unit time under particular environmental conditions. In an idealized, perfectly stable, and infinitely productive environment, a utilization curve might become a spike with no variance, with the organism using only its most optimal prey resource type. In actuality, limited and changing availabilities of resources, both in time and in space, result in utilization curves with breadth as well as height. In terms of the principle of allocation, an individual with a generalized diet adapted to eat prey of a wide range of available sizes presumably is not as effective at exploiting prey of intermediate size as another, more specialized, feeder. In other words, a jack-of-many-trades is a master of none. We consider this subject in more detail in Chapter 7 (pp. 257–260).

Adaptation and Deterioration of Environment

Organisms are adapted to their environments in that, to survive and reproduce, they must meet their environment's conditions for existence. Adaptation can be defined as *conformity between the organism and its environment*. Plants and animals have adapted to their environments both genetically and by means of physiological, behavioral, and/or developmental flexibility. The former includes instinctive behavior and the latter learning. Adaptation has many dimensions in that most organisms must conform simultaneously to numerous different aspects of their environments. Thus an organism, to be adapted, must cope not only with various aspects of its physical environment, such as temperature and humidity conditions, but also with competitors, predators, and escape tactics of its prey. Conflicting demands of these various environmental components often require that an organism compromise in its adaptations to each. Conformity to any given component takes a certain amount of energy that is then no longer available for other adaptations. The presence of predators, for example, may require that an animal be wary, which in turn is likely to reduce its foraging efficiency and hence its competitive ability.

Organisms can conform to and cope with highly predictable environments relatively easily, even when they change in a regular way, as long as they are not too extreme. Adaptation to an unpredictable environment may usually be much more difficult; adapting to extremely erratic environments may even prove impossible. Many organisms have evolved dormant stages that allow them to survive unfavorable periods, both predictable and unpredictable. Annual plants everywhere and brine shrimp in deserts are good examples. Brine shrimp eggs survive for years in the salty crust of dry desert lakes; when a rare desert rain fills one of these lakes, the eggs hatch, the shrimp grow rapidly to adults, and they produce many eggs. Some seeds known to be hundreds of years old are still viable and have been germinated. Changes in the environment that reduce overall adaptation are collectively termed the "deterioration of environment"; such changes cause directional selection resulting in accomodation to the new environment.

A simple model of adaptation and undirected environmental deterioration was developed by Fisher (1930). He reasoned that no organism is "perfectly adapted," but that all fail to conform to their environments in some ways and to differing degrees. However, a hypothetical, perfectly adapted organism can always be imagined (actually this reflects the environment), against which existing organisms may be compared. Fisher's mathematical argument is phrased in terms of an infinite number of "dimensions" for adaptation (only three are used here for ease of illustration). Imagine an adaptational space of three coordinates representing, respectively, the competitive, predatory, and physical environments (Figure 4.8). An ideal "perfectly adapted" or-

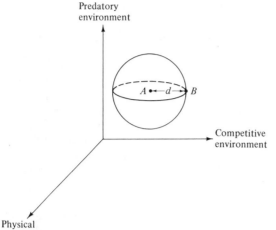

Predatory
environment

Competitive
environment

Physical
environment

Figure 4.8 Fisher's model of adaptation and
deterioration of environment. Point *A* represents a
hypothetical "perfectly adapted" organism; an actual
organism (point *B*) is never perfectly adapted and thus lies
at some distance, *d*, from point *A*. The surface of the
sphere represents all possible points with a level of
adaptation equal to the organism under consideration.
Very small undirected changes in either organism (point
B) or environment (point *A*) are equally likely to increase
or to decrease the level of adaptation, *d*.

ganism lies at a particular point (say *A*) in this space, but any given real
organism is at another point (say *B*), some distance, *d*, away from the
point of perfect adaptation. Changes in the position of *A* correspond to
environmental changes making the optimally adapted organism different; changes in *B* represent changes in the organisms concerned, such
as mutations. The distance between the two points, *d*, represents the
degree of conformity between organism and environment, or the level
of adaptation. Fisher noted that very small *undirected* changes in
either organism or environment have a 50:50 chance of being to the
organism's advantage (that is, of reducing the distance between *A* and
B). The probability of such improvement is inversely related to the
magnitude of the change (Figure 4.9). Very great changes in either
organism or environment are always maladaptive because even if they
are in the correct direction, they "overshoot" points of closer adaptation. (Of course, it is remotely possible that such major environmental
changes or "macromutations" could put an organism into a completely
new adaptive realm and thereby improve its overall level of adaptation.) Fisher makes an analogy with focusing a microscope: very fine
changes are as likely as not to improve the focus, but gross changes will
almost invariably throw the machine further out of focus. Organisms

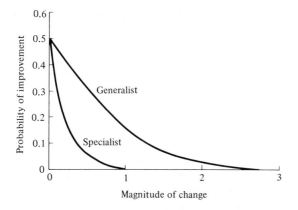

Figure 4.9   The probability of improvement of the level of adaptation (that is, of reducing d) is plotted against the magnitude of an undirected change when the number of dimensions is large. Two hypothetical organisms are shown, one highly adapted such as a specialist with narrow tolerance limits and one less highly adapted such as a generalist with broader tolerance limits and/or a greater number of niche dimensions. A random change of a given magnitude is more likely to improve the level of adaptation of the generalist than the specialist. [Partially adapted from Fisher (1958a).]

may be thought of as "tracking" their environments in both ecological and evolutionary time, changing as their environments change; thus, as point A shifts because of daily, seasonal, and long-term environmental fluctuations, point B follows it. Such environmental tracking may be physiological (as in acclimation), behavioral (including learning), and/or genetic (evolutionary), depending on the time scale of environmental change.

Individual organisms with narrow tolerance limits, such as highly adapted specialists, generally suffer greater losses in fitness due to a unit of environmental deterioration than do generalized organisms with more versatile requirements, all else being equal. Thus more specialized organisms and/or those with restricted homeostatic abilities cannot tolerate as much environmental change as generalists or organisms with better developed homeostasis (Figure 4.9). Better adapted organisms (smaller d) and/or those that are adapted to their environments in a greater number of dimensions can tolerate greater changes than those which are less well adapted (greater d) and/or those adapted to fewer dimensions. Fisher's model applies only to *non-directed* changes in either party of the adaptational complex—such as mutations and perhaps certain climatic fluctuations, or other random events. However, many environmental changes are probably nonrandom. Changes in other associated organisms, especially predators and prey, are invariably directed so as to reduce an organism's degree of conformity to its environment; thus they constitute a deterioration of that organism's environment (Chapter 6). Directed changes in competitors can either increase or decrease an organism's level of adaptation, depending upon whether they involve avoidance of competition or improvements in competitive ability per se.

Heat Budgets and Thermal Ecology

When averaged over a long enough period of time, heat gained by an organism must be exactly balanced by heat lost to its environment; otherwise the plant or animal would either warm up or cool off. Many different pathways of heat gains and heat losses exist (Figure 4.10).

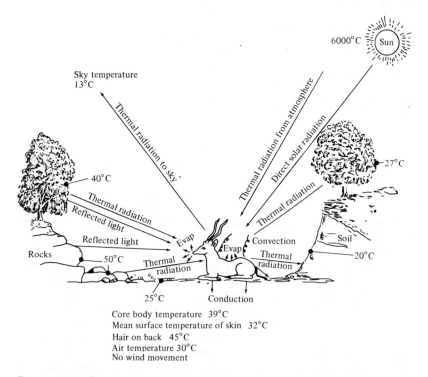

Figure 4.10   Diagrammatic representation of various pathways of heat gains and losses between an animal and its environment under moderately warm conditions. Heat exchange between an animal and its environment is roughly proportional to its body surface: because small animals have a larger surface area relative to their weight than larger ones, the former gain or lose relatively more heat than the latter. [From Bartholomew (1972). In M. S. Gordon (ed.), *Animal Physiology: Principles and Adaptations.* © 1972 by The Macmillan Co., Inc.]

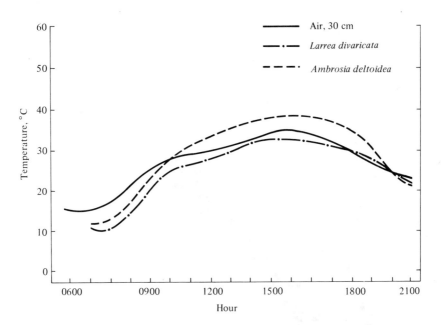

Figure 4.11   Average temperatures of leaves of *Larrea divaricata* and *Ambrosia deltoidea* during three sunny days in June with slight breezes. [From Patten and Smith (1975).]

The notion of a heat budget is closely related to the concept of an energy budget; balancing a heat budget requires very different adaptations under varying environmental conditions. At different times of day, ambient thermal conditions may change from being too cold to being too warm for a particular organism's optimal performance. Organisms living in hot deserts must avoid overheating by being able to minimize heat loads and to dissipate heat efficiently; in contrast, those that live in colder places such as at high altitudes or in polar regions must avoid overcooling—and they have hence evolved efficient means of heat retention, such as insulation by blubber, feathers, or fur which reduces the rate of heat exchange with the external environment.

As we have seen in previous chapters, environmental temperatures fluctuate in characteristic ways at different places over Earth's surface, both daily and seasonally. In the absence of a long-term warming or cooling trend, environmental temperatures at any given spot remain roughly constant when averaged over an entire annual cycle. Recall that the range in temperature within a year is much greater at high latitudes than it is nearer the equator (see Table 2.1). An organism could balance its annual heat budget by being entirely passive and simply allowing its temperature to mirror that of its environment. Such a passive thermoregulator is known as a thermoconformer (Figures 4.11 and 4.12). Of course it is also an ectotherm. Another extreme

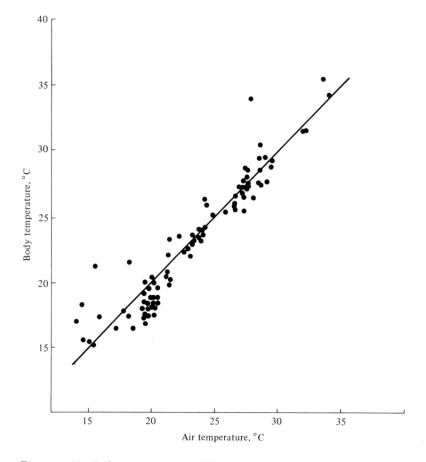

Figure 4.12   Body temperatures of 86 active Australian nocturnal gekkonid lizards *(Nephrurus laevissimus)* plotted against ambient air temperature. The line represents body temperatures equal to air temperature.

would be to maintain an absolutely constant body temperature by physiological and/or behavioral means, dissipating (or avoiding) excess bodily heat during warm periods but retaining (or gaining) heat during cooler periods (in endotherms, energy intake is often increased during cold periods and more metabolic heat is produced to offset the increased heat losses). Organisms that carefully regulate their internal temperatures are called regulators, or homeotherms. (Recall that both endotherms and ectotherms may regulate their body temperatures.) There is of course a continuum between the two extremes of perfect conformity and perfect regulation. Homeostasis, remember, is never perfect. Because regulation clearly has costs and risks as well as profits, an emerging conceptual framework envisions an optimal level of regulation that depends on the precise form of the constraints and interactions among costs and benefits arising from a particular ecological situation (Huey and Slatkin, 1976). Thermoregulation often involves both physiological and behavioral adjustments; as an example of the latter, consider a typical terrestrial diurnal desert lizard. During the early morning, when ambient temperatures are low, such a lizard locates itself in warmer microclimates of the environmental thermal mosaic (e.g., small depressions in the open or tree trunks), basking in the sun with its body as perpendicular as possible to the sun's rays and thereby maximizing heat gained. With the daily march of temperature, ambient thermal conditions quickly rise and the lizard seeks cooler shady microhabitats. Individuals of some species retreat into burrows as temperatures rise; others climb up off the ground into cooler air and orient themselves facing into the sun's rays, thereby reducing heat load. Such behavioral adjustments allow individual lizards to be active over a longer period of time than they could be if they conformed passively to ambient thermal conditions; presumably they are also more effective competitors and are better able to elude predators.

Hot arid regions typically support rich lizard faunas, whereas cooler forested areas have many fewer lizard species and individuals. By facilitating metabolic inactivity on both a daily and a seasonal basis, poikilothermy allows lizards to capitalize on unpredictable food supplies. Moreover, most endothermic diurnal birds and mammals must wait out the hot midday period at considerable metabolic cost, whereas lizards can effectively reduce temporal heterogeneity by retreating underground, becoming inactive, and lowering their metabolic rate during harsh periods (some desert rodents estivate when food and/or water is in short supply). Poikilothermy may well contribute to the apparent relative success of lizards over birds and mammals in arid regions. Forests and grasslands are probably too shady and too cold for ectothermic lizards to be very successful because these animals depend on basking to reach body temperatures high enough for activity; birds and mammals, in contrast, do quite well in such areas partly because of their endothermy.

Water Economy in Desert Organisms

Because water conservation is a major problem for desert organisms, their physiological and behavioral adaptations for acquisition of water and for economy of its use have been well studied. These interesting adaptations are quite varied. Like energy and heat budgets, water budgets must balance: losses must be replaced by gains. For examples of water acquisition mechanisms, consider rooting strategies. Desert plants may usually invest considerably more in root systems than plants from wetter areas; one study showed that perennial shrubs in the Great Basin desert allocate nearly 90 percent of their biomass to underground tissues (Caldwell and Fernandez, 1975), whereas roots apparently represent a much smaller fraction (only about 10 percent) of the standing crop biomass of a mesic hardwood forest. The creosote bush *Larrea divaricata* has both a surface root system and an extremely deep tap root that often reaches all the way to the water table. This long tap root provides *Larrea* with water even during long dry spells when surface soils contain little moisture. Cacti, in contrast, have an extensive but relatively shallow root system and rely on water storage to survive drought. Many such desert plants have tough sclerophyllous xerophytic leaves which do not allow much water to escape (they also photosynthesize at a low rate as a consequence). Mesophytic plants occur in deserts, too, but photosynthesize rapidly and grow only during periods when moisture is relatively available; they drop their leaves and become inactive during droughts (see pp. 54–55). Plants also reduce water losses during the heat of midday by closing their stomata and drooping their leaves (wilting). Many desert plants and animals absorb and use atmospheric and/or substrate moisture; most can also tolerate extreme desiccation. Camels do not rely on water storage to survive water deprivation, as is commonly thought, but can lose as much as a quarter of their body weight, primarily as water loss (Schmidt-Nielsen, 1964). Like many desert organisms, camels also conserve water by allowing their temperature to rise during midday. In deserts, small mammals like kangaroo rats survive without drinking by relying on metabolic water derived from the oxidation of their food (here, then, is an interface between energy budgets and water budgets). Most desert rodents are nocturnal and avoid the use of water for heat regulation by spending hot daytime hours underground in cool burrows with high relative humidity, thereby minimizing losses to evaporation (most desert organisms resort to evaporative cooling mechanisms such as panting only in emergencies). Urine of kangaroo rats is extremely concentrated, and their feces contain little water (Schmidt-Nielsen, 1964). Most other desert animals minimize water losses in excretion too. Birds and lizards produce solid uric acid wastes rather than urea, thereby requiring little water for excretion. Desert lizards also conserve water by retreating to burrows and lowering their metabolic rate during the heat of the day.

Other Limiting Materials

Numerous other materials, including calcium, chloride, magnesium, nitrogen, potassium, and sodium, may be in short supply for particular organisms and must therefore be budgeted. Neural mechanisms of animals depend on sodium, potassium, and chloride ions, which are sometimes available in limited quantities. Because many herbivorous mammals obtain little sodium from their plant foods (plants lack nerves and sodium is not essential to their physiology), these animals must conserve sodium and/or find supplemental sources at salt licks— indeed, Feeny (1975) suggests that plants may actually withhold sodium as an antiherbivore tactic! Similarly, amino acids are in short supply for many insects, as in *Heliconius* butterflies, which supplement their diets with protein-rich pollen (Gilbert, 1972).

An organism's nutrient and vitamin requirements are strongly influenced by the evolution of its metabolic pathways; likewise, these same pathways may themselves determine certain of the organism's nutritional needs. To illustrate: The vast majority of species of vertebrates synthesize their own ascorbic acid, but humans and several other primate species that have been tested cannot; they require a dietary supplement of ascorbic acid, known as vitamin C. Of thousands of other species of mammals, only two—the guinea pig and an Indian fruit-eating bat—are known to have lost the ability to synthesize their own ascorbic acid. A few species of birds must supplement their diets with ascorbic acid, too. Thus a frog, a lizard, a sparrow, or a rat can make its own ascorbic acid, but we cannot. Why should natural selection favor the loss of the ability to produce a vital material? Pauling (1970) suggests that species of animals which have lost this capacity evolved in environments with ample supplies of ascorbic acid in available foods. It might actually be advantageous to dismantle a biochemical pathway in favor of another once it becomes redundant. Conversely, natural selection should favor evolution of the ability to synthesize any necessary materials that cannot be predictably obtained from available foods where this is possible (clearly organisms cannot synthesize elements: herbivores cannot make sodium).

Sensory Capacities and Environmental Cues

Animals vary tremendously in their perceptive abilities. Most (except some cave dwellers and deep sea forms) use light to perceive their environments. But visual spectra and acuity vary greatly. Some, such as insects, lizards, and birds, have color vision, while others (most mammals, except squirrels and primates) do not. Ants, bees, and some birds can detect polarized light and navigate by the sun's position; pigeons have poorly understood back-up systems that allow them to return home remarkably well even when their vision is severely impaired with opaque contact lenses. Many temperate zone species of plants and animals rely on changes in daylength to anticipate seasonal changes in climatic conditions (some may also use barometric pressure). Certain snakes, such as pit vipers and boids, have infrared receptors which allow them to locate and capture endothermic prey in total darkness. Most animals can hear, of course, although response to different frequencies varies considerably (some actually perceive ultrasonic sounds). Bats and porpoises emit and exploit sonar signals to navigate by echolocation. Similarly, nocturnal electric fish perceive their immediate environments by means of self-generated electrical fields. Recent work even suggests that some birds can detect magnetic fields. Although a few animals have only a relatively feeble sense of smell (birds and men, for example), most have keen chemoreceptors and/or olfactory abilities. Certain male moths can detect dilute pheromones released by a female a full kilometer upwind, allowing these males to find females at considerable distances. Similarly, dung beetles use a zigzag flight to "home in" on upwind fecal material with remarkable precision.

Various environmental cues clearly provide particular animals with qualitatively and quantitatively different kinds and amounts of information. Certain environmental cues are useful in the context of capturing prey and escaping predators; others may facilitate timing of reproduction to coincide with good conditions for raising young. Some environmental cues are noisier and less dependable than others. Moreover, the ability to process information received from the environment is limited by a finite neural capacity. The principle of allocation and the notion of tradeoffs dictates that an individual cannot perceive all environmental cues with high efficiency. If ability to perceive a broad range of environmental stimuli actually requires lowered levels of performance along each perceptual dimension, natural selection should improve perceptual abilities along certain critical dimensions at the expense of other less useful ones. Clearly, echolocation has tremendous utility for a nocturnal bat, whereas vision is relatively much less useful. In contrast, the values of these two senses are reversed for a diurnal arboreal squirrel. Within phylogenetic constraints imposed by its evolutionary history, an animal's sensory capacities can be viewed as a bioassay of the importance of particular perceptual dimensions and environmental cues in that animal's ecology.

Adaptive Suites

A basic point of this chapter is that any given organism possesses a unique coadapted complex of physiological, behavioral and ecological traits, whose functions complement one another and enhance that organism's reproductive success. Such a constellation of adaptations has been called an optimal design (Rosen, 1967) or an *adaptive suite* (Bartholomew, 1972).

Consider the desert horned lizard *Phrynosoma platyrhinos*. Various features of its anatomy, behavior, diet, temporal pattern of activity, thermoregulation, and reproductive tactics can be profitably interrelated and interpreted to provide an integrated view of the ecology of this interesting animal (Pianka and Parker, 1975b). Horned lizards are ant specialists and usually eat essentially nothing else. Ants are small and contain much undigestible chitin, so that large numbers must be consumed. Hence an ant specialist must possess a large stomach for its body size. When expressed as a proportion of total body weight, the stomach of this horned lizard occupies a considerably larger fraction of the animal's overall body mass (about 13 percent) than do stomachs of all other sympatric desert lizard species, including the herbivorous desert iguana *Dipsosaurus dorsalis* (herbivores typically have lower assimilation rates and larger stomachs than carnivores). Possession of such a large gut necessitates a tanklike body form, reducing speed and decreasing the lizard's ability to escape from predators by movement. As a result, natural selection has favored a spiny body form and cryptic behavior rather than a sleek body and rapid movement to cover (as in the majority of other species of lizards). Risks of predation are likely to be increased during long periods of exposure while foraging in the open. A reluctance to move, even when actually threatened by a potential predator, could well be advantageous: movement might attract the predator's attention and negate the advantage of concealing coloration and contour. Such decreased movement doubtless contributes to the observed high variance in body temperature of *Phrynosoma*, which is significantly greater than that of all other species of sympatric lizards.

Phrynosoma platyrhinos are also active over a longer time interval than any sympatric lizard species. Wide fluctuations in horned lizard body temperatures under natural conditions presumably reflect both the long activity period and perhaps their reduced movements into or out of the sun and shade (the vast majority of the animals are in the open sun when first sighted). More time is thus made available for activities such as feeding. A foraging ant eater must spend considerable time feeding. Food specialization on ants is economically feasible only because these insects usually occur in a clumped spatial distribution and hence constitute a concentrated food supply. To make use of this patchy and spatially concentrated, but at the same time not overly nutritious, food supply, *P. platyrhinos* has evolved a unique constellation of adaptations that include a large stomach, spiny body form, an expanded period of activity, and "relaxed" thermoregulation (eurythermy). The high reproductive investment of adult horned lizards is probably also a simple and direct consequence of their robust body form. Lizards that must be able to move rapidly to escape predators, such as racerunners *(Cnemidophorus)*, would hardly be expected to weight themselves down with eggs to the same extent as animals like horned lizards that rely almost entirely upon spines and camouflage to avoid their enemies.

Energetics of metabolism of weasels provide another, somewhat more physiological, example of a suite of adaptations (Brown and Lasiewski, 1972). Due to their long thin body shape, weasels have a higher surface-to-volume ratio than mammals with a more standard

shape, and as a consequence, they have an increased energy require-
ment. Presumably benefits of the elongate body form more than outweigh associated costs; otherwise natural selection would not have
favored evolution of the weasel body shape. Brown and Lasiewski
speculate that a major advantage of the elongate form is the ability to
enter burrows of small mammals (weasel prey), which results in increased hunting success and thus allows weasels to balance their energy budgets (Figure 4.13). A further spin-off of the elongate shape is
the evolution of a pronounced sexual dimorphism in body size, which
allows male and female weasels to exploit prey of different sizes and
hence reduces competition between the sexes (related mustelids such

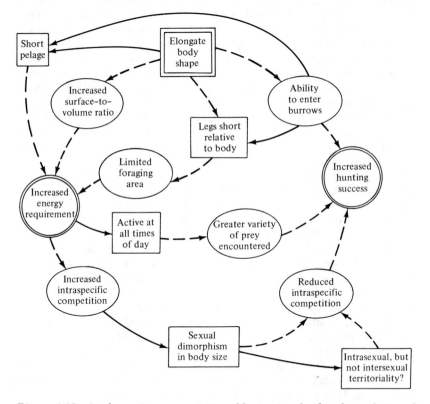

Figure 4.13 A schematic representation of factors involved in the evolution of
elongate body shape in weasels. Circles indicate primary consequences of
evolving a long, thin body configuration; ellipses show secondary
consequences; rectangles indicate phenotypic characteristics of weasels
affected by evolution of this body shape. Unbroken lines indicate selective
pressures and dashed lines show causal sequences. Changes proceed in the
direction of the arrows as long as selection favors a more elongate shape.
[From Brown and Lasiewski (1972). Copyright by the Ecological Society of
America.]

as skunks and badgers do not have the marked sexual size dimorphism characteristic of weasels).

Selected References

Bligh (1973); Cloudsley-Thompson (1971); Florey (1966); Folk (1974); Gates and Schmerl (1975); Gordon (1972); Guyton and Horrobin (1974); Hadley (1975); Hochachka and Somero (1973); Levitt (1972); Prosser (1973); Schmidt-Nielsen (1964, 1975); Vernberg and Vernberg (1974); Vernberg (1975); Wieser (1973); Yousef, Horvath, and Bullard (1972).

Physiological Optima and Tolerance Curves

Brown and Feldmeth (1971); Ruibal and Philibosian (1970); Schmidt-Nielsen (1975); Shelford (1913b).

Energetics of Metabolism and Movement

McNab (1963); Pearson (1948); Schmidt-Nielsen (1972, 1975); Schoener (1968b); Tucker (1975); Turner, Jennrich, and Weintraub (1969).

Energy Budgets and the Principle of Allocation

Fitzpatrick (1973); Levins (1968); Randolph, Randolph, and Barlow (1975).

Adaptation and Deterioration of Environment

Fisher (1930, 1958a, 1958b); Henderson (1913); Maynard Smith (1976); Van Valen (1973, 1974).

Heat Budgets and Thermal Ecology

Bartholomew (1972); Bartlett and Gates (1967); Brown and Feldmeth (1971); Brown and Lasiewski (1972); Cowles and Bogert (1944); Dawson (1975); Gates (1962); Hamilton (1973); Heinrich (1975); Huey and Slatkin (1976); Porter and Gates (1969); Porter *et al.* (1973); Ruibal (1961); Ruibal and Philibosian (1970); Schmidt-Nielsen (1964); Schmidt-Nielsen and Dawson (1964); Whittow (1970); Wieser (1973).

Water Economy in Desert Organisms

Cloudsley-Thompson (1971); Folk (1974); Gindell (1973); Hadley (1975); Main (1976); Schmidt-Nielsen (1964, 1975).

Other Limiting Materials

Feeny (1974); Gilbert (1972); Pauling (1970).

Sensory Capacities and Environmental Cues

Griffin (1958); Machin and Lissmann (1960); Schmidt-Nielsen (1975).

Adaptive Suites

Bartholomew (1972); Brown and Lasiewski (1972); Frazzetta (1975); Pianka and Parker (1975b); Rosen (1967); Wilbur (1977).

Principles of
Population Ecology

5

Introduction

Each generation, sexually reproducing organisms mix their genetic materials. Such shared genetic material is called a *gene pool*, and all the organisms involved in a gene pool are collectively termed a Mendelian *population*. Populations are more abstract conceptual entities than cells or organisms and are somewhat more elusive, but they are nonetheless real. A gene pool has continuity in both space and time, and organisms belonging to a given population either have a common immediate ancestry or are potentially able to interbreed. Alternatively, a population may be defined as a cluster of individuals with a high probability of mating with each other compared to their probability of mating with a member of some other population. As such, Mendelian populations are groups of organisms with a substantial amount of genetic exchange. Such populations are also called *demes*, and the study of their vital statistics is termed *demography*.

In practice it is often extremely difficult to draw boundaries between populations except in most unusual circumstances. Certainly the English sparrows introduced into Australia are no longer exchanging genes with those introduced into North America, and so each represents a functionally distinct population. Although they might well be potentially able to interbreed, the probability of their doing so is remote because of geographical separation. Such differences also exist at a much finer and more local level, both between and within habitats. For example, English sparrows in eastern Australia are separated from those in western Australia by a desert that is uninhabitable to these birds, so they form different populations.

By the above definition, organisms reproducing asexually (e.g., a plant that buds off another individual) strictly speaking do not form true populations; there is no gene pool, no interbreeding, and all offspring are essentially identical genetically. However, even such non-interbreeding plants and animals often form collections of organisms

97

(a) Frequency distribution of age at death

(b) Force of mortality in various age groups

(c) Survivorship of various age groups

Figure 5.1 Hypothetical death data of a life insurance actuary (who would treat the sexes separately). (*a*) The "raw" data consist of a frequency distribution of the age of death of, say, 1000 individuals. For convenience, data are lumped into five-year age intervals. (*b*) Death rate at age x, or q_x, expressed as the percentage of each age group that die during that age interval. (The age

with many of the populational attributes of true sexual populations. Most of the animal and many of the plant species that have been studied resort to sexual reproduction at least periodically, and therefore mix up their genes to some extent and form true Mendelian populations. Populations vary in size from very small (a few individuals on a newly colonized island) to very large, such as some wide ranging and common small insects with populations in the millions. Populations are more usually in the hundreds or thousands. Consideration of both asexual and sexual organisms at the population level often allows us to extend our insight into the activities of individuals in remarkable ways.

An individual's ability to perpetuate its genes in the gene pool of its population, or its reproductive success, represents that individual's fitness. Each member of a population has its own relative fitness within that population, which determines in part the fitness of other members of its population; likewise, every individual's fitness is influenced by all other members of its population. Fitness can be defined and understood only in the context of an organism's total environment.

If we consider any continuously varying measurable characteristic, such as height or weight, a population under consideration has a *mean* (average) and a *variance* (a statistical measure of dispersion based on the average of the squared deviations from the mean). Any one individual has only a single value, but the population of individuals has both a mean and a variance. (In fact we usually estimate true values with the mean and variance of a sample.) These are *population parameters*, characteristics of the population concerned, and they are impossible to define unless we consider a population. Populations also have birth rates, death rates, age structures, sex ratios, gene frequencies, genetic variability, growth rates and growth forms, densities, and so on. Here we examine a variety of such populational characteristics.

Life Tables and Tables of Reproduction

Insurance companies employ actuaries to calculate insurance risks. An actuary obtains a large sample of data on some past event and uses them to estimate the average rate of occurrence of a phenomenon; the company then allows itself a suitable profit and margin of safety and sells insurance on the event concerned. Let us consider how an actuary calculates life insurance risks. The raw data consist simply of the average number of deaths at every age in a population: that is, a frequency distribution of deaths by ages (Figure 5.1*a*). From these values and the

---

distribution of the population at large is needed to compute such age-specific mortality rates.) Values are high in older age groups because they contain relatively few individuals, many of whom die during that age interval. (*c*) Percentage surviving from an initial cohort of individuals with the above age-specific death rates. When divided by 100, these values give the probability that an average newborn will survive to age *x*.

age distribution of the population, the actuary calculates age-specific death rates, which are simply the percentages of individuals of any age group who die during that age period (Figure 5.1b). Death rate at age x is designated by q_x, which is sometimes called the "force of mortality" or the *age-specific death rate*. In Figure 5.1 deaths are combined into age classes covering a five-year period. If the population is large and age groups are fine (consisting of, for instance, only individuals born on a given day) these curves would be much smoother or more nearly continuous (the distinction between *discrete* and *continuous* events or characteristics will be made repeatedly in this chapter). Demographers begin with discrete age intervals and use the methods of calculus to fit continuous functions to them for estimates at various points within age intervals.

Still another useful way of manipulating life tables is to calculate age-specific percentage survival (Figure 5.1c). Starting with an initial number or *cohort* of newborn individuals, one calculates the percentage of this initial population alive at every age by sequentially subtracting the percentage of deaths at each age. A smoothed continuous version of survivorship (as in Figures 5.2 and 5.3) is called a *survivorship curve*. The fraction surviving at age x gives the probability that an average newborn will survive to that age, which is usually designated l_x.

Finally, our actuary is interested in estimating the expectation of further life. How long, on the average, will someone of age x live? For newborn individuals (age 0) average life expectancy is equal to the mean length of life of the cohort. In general, expectation of life at any age x is simply the mean life span remaining to those individuals attaining age x. In symbols,

$$E_x = \frac{\sum_{y=x}^{\infty} l_y}{l_x} \quad \text{or} \quad E_x = \frac{\int_x^{\infty} l_y dy}{l_x} \tag{1}$$

where E_x is the expectation of life at age x, and y subscripts age. The left-hand equation is the discrete version, the right-hand one a continuous version using the symbolism of integral calculus. Calculation of E_x is illustrated in Table 5.1.

Life insurance premiums for men are higher than they are for women because males have a steeper survivorship curve and therefore, at a given age, a shorter life expectancy than females. Figures 5.2 and 5.3 show a variety of survivorship curves, representing the great range they take in natural populations. Rectangular survivorship on a semilogarithmic plot—that is, little mortality until some age and then fairly steep mortality thereafter, as in the lizards *Xantusia vigilis* and *Scincella laterale*, dall mountain sheep, most African ungulates, humans, and perhaps most mammals (Caughley, 1966)—has been called Type I survivorship (Pearl, 1928). Relatively constant death rates with

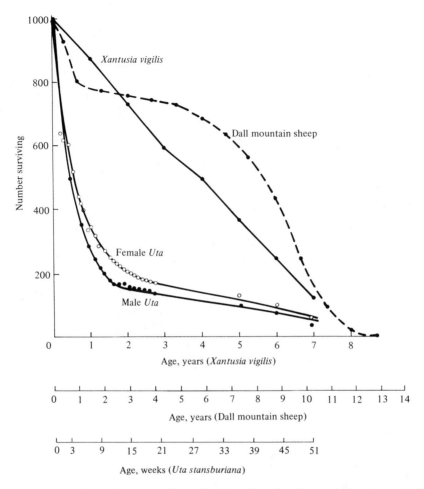

Figure 5.2   Several survivorship schedules plotted with an arithmetic vertical axis. (Compare with the more rectangular semilogarithmic plots of Figure 5.3.) Although both types of plots are in common use, logarithmic ones are preferable. In a logarithmic plot, survivorship of the lizard *Xantusia* becomes rectangular (rather than diagonal), whereas that of another lizard, *Uta*, is diagonal (rather than inversely hyperbolic). [After Deevey (1947), Tinkle (1967), and Zweifel and Lowe (1966).]

age produce diagonal survivorship curves on semilogarithmic plots as in the lizards *Uta stansburiana* and *Eumeces fasciatus*, the warthog, and most birds; these are classified as Type II curves [actually there are two kinds of Type II curves, representing, respectively, constant risk of death per unit time and constant numbers of deaths per unit time (Slobodkin, 1962)]. Many fish, marine invertebrates, most insects, and

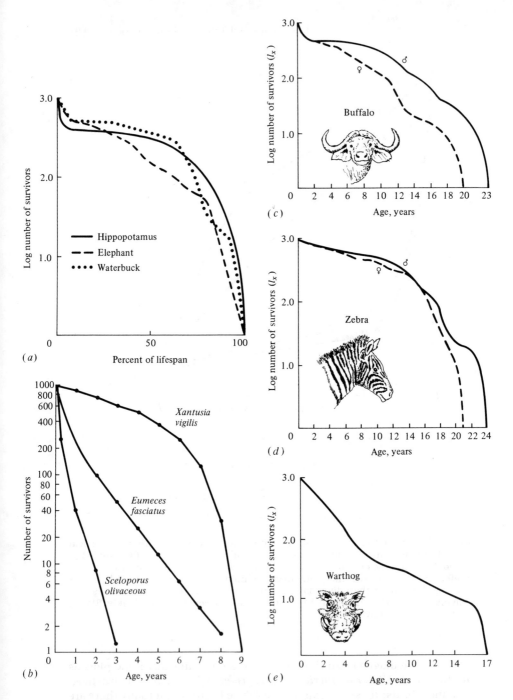

Figure 5.3 Semilogarithmic plots of some survivorship schedules. Compare *Xantusia* survivorship in (*b*) with Figure 5.2 plotted from the same data. The virtue of a semilogarithmic plot is that a straight line implies equal mortality rates with respect to age. [After Zweifel and Lowe (1966) and Spinage (1972). Spinage by permission of Duke University Press.]

Table 5.1   Illustration of the Calculation of E_x, T, R_0, and v_x in a Hypothetical Stable Population with Discrete Age Classes, Using Equations (1) to (4)

Age (x)	l_x	m_x	$l_x m_x$	$x l_x m_x$	E_x (see below)	v_x (see below)
0	1.0	0.0	0.00	0.00	3.40	1.00
1	0.8	0.2	0.16	0.16	3.00	1.25
2	0.6	0.3	0.18	0.36	2.67	1.40
3	0.4	1.0	0.40	1.20	2.50	1.65
4	0.4	0.6	0.24	0.96	1.50	0.65
5	0.2	0.1	0.02	0.10	1.00	0.10
6	0.0	0.0	0.00	0.00	0.00	0.0(
Sums		2.2 (GRR)	1.00 (R_0)	2.78 (T)		

Expectation of life:
$E_0 = (l_0 + l_1 + l_2 + l_3 + l_4 + l_5)/l_0 = (1.0 + 0.8 + 0.6 + 0.4 + 0.4 + 0.2)/1.0 = 3.4/1.0$
$E_1 = (l_1 + l_2 + l_3 + l_4 + l_5)/l_1 = (0.8 + 0.6 + 0.4 + 0.4 + 0.2)/0.8 = 2.4/0.8 = 3.0$
$E_2 = (l_2 + l_3 + l_4 + l_5)/l_2 = (0.6 + 0.4 + 0.4 + 0.2)/0.6 = 1.6/0.6 = 2.67$
$E_3 = (l_3 + l_4 + l_5)/l_3 = (0.4 + 0.4 + 0.2)/0.4 = 1.0/0.4 = 2.5$
$E_4 = (l_4 + l_5)/l_4 = (0.4 + 0.2)/0.4 = 0.6/0.4 = 1.5$
$E_5 = l_5/l_5 = 0.2/0.2 = 1.0$

Reproductive value:
$$v_0 = \frac{l_0}{l_0} m_0 + \frac{l_1}{l_0} m_1 + \frac{l_2}{l_0} m_2 + \frac{l_3}{l_0} m_3 + \frac{l_4}{l_0} m_4 + \frac{l_5}{l_0} m_5 = 0.0 + 0.16 + 0.18 + 0.40 + 0.24 + 0.02 = 1.00$$
$$v_1 = \frac{l_1}{l_1} m_1 + \frac{l_2}{l_1} m_2 + \frac{l_3}{l_1} m_3 + \frac{l_4}{l_1} m_4 + \frac{l_5}{l_1} m_5 = 0.20 + 0.225 + 0.50 + 0.30 + 0.025 = 1.25$$
$$v_2 = \frac{l_2}{l_2} m_2 + \frac{l_3}{l_2} m_3 + \frac{l_4}{l_2} m_4 + \frac{l_5}{l_2} m_5 = 0.30 + 0.67 + 0.40 + 0.03 = 1.40$$
$$v_3 = \frac{l_3}{l_3} m_3 + \frac{l_4}{l_3} m_4 + \frac{l_5}{l_3} m_5 = 1.0 + 0.6 + 0.05 = 1.65$$
$$v_4 = \frac{l_4}{l_4} m_4 + \frac{l_5}{l_4} m_5 = 0.60 + 0.05 = 0.65$$
$$v_5 = \frac{l_5}{l_5} m_5 = 0.10$$

many plants have extremely steep juvenile mortality and relatively high survivorship afterward—that is, inverse hyperbolic or Type III survivorship. Of course, nature does not fall into three or four convenient categories, and many real survivorship curves are intermediate between the various "types" categorized above. Moreover, survivorship schedules are not constant but change with immediate environmental conditions. Later we examine evolution of death rates and old age, but first we must consider the other important populational phenomenon—reproduction.

The number of offspring produced by an average organism of age x during that age period is designated m_x; only those progeny that enter age class zero are counted (age class zero is arbitrary—we could begin a life table at conception, birth, or the age of independence from parental care, depending on which was most convenient and appropriate).

Males and females are each credited with one-half of one reproduction for every such offspring produced, so an organism must have two progeny to replace itself. (This procedure makes biological sense in that a sexually reproducing organism passes only half its genome to each of its progeny.) The sum of m_x over all ages, or the total number of offspring that would be produced by an average organism in the absence of mortality, is termed the *gross reproductive rate* (GRR). Like survivorship, patterns of reproduction and fecundities, or m_x schedules, vary widely both with environmental conditions and among different species of organisms. Some, such as annual plants and many insects, breed only once during their lifetime. Others, such as perennial plants and many vertebrates, breed repeatedly. The number of eggs produced, and their size relative to the parent, also vary over many orders of magnitude. Litter size (usually designated by *B*) refers to the number of young produced during each act of reproduction and is discussed further on pages 125–138.

Reproduction may be delayed until fairly late in life, or reproductive activities may begin almost immediately after hatching or birth. The age of first reproduction is usually termed α, and the age of last reproduction, ω. For an organism that breeds only once, the average time from egg to egg, or the time between generations, termed *generation time (T)*, is simply equal to α. But generation time in animals that breed repeatedly is somewhat more complicated. The average time between generations of repeated reproducers can be roughly estimated as $\overline{T} = (\alpha + \omega)/2$. A more accurate calculation of T is possible by weighing each age by its total realized fecundity, $l_x m_x$, using the following equations (see also Table 5.1)

$$T = \sum_{x=\alpha}^{\omega} x l_x m_x \quad \text{or} \quad T = \int_{\alpha}^{\omega} x l_x m_x \, dx \qquad (2)$$

[These equations apply only in a nongrowing population; if a population is expanding or contracting the right-hand side must be divided by the net reproductive rate, R_0 (below), to standardize for the average number of successful offspring per individual.] Mean generation time is thus the average age of parenthood, or the *average parental age at which all offspring are born*.

Net Reproductive Rate and Reproductive Value

Clearly not many organisms live to realize their full potential for reproduction, and we need an estimate of the number of offspring produced by an organism that suffers average mortality. Thus the *net reproductive rate (R$_0$)* is defined as *the average number of age class zero offspring produced by an average newborn organism during its entire lifetime*. Mathematically, R_0 is simply the product of the age-specific survivorship and fecundity schedules, over all ages at which reproduction occurs:

$$R_0 = \sum_{x=0}^{\infty} l_x m_x \quad \text{or} \quad R_0 = \int_0^{\infty} l_x m_x \, dx \tag{3}$$

Alternatively, α and ω could be substituted for the 0 and ∞ limits since $l_x m_x$ is zero at all nonreproductive ages. Once again, the equation on the left is for discrete age groups, while that on the right is for continuous ones. Table 5.1 illustrates calculation of R_0 from a pair of discrete l_x and m_x schedules, and Figure 5.4 diagrams its calculation for continuous ones.

When R_0 is greater than 1 the population is increasing, when R_0 equals 1 it is stable, and when R_0 is less than 1 the population is decreasing. Because of this, the net reproductive rate has also been called the *replacement rate* of the population. A stable population, at equilibrium, with a steep l_x curve must have a correspondingly high m_x curve in order to replace itself (when death rate is high, birth rate must also be high). Conversely, when l_x is high, m_x must be low in order for R_0 to equal unity.

Another important concept, first elaborated by Fisher (1930), is reproductive value. To what extent, on the average, do members of a

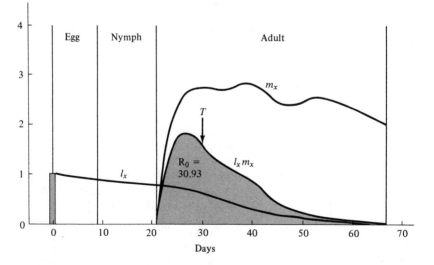

Figure 5.4 Continuous l_x and m_x schedules are multiplied together over all ages to obtain the area (shaded) under the $l_x m_x$ product curve of realized fecundity, which is equal to the net reproductive rate, R_0. [From F. E. Smith in *Dynamics of Growth Processes*, ed. E. J. Boell (copyright 1954 by Princeton University Press), fig. 1, p. 278, after Evans and Smith (1952) from data for the human louse, *Pediculus humanus*. Reprinted by permission of Princeton University Press.]

given age group contribute to the next generation between that age and death? In a stable population that is neither increasing nor decreasing *reproductive value* (v_x) is defined as the *age-specific expectation of future offspring*. Its mathematical definition in a stable population at equilibrium is:

$$v_x = \sum_{t=x}^{\infty} \frac{l_t}{l_x} m_t \quad \text{or} \quad v_x = \int_x^{\infty} \frac{l_t}{l_x} m_t dt \tag{4}$$

As before, the equation on the left is for discrete age groups and that on the right for continuous age groups. The term l_t/l_x represents the probability of living from age x to age t, and m_t is the average reproductive success of an individual at age t. Clearly for newborn individuals in a stable population, v_0 is exactly equal to the net reproductive rate, R_0. A postreproductive individual has a reproductive value of zero because it can no longer expect to produce offspring; moreover, since natural selection operates only by differential reproductive success (Chapter 1), such a postreproductive organism is no longer subject to the direct effects of natural selection (see also pp. 139–141). Under many l_x and m_x schedules, reproductive value is maximal around the onset of reproduction and falls off after that because fecundity often decreases with age (but fecundity is subject to natural selection too—see pp. 125–138). Table 5.1 illustrates calculation of reproductive value in a stable population. Figure 5.5 shows how reproductive value changes with age in a variety of populations.

In populations that are changing in size, the definition of reproductive value is the *present value of future offspring*. Basically it represents the number of progeny that an organism dying before it reaches age $x + 1$ would have to produce at age x in order to leave as many descendants as it would if it had instead survived to age $x + 1$ and enjoyed average survivorship and fecundity thereafter. In an expanding population (Figure 5.5a), reproductive value of very young individuals is low for two reasons: (1) there is a finite probability of death before reproduction and (2) because the future breeding population will be larger, offspring to be produced later will contribute less to the total gene pool than offspring currently being born (similarly, in a declining population, offspring expected at some future date are worth relatively more than current progeny because the total future population will be smaller). Thus present progeny are worth more than future offspring in a growing population, whereas future progeny are more valuable than present offspring in a declining population. This component of reproductive value is tedious to calculate and applies only to populations changing in size.

The general equations for reproductive value in any population, either stable or changing, are:

$$\frac{v_x}{v_0} = \frac{e^{rx}}{l_x} \sum_{t=x}^{\infty} e^{-rt} l_t m_t \quad \text{or} \quad \frac{v_x}{v_0} = \frac{e^{rx}}{l_x} \int_x^{\infty} e^{-rt} l_t m_t \, dt \tag{5}$$

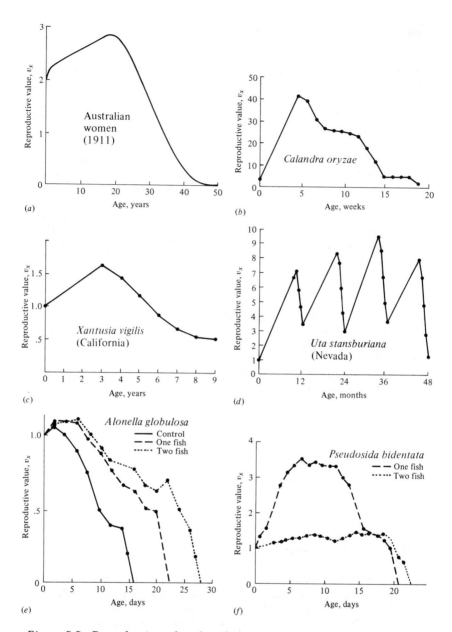

Figure 5.5 Reproductive value plotted against age for a variety of populations. *(a)* Australian women, about 1911. [From Fisher (1958a).] *(b) Calandra oryzae,* a beetle, in the laboratory. [From data of Birch (1948).] *(c) Xantusia vigilis,* a viviparous lizard with a single litter each year. [From data of Zweifel and Lowe (1966).] *(d) Uta stansburiana,* an egg-laying lizard that lays several clutches each reproductive season. [From data of Turner *et al.* (1970).] *(e) Alonella globulosa,* a microscopic aquatic crustacean, under three different competitive and predatory regimes in laboratory microcosms. *(f) Pseudosida bidentata,* a microscopic aquatic crustacean, under two different laboratory situations. [e, f after Neill (1972).]

where e is the base of the natural logarithms and r is the instantaneous rate of increase per individual (see also below). The exponentials, e^{rx} and e^{-rt}, weight offspring according to the direction in which the population is changing. In a stable population, r is zero. Remembering that e^0 and e^{-0} equal unity, and that v_0 in a stable population is also 1, the reader can verify that (5) reduces to (4) when r is zero. Reproductive value does not directly take into account social phenomena such as parental care* or a grandmother's caring for her grandchildren and thereby increasing their probability of survival and successful reproduction.

It is sometimes useful to partition reproductive value into two components—progeny expected in the immediate future versus those expected in the more distant future. For a nongrowing population:

$$v_x = m_x + \sum_{t=x+1}^{\omega} \frac{l_t}{l_x} m_t \tag{6}$$

The second term on the right-hand side represents expectation of offspring of an organism at age x in the distant future (at age $x + 1$ and beyond) and is known as its *residual reproductive value* (Williams, 1966b). Rearranging (6) shows that residual reproductive value (v_x^*) is equal to an organism's reproductive value in the next age interval, v_{x+1}, multiplied by the probability of surviving from age x to age $x + 1$, or l_{x+1}/l_x.

$$v_x^* = \frac{l_{x+1}}{l_x} v_{x+1} \tag{7}$$

In summary, any pair of age-specific mortality and fecundity schedules has its own implicit T, R_0, r (see below), as well as v_x and v_x^* curves.

Stable Age Distribution

Another important aspect of a population's structure is its age distribution (Figure 5.6), indicating the proportions of its members belonging to each age class. Two populations with identical l_x and m_x schedules, but with different age distributions, will behave differently and may even grow at different rates if one population has a higher proportion of reproductive members. Lotka (1922) proved that any pair of unchanging l_x and m_x schedules eventually gives rise to a population with a *stable age distribution*. When a population reaches this equilibrium age distribution, the percentage of organisms in each age group re-

---

* Defining the age of independence from parental care as age class zero neatly circumvents this problem.

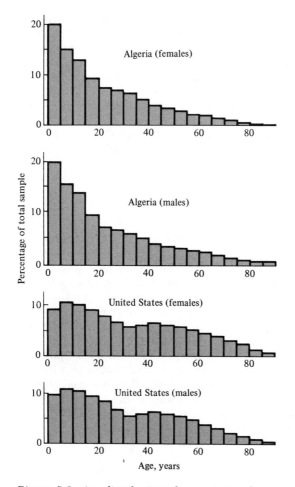

Figure 5.6 Age distributions, by sex, in two human
populations. The Algerian population is increasing
rapidly; the United States population is growing more
slowly. [After Krebs (1972) after 1968 Demographic
Yearbook of the United Nations. Copyright, United
Nations (1969). Reproduced by permission.]

mains constant. Recruitment into every age class is exactly balanced
by its losses due to mortality and aging. Provided l_x and m_x schedules
are not changing, a kind of stable age distribution is quickly reached
even in an expanding population, in which the per capita rate of growth
of each age class is the same (equal to the intrinsic rate of increase per
head, r, below), with the consequence that proportions of various age
groups also stay constant. Equations (2) through (9) and (12) through

(15) assume that a population is in its stable age distribution. However, the intrinsic rate of increase (r), generation time (T), and reproductive value (v_x) are conceptually *independent* of these specific equations, since these concepts can be defined equally well in terms of the age distribution of a population, although they are somewhat more complex mathematically (Vandermeer, 1968). The stable age distribution of a stable population with a net reproductive rate equal to 1 is called the "stationary age distribution." Computation of the stable age distribution is somewhat tedious, and so no more will be said about it here. The interested reader is referred to references at the end of the chapter.

Intrinsic Rate of Natural Increase

Still another useful parameter implicit in every pair of schedules of births and deaths is the *intrinsic rate of natural increase,* sometimes called the Malthusian parameter. Usually designated by r, it is a measure of the instantaneous rate of change of population size (per individual): r is expressed in numbers per unit time per individual and has the units of 1/time. In a closed population the intrinsic rate of increase is defined as the instantaneous birth rate (per individual), b, minus the instantaneous death rate (per individual), d. [Similarly, in an open population r is equal to (births + immigration) − (deaths + emigration).] When per capita births exceed per capita deaths $(b > d)$ the population is increasing and r is positive; when deaths exceed births $(b < d)$, r is negative and the population is decreasing.

In practice, r is somewhat tedious to calculate, as its value must be determined by iteration using Euler's implicit equation

$$\sum_x e^{-rx} l_x m_x = 1 \tag{8}$$

where e is the base of the natural logarithms and x subscripts age (derivations of this equation may be found in Mertz, 1970, or Emlen, 1973). Provided that the net reproductive rate, R_0, is near one, r can be estimated using the approximate formula

$$r \cong \frac{\log_e R_0}{T} \tag{9}$$

where T is generation time computed with equation (2) (see also May, 1976). From (9) we see that r is positive when R_0 is greater than 1, and negative when R_0 is less than 1. Because $\log_e 1$ is zero, an R_0 of unity corresponds to an r of zero. Under optimal conditions, when R_0 is as high as possible, the maximal rate of natural increase is realized and is designated by r_{max}. Note that the intrinsic rate of increase is inversely related to generation time, T (see also Figure 5.27, p. 143).

The maximal instantaneous rate of increase per head, r_{max}, varies among animals by several orders of magnitude (Table 5.2). Small short-lived organisms such as the common human intestinal bacterium *Escherichia coli* have a relatively high r_{max} value, whereas larger and longer-lived organisms such as man have, comparatively, very low r_{max} values. The components of r_{max} are the instantaneous birth rate per head, b, and the instantaneous death rate per head, d, under optimal environmental conditions. The evolution of rates of reproduction and death rates are taken up later in this chapter.

A population whose size increases linearly in time would have a constant populational growth rate given by

$$\begin{matrix} \text{growth rate} \\ \text{of population} \end{matrix} = \frac{N_t - N_0}{t - t_0} = \frac{\Delta N}{\Delta t} = \text{constant} \qquad \textbf{(10)}$$

where N_t is the number at time t, N_0 the initial number, and t_0 the initial time. But at any fixed positive value of r, the per capita rate of increase is constant, and a population grows exponentially (Figure 5.7). Its growth rate is a function of population size, with the population growing faster as N becomes larger. Suppose you wanted to estimate the rate of change of the population shown in Figure 5.7 at an instant in time, say at time t. As a first approximation, you might look at N immediately before and immediately after time t, say 1 hour before and 1 hour after, and apply the above $\Delta N / \Delta t$ equation. But examination of

Table 5.2   Estimated Maximal Instantaneous Rates of Increase (r_{max}, Per Capita Per Day) and Mean Generation Times (in Days) for a Variety of Organisms

Taxon	Species	r_{max}	Generation Time (T)
Bacterium	*Escherichia coli*	ca. 60.0	0.014
Protozoa	*Paramecium aurelia*	1.24	0.33–0.50
Protozoa	*Paramecium caudatum*	0.94	0.10–0.50
Insect	*Tribolium confusum*	0.120	ca. 80
Insect	*Calandra oryzae*	0.110(.08–.11)	58
Insect	*Rhizopertha dominica*	0.085(.07–.10)	ca. 100
Insect	*Ptinus tectus*	0.057	102
Insect	*Gibbium psylloides*	0.034	129
Insect	*Trigonogenius globulus*	0.032	119
Insect	*Stethomezium squamosum*	0.025	147
Insect	*Mezium affine*	0.022	183
Insect	*Ptinus fur*	0.014	179
Insect	*Eurostus hilleri*	0.010	110
Insect	*Ptinus sexpunctatus*	0.006	215
Insect	*Niptus hololeucus*	0.006	154
Mammal	*Rattus norwegicus*	0.015	150
Mammal	*Microtus aggrestis*	0.013	171
Mammal	*Canis domesticus*	0.009	ca. 1000
Insect	*Magicicada septendecim*	0.001	6050
Mammal	*Homo sapiens*	0.0003	ca. 7000

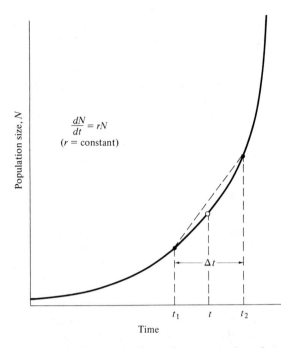

Figure 5.7 Exponential population growth under the assumption that the rate of increase per individual, r, remains constant with changes in population density. Note that a straight-line estimate of the rate of population growth at time t becomes more and more accurate as t_1 and t_2 converge; in the limit, as t_1 and t_2 approach t, or $\Delta t \to 0$, the rate of population growth equals the slope of a line tangent to the curve at time t (open circle).

Figure 5.7 reveals that at time t_1 (say $t - 1$ hr) the true rate is less than, and at time t_2 (say $t + 1$hr) greater than, your straight-line estimate. Differential calculus was developed to handle just such cases, and it allows us to calculate the rate of change at an *instant* in time. As ΔN and Δt are made smaller and smaller, $\Delta N / \Delta t$ gets closer and closer to the true rate of change at time t (Figure 5.7). In the limit, as the Δ's approach zero, $\Delta N / \Delta t$ is written as dN/dt, which is calculus shorthand for the instantaneous rate of change of N at t. Exponential population growth is described by the simple differential equation

$$\frac{dN}{dt} = bN - dN = (b - d)N = rN \tag{11}$$

where, again, b is the instantaneous birth rate per individual and d the instantaneous death rate per individual (remember that $r = b - d$).

(Note that the ds in dN/dt do not refer to death rate, but the entire term represents the instantaneous rate of change in population density, N.)

Using calculus to integrate (11) shows that the number of organisms at some time t, N_t, under exponential growth is a function of the initial number at time zero, N_0, r, and the time available for growth since time zero, t:

$$N_t = N_0 e^{rt} \tag{12}$$

Here again, e is the base of the natural logarithms. By taking logarithms of (12), which is simply an integrated version of (11), we get

$$\log_e N_t = \log_e N_0 + \log_e e^{rt} = \log_e N_0 + rt \tag{13}$$

This equation indicates that $\log_e N$ changes linearly in time; that is, a semilog plot of $\log_e N$ against t gives a straight line with a slope of r and a y-intercept of $\log_e N_0$.

Setting N_0 equal to 1 (i.e., a population initiated with a single organism), after one generation, T, the number of organisms in the population is equal to the net reproductive rate of that individual, or R_0. Substituting these values in (13):

$$\log_e R_0 = \log_e 1 + rT \tag{14}$$

Since $\log_e 1$ is zero, (14) is identical with (9).

Another population parameter closely related to the net reproductive rate and the intrinsic rate of increase is the so-called *finite rate of increase*, λ, defined as the rate of increase per individual per unit time. The finite rate of increase is measured in the same time units as the instantaneous rate of increase, and

$$r = \log_e \lambda \text{ or } \lambda = e^r \tag{15}$$

In a population without age structure, λ is thus identical with R_0 [T equal to 1 in (9) and (14)].

Population Growth and Regulation

In a finite world, no population can grow exponentially for very long (Figure 5.8). Sooner or later every population must encounter either difficult environmental conditions or shortages of its requisites for reproduction. Over a long period of time, unless the average actual rate of increase is zero, a population either decreases to extinction or increases to the extinction of other populations.

So far our populations have had fixed age-specific parameters, such as their l_x and m_x schedules. In this section we ignore age specificity and instead allow R_0 and r to vary with population density. To do this, we define *carrying capacity*, K, as the density of organisms (i.e., the number per unit area) at which the net reproductive rate (R_0)

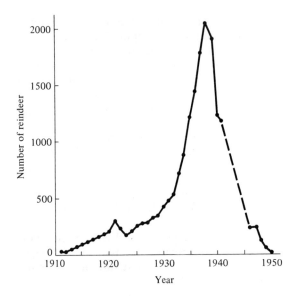

Figure 5.8   In 1911, 25 reindeer were introduced on
Saint Paul Island in the Pribolofs off Alaska. The
population grew rapidly and nearly exponentially until
about 1938, when there were over 2000 animals in the
41-square-mile island. The reindeer badly overgrazed
their food supply (primarily lichens) and the population
"crashed." Only 8 animals could be found in 1950. A
similar sequence of events occurred on Saint Matthew
Island from 1944 through 1966. [After Krebs (1972)
after Scheffer.]

equals unity and the intrinsic rate of increase (r) is zero. At "zero
density" (only one organism, or a perfect competitive vacuum), R_0 is
maximal and r becomes r_{max}. For any given density above zero density,
both R_0 and r decrease until, at K, the population ceases to grow. A
population initiated at a density above K decreases until it reaches the
steady state at K (Figure 5.9). Thus we define r_a (or $dN/dt \cdot 1/N$) as the
actual instantaneous rate of increase; it is zero at K, negative above K,
and positive when the population is below K.

The simplest assumption we can make is that r_a decreases linearly
with N and becomes zero at an N equal to K (Figure 5.9); this assumption leads to the classical Verhulst-Pearl logistic equation

$$\frac{dN}{dt} = rN - rN\left(\frac{N}{K}\right) = rN - \frac{rN^2}{K} \tag{16}$$

Alternatively, by factoring out an rN, (16) can be written

$$\frac{dN}{dt} = rN\left(1 - \frac{N}{K}\right) = rN\left(\frac{K-N}{K}\right) \tag{17}$$

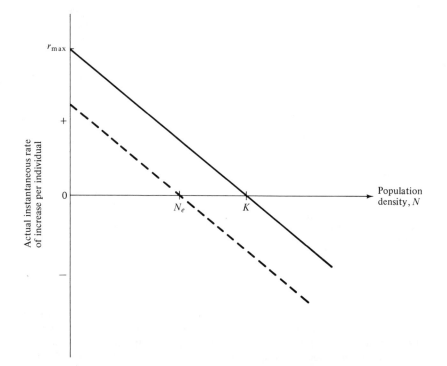

Figure 5.9   The actual instantaneous rate of increase per individual, r_a, decreases linearly with increasing population density under the assumptions of the Pearl-Verhulst logistic equation. The solid line depicts conditions in an optimal environment in which the difference between b and d is maximal. The dashed line shows how the actual rate of increase decreases with N when the death rate per head, d, is higher; equilibrium population size, N_e, is then less than carrying capacity, K. (See also Figure 5.12.)

Or, simplifying by setting r/K in (16) equal to z,

$$\frac{dN}{dt} = rN - zN^2 \tag{18}$$

The term $rN(N/K)$ in (16) and the term zN^2 in (18) represent the density-dependent reduction in the rate of population increase. Thus, at N equal to unity (an ecologic vacuum), dN/dt is nearly exponential, while at N equal to K, dN/dt is zero and the population is in a steady state at its carrying capacity. Logistic equations (there are many more besides the Verhulst-Pearl) generate so-called sigmoid (s-shaped) population growth curves (Figure 5.10). Implicit in the Verhulst-Pearl logistic equation are three assumptions: (1) that all individuals are equivalent—that is, that the addition of every new individual reduces the actual rate of increase by the same fraction, $1/K$, at every density

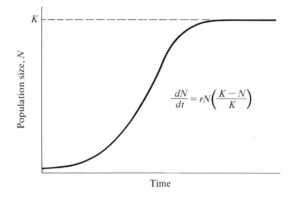

$$\frac{dN}{dt} = rN\left(\frac{K-N}{K}\right)$$

Figure 5.10   Population growth under the
Pearl-Verhulst logistic equation is sigmoidal
(s-shaped), reaching an upper limit termed the
carrying capacity, K. Populations initiated at densities
above K decline exponentially until they reach K,
which represents the only stable equilibrium.

(Figure 5.9); (2) that r_{max} and K are immutable constants; and (3) that
there is no time lag in the response of the actual rate of increase per
individual to changes in N. All three assumptions are unrealistic, so
the logistic has been strongly criticized (Allee *et al.*, 1949; Smith,
1952, 1963a; Slobodkin, 1962).

More plausible curvilinear relationships between the rate of increase and population density are shown in Figure 5.11. Note that

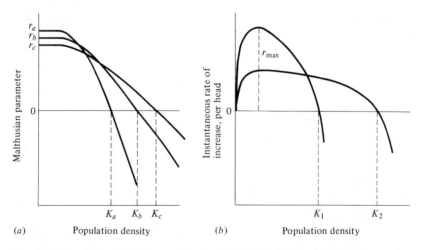

Figure 5.11   Hypothetical curvilinear relationships between instantaneous
rates of increase and population density. Concave upward curves have also been
postulated. [From Gadgil and Bossert (1970) and Pianka (1972).]

density-dependent effects on birth rate and death rate are combined by the use of r (these effects are separated later in this section). Carrying capacity is also an extremely complicated and confounded quantity, for it necessarily includes both renewable and nonrenewable resources, which are variables themselves. Carrying capacity almost certainly varies a great deal from place to place and from time to time for the majority of organisms. There is also some inevitable lag in feedback between population density and the actual instantaneous rate of increase. All these assumptions can be relaxed and more realistic equations developed, but the mathematics quickly become extremely complex and unmanageable. Nevertheless a number of populational phenomena can be nicely illustrated using the simple Verhulst-Pearl logistic, and a thorough understanding of it is a necessary prelude to the equally simple Lotka-Volterra competition equations, which are taken up in Chapter 6. However, the numerous flaws of the logistic must be recognized, and it should be taken only as a first approximation for small changes in population growth, most likely to be valid near equilibrium and over short time periods (i.e., situations in which linearity should be approximated).

Notice that r in the logistic equation (16) is actually r_{max}. The equation can be solved for the *actual* rate of increase, r_a, which is a variable and a function of r, N, and K, by simply factoring out an N:

$$r_a = \frac{dN}{Ndt} = r \left(\frac{K - N}{K} \right) = r - \left(\frac{N}{K} \right) r \qquad (19)$$

The actual instantaneous rate of increase per individual, r_a, is always less than or equal to r_{max} (r in the logistic); Equation (19) and Figure 5.9 show how r_a decreases linearly with increasing density under the assumptions of the Verhulst-Pearl logistic equation.

The two components of the actual instantaneous rate of increase per individual, r_a, are the actual instantaneous birth rate per individual, b, and the actual instantaneous death rate per individual, d. The difference between b and d (i.e., $b - d$) is r_a. Under theoretical ideal conditions when b is maximal and d is minimal, r_a is maximized at r_{max}. In the logistic, this is realized at a minimal density, or a perfect competitive vacuum. To be more precise, we subscript b and d, which are functions of density. Thus $b_N - d_N = r_N$ (which is r_a at density N), and $b_0 - d_0 = r_{max}$. When $b_N = d_N$, r_a and dN/dt are zero and the population is at equilibrium. Figure 5.12 diagrams the way in which b and d vary linearly with N under the logistic. At any given density, b_N and d_N are given by linear equations

$$b_N = b_0 - xN \qquad (20)$$

$$d_N = d_0 + yN \qquad (21)$$

where x and y represent, respectively, the slopes of the lines plotted in Figure 5.12 (see also Bartlett, 1960, and Wilson and Bossert,

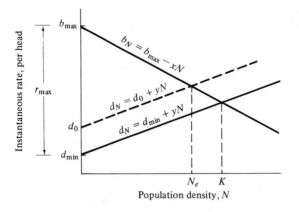

Figure 5.12 The instantaneous birth rate per individual decreases linearly with population density under the logistic equation, whereas the instantaneous death rate per head rises linearly as population density increases. Two death-rate lines are plotted, one with a high death rate (dashed line) and one with a lower death rate (solid line). Equilibrium population density, N_e, is lowered by an increased death rate or by a reduced birth rate.

1971). The instantaneous death rate, d_N, clearly has both density-dependent and density-independent components; in (21) and Figure 5.12, yN measures the density-dependent component of d_N while d_0 determines the density-independent component.

At equilibrium, b_N must equal d_N, or

$$b_0 - xN = d_0 + yN \qquad (22)$$

substituting N_e for N at equilibrium, r for $(b_0 - d_0)$, and rearranging terms

$$r = (x + y)N_e \qquad (23)$$

or

$$N_e = \frac{r}{x + y} \qquad (24)$$

Note that the sum of the slopes of the birth and death rates $(x + y)$ is equal to z, or r/K. Clearly z is the density-dependent constant which is analogous to the density-independent constant r_{max}.

Density Dependence and Density Independence

Various factors can influence populations in two fundamentally different ways. If their effects on a population do not vary with population

Table 5.3   Fish Kills Following Sudden, Severe Cold
Weather on the Texas Gulf Coast in the
Winter of 1940, to Illustrate Density-
Independent Mortality

Locality	Commerical Catch		Decline (%)
	before	after	
Matagorda	16,919	1,089	93.6
Aransas	55,224	2,552	95.4
Laguna Madre	2,016	149	92.6

Source: After Odum (1959) after Gunter.

density, but the same *proportion* of organisms are affected at any density, factors are said to be *density-independent*. Climatic factors often, though by no means always, affect populations in this manner
(Table 5.3). If, on the other hand, a factor's effects vary with population
density so that the proportion of organisms influenced actually changes
with density, that factor is said to be *density-dependent*. Density-
dependent factors and events can be either positive or negative. Death
rate, which presumably often increases with increasing density, is an
example of positive or direct density dependence (Figure 5.12); birth
rate, which normally decreases with increasing density, is an example
of negative or inverse density dependence (Figure 5.13). Density-
dependent influences on populations frequently result in an equilibrium density at which the population ceases to grow. Biotic factors,

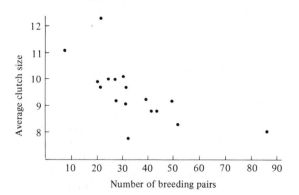

Figure 5.13   A plot of average clutch size against the
density of breeding pairs of English great tits (birds) in
a particular woods in a series of years over a 17-year
period. [After Perrins (1965).]

such as competition, predation, and pathogens, often (though not always) act in this way.

Ecologists are divided in their opinions as to the relative importance of density dependence and density independence in natural populations (Andrewartha and Birch, 1954; Lack, 1954, 1966; Nicholson, 1957; Orians, 1962; McLaren, 1972; Ehrlich *et al.*, 1972). A few even go so far as to deny categorically the existence of one or the other of these types of influences on populations.

Detection of density dependence can be difficult. In their studies on the population dynamics of *Thrips imaginis* (a small herbivorous insect), Davidson and Andrewartha (1948) found that they could predict population sizes of the insects fairly accurately using only past population sizes and recent climatic conditions. These workers could find no evidence of any density effects; they therefore interpreted their data to mean that the populations of *Thrips* were controlled primarily by density-independent climatic factors. However, Smith (1961) reanalyzed their data and discovered pronounced density-dependent effects at high densities. He found a strong inverse correlation between population change and population size, which strongly suggests density dependence. Smith also demonstrated a rapidly decreasing variance in population size during the later portion of the spring population increase. Further, Smith showed that these patterns persisted even after partial correlation analysis, which held constant the very climatic variables that Davidson and Andrewartha considered to be so important. This example illustrates the great difficulty ecologists frequently encounter in distinguishing cause from effect. There is now little real doubt that both density-dependent and density-independent events occur; their relative importance, however, may vary by many orders of magnitude from population to population—and even within the same population from time to time as the size of the population changes (Horn, 1968a; McLaren, 1971).

Opportunistic versus Equilibrium Populations

Periodic disturbances, including fires, floods, hurricanes, and droughts, often result in catastrophic density-independent mortality, suddenly reducing population densities well below the maximal sustainable level for a particular habitat. Populations of annual plants and insects typically grow rapidly during the spring and summer but are greatly reduced at the onset of cold weather. Because populations subjected to such forces grow in erratic or regular bursts (Figure 5.14), they have been termed *opportunistic populations*. In contrast, populations such as those of many vertebrates may usually be closer to an equilibrium with their resources and generally exist at much more stable densities (provided that their resources do not fluctuate); such

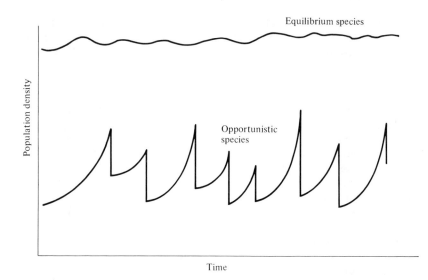

Figure 5.14 Population growth in an equilibrium species versus that in an opportunistic species subjected to irregular catastrophic mortality.

populations are called *equilibrium populations*. Clearly these two sorts of populations represent endpoints of a continuum; however, the dichotomy is useful in comparing different populations. The significance of opportunistic versus equilibrium populations is that density-independent and density-dependent factors and events differ in their effects on natural selection and on populations. In highly variable and/or unpredictable environments, catastrophic mass mortality (such as that illustrated in Table 5.3) presumably often has relatively little to do with the genotypes and phenotypes of the organisms concerned or with the size of their populations. (Some degree of selective death and stabilizing selection has, however, been demonstrated in winter kills of certain bird flocks.) By way of contrast, under more stable and/or predictable environmental regimes, population densities fluctuate less and much mortality is more directed, favoring individuals that are better able to cope with high densities and strong competition. Organisms in highly rarefied environments seldom deplete their resources to levels as low as do organisms living under less rarefied situations; as a result, the former usually do not encounter such intense competition. In a "competitive vacuum" (or an extensively rarefied environment) the best reproductive strategy is often to put maximal amounts of matter and energy into reproduction and to produce as many total progeny as possible as soon as possible. Because there is little competition, these offspring often can thrive even if they are quite small and therefore energetically inexpensive to produce. However, in a "saturated" environment, where density effects are pronounced and competition is keen, the best strategy may often be to put more energy into competition and maintenance and to produce offspring with more substantial competitive abilities. This usually requires larger offspring; and, since they are energetically more expensive, it means that fewer can be produced.

MacArthur and Wilson (1967) designate these two opposing selective forces as r selection and K selection, after the two terms in the logistic equation. Of course things are seldom so black and white, but there are usually all shades of gray. No organism is completely r selected or completely K selected; rather all must reach some compromise between the two extremes. Indeed, one can think of a given organism as an "r-strategist" or a "K-strategist" only relative to some other organism; thus statements about r and K selection are invariably comparative. We think of an $r \rightarrow K$ selection continuum, and an organism's position along it in a particular environment at a given instant in time (Pianka, 1970, 1972; see also pp. 60–64). Table 5.4 lists a variety of correlates of r and K selection.

An interesting special case of an opportunistic species is the *fugitive* species, envisioned as a predictably inferior competitor which is

Table 5.4 Some of the Correlates of r and K Selection

	r Selection	K Selection
Climate	Variable and/or unpredictable; uncertain	Fairly constant and/or predictable; more certain
Mortality	Often catastrophic, nondirected, density independent	More directed, density dependent
Survivorship	Often Type III	Usually Types I and II
Population size	Variable in time, nonequilibrium; usually well below carrying capacity of environment; unsaturated communities or portions thereof; ecologic vacuums; recolonization each year	Fairly constant in time, equilibrium, at or near carrying capacity of the environment; saturated communities; no recolonization necessary
Intra- and interspecific competition	Variable, often lax	Usually keen
Selection favors	1 Rapid development 2 High maximal rate of increase, r_{max} 3 Early reproduction 4 Small body size 5 Single reproduction 6 Many small offspring	1 Slower development 2 Greater competitive ability 3 Delayed reproduction 4 Larger body size 5 Repeated reproduction 6 Fewer larger progeny
Length of life	Short, usually less than 1 year	Longer, usually more than 1 year
Leads to	Productivity	Efficiency
Stage in succession	Early	Late, climax

Source: After Pianka (1970).

always excluded locally by interspecific competition but which persists in newly disturbed regions by virtue of a high dispersal ability (Hutchinson, 1951). Such a colonizing species can persist in spite of pressures from competitively superior species in a continually changing *patchy* environment. Hutchinson (1961) used another argument to explain the apparent "paradox of the plankton," the coexistence of many species in diverse planktonic communities under relatively homogeneous physical conditions, with limited possibilities for ecological separation. He suggested that temporally changing environments may promote diversity by periodically altering relative competitive abilities of component species, thereby allowing their coexistence.

Population "Cycles": Cause and Effect

Ecologists have long been intrigued by the regularity of certain population fluctuations, such as those of the snowshoe hare, the Canadian lynx, the ruffed grouse, and many microtine rodents (voles and lemmings) as well as their predators, including the arctic fox and the snowy owl (Keith, 1963; Elton, 1942). These population fluctuations (sometimes called "cycles") are of two types: voles, lemmings, and their predators display roughly a four-year periodicity; hare, lynx, and grouse have approximately a ten-year cycling time. Lemming population eruptions and the fabled, but very rare, suicidal marches of these rodents into the sea have frequently been popularized and are well known to the layman.

The tantilizing regularity of these fluctuations in population density presents ecologists with a "natural experiment"—hopefully one that can provide some general insights into factors influencing population densities. Many different hypotheses for the explanation of population cycles have been offered, and the literature on them is extensive (see references at end of chapter). Here, as elsewhere in ecology, it is often extremely difficult or even impossible to devise tests that separate cause from effect, and many of the putative causes of population cycles may in fact merely be side effects of the cyclical changes in the populations concerned. Several of the currently more popular hypotheses, which are not necessarily mutually exclusive, are outlined below. Descriptions and discussion of others, such as the "sunspot" and the "random peaks" hypotheses, can be found in the references. Bear in mind that two or more of these hypothetical mechanisms could act together in any given situation.

STRESS PHENOMENA HYPOTHESIS. At the extremely high densities that occur during population peaks of voles, a great deal of fighting occurs among these rodents. The so-called stress syndrome is manifested by the animals, their adrenal weights increase, and they become extremely aggressive—so much so that successful reproduction is almost completely curtailed. Eventually "shock disease" may set in and large numbers of animals may die off, apparently because of the physiological stresses on them. Christian and Davis (1964) review evidence pertaining to this hypothesis.

PREDATOR–PREY OSCILLATION HYPOTHESIS. In simple ecological systems, predator and prey populations can oscillate because of the interaction between them (see also Chapter 6). When the predator population is low the prey increase, which then allows the predators to increase, although this increase lags behind that of the prey. Eventually predators overeat their prey and the prey population begins to decline; but, because of time lag effects, the predator population continues to increase for a period, driving the prey to an even lower density. Finally at low enough prey densities many predators starve and the cycle repeats itself. However, prey populations oscillating for reasons other than predation pressures obviously constitute cyclical food supplies for their predators, which should in turn lag behind and oscillate with prey availability; short of a predator removal experiment, it is thus extremely difficult to determine whether or not changes in prey populations are causally related to changes in predator population density.

NUTRIENT RECOVERY HYPOTHESIS. According to this hypothesis (Pitelka, 1964; Schultz, 1964, 1969), one reason for the periodic decline of rodent populations (especially lemmings) is that the quality of their plant food changes in a cyclical way. During a lemming "high" the ground is blanketed with lemming fecal pellets, and many important chemical elements such as nitrogen and phosphorus are tied up and unavailable to growing plants. In the cold arctic tundra, decomposition of fecal materials takes a long time. During this period lemmings decline due to inadequate nourishment. After a lapse of a few years, feces are decomposed and their nutrients recycled once again and taken up by plants. Because their plant food is now especially nutritious, lemmings increase in numbers and the cycle repeats. Schultz (1969) has evidence of such cyclical changes, but whether they are causing, or merely effects of, the lemming population fluctuations has not been definitely established. The nutrient recovery mechanism does not apply to microtine cycles in general, for Krebs and DeLong (1965) provided supplemental food to a declining population of *Microtus* but failed to reverse the decline. More unambiguous experiments like this one are needed to assess the importance of various mechanisms of population control.

Freeland (1974) proposed a related hypothesis based on changes in the relative abundances of toxic food plants versus competitively superior palatable food plants due to preferential grazing pressures by voles.

FOOD QUANTITY HYPOTHESIS. Arctic hare populations sometimes oscillate without lynx, perhaps due to a predator–prey "cycle" involving themselves as predators and their own food plants as prey. Under this hypothesis, dense hare populations decrease the quantity of suitable foods which in turn causes a decline in the hare population. In time the plants recover, and after a lag period hares again increase. Clearly, lynx could entrain to this cycle.

GENETIC CONTROL HYPOTHESIS. This hypothesis, credited to Chitty (1960, 1967a), explains population fluctuations in terms of changing genetic composition of the population concerned. During troughs the animals experience little competition and are relatively r-selected, whereas at peaks competition is intense and they are more K-selected. Thus directional selection, related to population density, is always occurring; modal phenotypes are never the most fit individuals in the population, and each generation the gene pool changes. The population always lags somewhat behind the changing selective pressures and so no stable equilibrium exists. Some evidence exists for such genetic changes in populations of *Microtus* (Tamarin and Krebs, 1969), but here again cause and effect are extremely difficult to disentangle.

One must always be wary of oversimplification and "single-factor thinking"; most or even all of the above hypothetical mechanisms could work in concert to produce observed population "cycles." The extreme difficulty of separating cause from effect, illustrated above, plagues much of ecology. Simple tests, such as that of Krebs and De-Long (1965), which actually refute a hypothesis are badly needed. Indeed, for scientific understanding to progress rapidly and efficiently, a logical framework of *refutable* hypotheses, complete with alternatives, is absolutely essential (Platt, 1964).

Evolution of Reproductive Tactics

Natural selection recognizes only one currency: successful offspring. Yet even though all living organisms have presumably been selected to maximize their own lifetime reproductive success, they vary greatly in exact modes of reproduction. Some, such as most annual plants, a multitude of insects, and certain fish like the Pacific salmon, reproduce only once during their entire lifetime. These "big bang" or semelparous reproducers typically exert a tremendous effort in this one and only opportunity to reproduce (in fact their exceedingly high investment in reproduction may well itself contribute substantially to their own demise!). Many other organisms, including perennial plants and many vertebrates, do not engage in such suicidal bouts of reproduction but reproduce again and again during their lifetime. Such organisms have been called "iteroparous" (repeated parenthood). Even within organisms that use either the big bang or the iteroparous tactic, individuals and species differ greatly in numbers of progeny produced. Annual seed set of different species of trees ranges from a few hundred or a few thousand in many oaks (which produce relatively large seeds—acorns) to literally millions in the redwood (Harper and White, 1974). Seed production may vary greatly even among individual plants of the same species grown under different environmental conditions; an individual poppy *(Papaver rhoeas)* produces as few as four seeds under stress conditions, but as many as a third of a million seeds when grown under conditions of high fertility (Harper, 1966). Fecundity is equally variable among fish; a female cod lays millions of relatively tiny eggs, whereas most elasmobranchs produce considerably fewer but much larger offspring. Variability of clutch and/or litter size is not nearly so great among other classes of vertebrates, but it is still significant. Among lizards, for example, clutch size varies from a fixed clutch of one in some geckos and *Anolis* to as many as 40 in certain horned lizards *(Phrynosoma)* and the large *Iguana*. Timing of reproduction also varies considerably among organisms. Due to the finite chance of death, earlier reproduction is always advantageous, all else being equal. Nevertheless many organisms postpone reproduction. The century plant, an *Agave*, devotes years to vegetative growth before suddenly sending up its inflorescence (some related monocots bloom much sooner). Delayed reproduction also occurs in most perennial plants, many fish such as salmon, a few insects like cicadas, some lizards, and many mammals and birds, especially large seabirds.

Innumerable other examples of the diversity of existing reproductive tactics could be listed. Clearly natural selection has shaped observed reproductive tactics, with each presumably corresponding in some way to a local optimum that maximizes an individual's lifetime reproductive success in its particular environment. Population biologists would like to understand the factors that influence evolution of various modes of reproduction.

Reproductive Effort

How much should an organism invest in any given act of reproduction? R. A. Fisher (1930) anticipated this question years ago:

> It would be instructive to know not only by what *physiological mechanism* a just apportionment is made between the nutriment devoted to the gonads and that devoted to the rest of the parental organism, but also what *circumstances in the life history and environment* would render profitable the diversion of a greater or lesser share of the available resources towards reproduction. [My italics.]

Fisher clearly distinguished between the proximate factor (physiological mechanism) and the ultimate factors (circumstances in the life history and environment) that determine the allocation of resources into reproductive versus nonreproductive tissues and activities. Loosely defined as an organism's investment in any current act of reproduction, *reproductive effort* has played a central role in thinking about reproductive tactics. Although reproductive effort is conceptually quite useful, it has not yet been adequately quantified. Ideally, an operational measure of reproductive effort would include not only the direct material and energetic costs of reproduction but also risks associated with a given level of current reproduction. Another difficulty concerns the temporal patterns of collection and expenditure of matter and energy. Many organisms gather and store materials and energy during time periods that are unfavorable for successful reproduction but then expend these same resources on reproduction during a later, more suitable, time. The large first clutch of a fat female lizard that has just overwintered many actually represent a *smaller* investment in reproduction than her subsequent smaller clutches which must be produced with considerably diminished energy reserves. Reproductive effort could perhaps be best measured operationally in terms of the effects of various current levels of reproduction upon future reproductive success (see also below).

In spite of these rather severe difficulties, instantaneous ratios of reproductive tissues over total body tissue are sometimes used as a crude first approximation of an organism's reproductive effort (both weights and calories have been used). Thus measured, the proportion of the total resources available to an organism that is allocated to reproduction varies widely among organisms. Among different species of plants, energy expenditure on reproduction, integrated over a plant's lifetime, ranges from near zero to as much as 40 percent (Harper, Lovell, and Moore, 1970). Annual plants tend to expend more energy on reproduction than most perennials (about 14–30 percent versus 1–24 percent). An experimental study of the annual euphorb *Chamaesyce hirta* showed that calories allocated to reproduction varied directly with nutrient availability and inversely with plant density and competition (Snell and Burch, 1975).

Returning to Fisher's dichotomy for the apportionment of energy into reproductive versus nonreproductive (somatic) tissues, organs, and activities, we examine optimal reproductive effort. Somatic tissues are clearly necessary for acquisition of matter and energy; at the same time, an organism's soma is of no selective value except inasmuch as it contributes to that organism's lifelong production of successful offspring. Allocation of time, energy, and materials to reproduction in itself usually decreases growth of somatic tissues and often reduces future fecundity. Increased reproductive effort may also cost by reducing survivorship of the soma; this is easily seen in the extreme case of big bang reproduction, in which the organism puts everything available into a suicidal bout of reproduction and then dies. More subtle changes in survivorship also occur with minor alterations in reproductive effort (Figure 5.15).

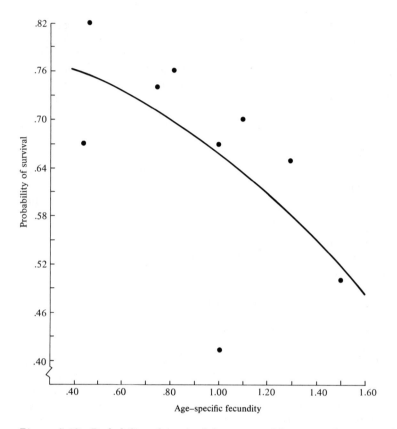

Figure 5.15   Probability of survival decreases with increased age-specific fecundity in the rotifer *Asplanchna*. [From Snell and King (1977).]

How great a risk should an optimal organism take with its soma in any given act of reproduction? To answer this question, we exploit the concept of residual reproductive value (see p. 108), which is simply age-specific expectation of all future offspring beyond those immediately at stake. To maximize its overall lifetime contribution to future generations, an optimal organism should weigh the profits of its immediate prospects of reproductive success against the costs to its long-term future prospects (Williams, 1966b). An individual with a high probability of future reproductive success should be more hesitant to risk its soma in present reproductive activities than another individual with a lower probability of reproducing successfully in the future. Moreover, to the extent that present reproduction decreases expectation of further life, it may reduce residual reproductive value directly. For both reasons current investment in reproduction should vary inversely with expectation of future offspring (Figure 5.16).

Several possible different forms for the inverse interaction between reproductive effort and residual reproductive value are depicted in Figure 5.17. Curves in this simple graphical model relate costs and profits in future offspring, respectively, to profits and costs associated with various levels of current reproduction, the latter measured in present progeny. Each curve describes all possible tactics available to a given organism at a particular instant, ranging from a current reproductive effort of zero to all-out big bang reproduction. In a stable population, immediate progeny and offspring in the more distant future are of equivalent value in perpetuation of an organism's genes (see also pp. 106–108); here a straight line with a slope of minus 45° represents equal lifetime production of offspring. A family of such lines (dashed) is plotted in Figure 5.17. An optimal reproductive tactic exists at the point of intersection of any given curve of possible tactics with the line of equivalent lifetime reproductive success that is farthest from the origin; this level of current reproduction maximizes both reproductive value at that age and total lifetime production of offspring (dots in Figure 5.17). For any given curve of possible tactics, all other tactics yield lower returns in lifetime reproductive success. The precise form

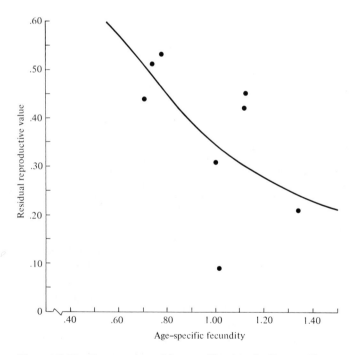

Figure 5.16 Expectation of future offspring declines with current investment in reproduction in laboratory populations of the rotifer *Asplanchna*. [From Snell and King (1977).]

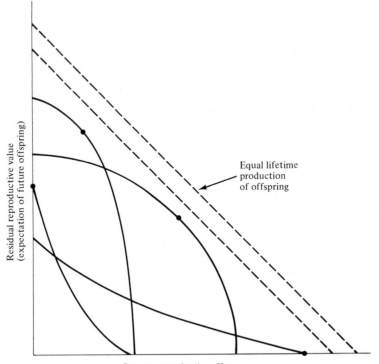

Figure 5.17 Tradeoffs between current reproductive effort and
expectation of future offspring at any particular instant (or age). Four
curves relate costs in future progeny to profits in present offspring (and
vice versa), with a dot marking the reproductive tactic that maximizes
total possible lifetime reproductive success. Concave upward curves
lead to all-or-none "big-bang" reproduction, while convex upward
curves result in repeated reproduction (iteroparity). Figures 5.18 and
5.19 depict these tradeoffs through the lifetime of a typical iteroparous
and a semelparous organism, respectively. [From Pianka (1976b).]

of the tradeoff between present progeny and expectation of future offspring thus determines the optimal current level of reproductive effort
at any given time. Note that concave upwards curves always lead to big
bang reproduction, whereas convex upwards curves result in iteroparity since reproductive value and lifetime reproductive success are
maximized at an intermediate current level of reproduction.

Probable tradeoffs between immediate reproduction and future reproductive success over the lifetime of an iteroparous organism are
depicted in Figure 5.18. The surface in this figure shows the effects of
different levels of current fecundity on future reproductive success;

the dark dots trace the optimal tactic that maximizes overall lifetime reproductive success. The shadow of this line on the age versus current fecundity plane represents the reproductive schedule one would presumably observe in a demographic study. In many organisms residual reproductive value first rises and then falls with age; optimal current level of reproduction will rise as expectation of future offspring declines. An analogous plot for a semelparous organism is shown in Figure 5.19; here current fecundity also increases as residual reproductive value falls, but the surface for a big bang reproducer is always concave upwards. Exact shapes of the surfaces depicted in these two figures depend upon the actual reproductive tactic taken by an organism as well as immediate environmental conditions for foraging, reproduction, and survival. The precise form of the tradeoffs between present progeny and expectation of future offspring is of course influenced by numerous factors including predator abundance, resource

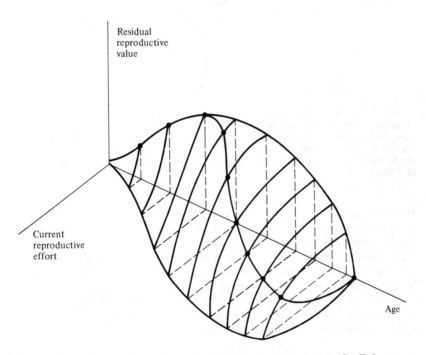

Figure 5.18   During the lifetime of an iteroparous organism, tradeoffs between current reproductive effort and future reproductive success might vary somewhat as illustrated, with the dark solid curve connecting dots tracing the optimal reproductive tactic that maximizes *total lifetime reproductive success*. The shape of this three-dimensional surface would vary with immediate environmental conditions for foraging, survival, and reproduction, as well as with the actual reproductive tactic taken by the organism concerned. [From Pianka (1976b).]

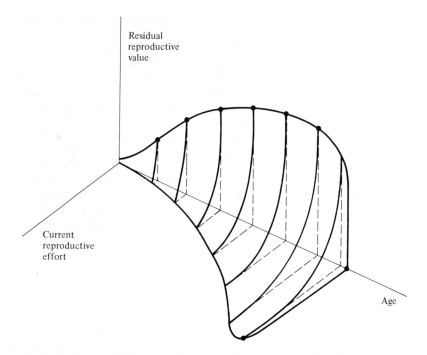

Figure 5.19   A plot (like that of Figure 5.18) for a typical semelparous or "big-bang" reproducer. The surface relating costs and profits in present versus future offspring is always concave upwards, and reproduction is all or none. Again, the actual shape of such a surface would reflect immediate environmental conditions as well as an organism's actual tactic. [From Pianka (1976b).]

availability, and numerous aspects of the physical environment. Unfavorable conditions for immediate reproduction decrease costs of allocating resources to somatic tissues and activities, resulting in lower reproductive effort. (Improved conditions for survivorship, such as good physical conditions or a decrease in predator abundance, would have a similar effect by increasing returns expected from investment in soma.) Conversely, good conditions for reproduction and/or poor conditions for survivorship result in greater current reproductive effort and decreased future reproductive success.

Expenditure per Progeny

Not all offspring are equivalent. Progeny produced late in a growing season often have lower probabilities of reaching adulthood than those produced earlier—hence they contribute less to enhancing parental fitness. Likewise, larger offspring may usually cost more to produce, but they are also "worth more." How much should a parent devote to any single progeny? For a fixed amount of reproductive effort, average fitness of individual progeny varies inversely with the total number produced. One extreme would be to invest everything in a single very large but extremely fit progeny. Another extreme would be to maximize the total number of offspring produced by devoting a minimal possible amount to each. Parental fitness is often maximized by producing an intermediate number of offspring of intermediate fitness: here the best reproductive tactic is a compromise between conflicting demands for production of the largest possible total number of progeny (*r* selection) and production of offspring of the highest possible individual fitness (*K* selection).

A simple graphical model illustrates this tradeoff between quantity and quality of offspring (Figures 5.20 and 5.21). In the rather unlikely

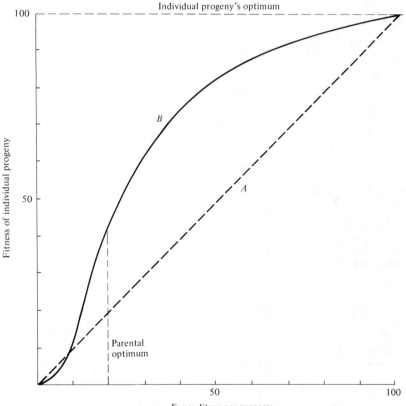

Figure 5.20   Fitness of an individual progeny should generally increase with parental expenditure. Because initial outlays on an offspring usually contribute more to its fitness than subsequent ones, curve *B* is biologically more realistic than line *A*. Note that the parental optimum differs from the optimum for individual progeny.

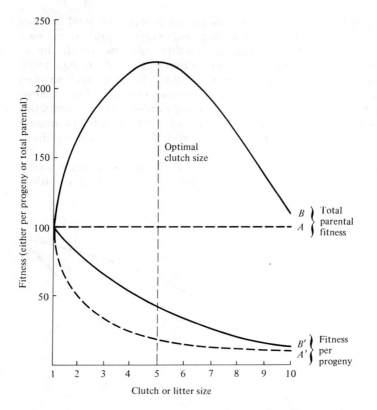

Figure 5.21 Fitness per progeny (*A'* and *B'*) and total parental fitness, the sum of the fitnesses of all offspring produced (*A* and *B*), plotted against clutch and litter size under the assumptions of Figure 5.20. Total investment in reproduction, or reproductive effort, is assumed to be constant. Note that parental fitness peaks at an intermediate clutch size under assumption *B*; optimal clutch size in this example is five.

event that progeny fitness increases linearly with parental expenditure (dashed line *A* in Figure 5.20), fitness of individual progeny decreases with increased clutch or litter size (the lowermost dashed curve *A* in Figure 5.21). However, because parental fitness—the total of the fitnesses of all progeny produced—is flat, no optimal clutch size exists from a parental viewpoint (upper dashed line *A* in Figure 5.21). If, however, the biologically plausible assumption is made that progeny fitness increases sigmoidally with parental investment* (curve *B* in

---

* Gains in progeny fitness per unit of parental investment are likely to be greater at lower expenditures per progeny than at higher ones because the proportional increase per unit of allocation is greater at low levels of investment; curves level off at higher expenditures due to the law of diminishing returns.

(Figure 5.20), there is an optimal parental clutch size (peak of upper-most curve *B* in Figure 5.21). In this hypothetical example, parents that allocate only 20 percent of their reproductive effort to each of five offspring gain a higher return on their investment than parents adopt-ing any other clutch size (Figure 5.21). While this tactic is optimal for parents, it is not the optimum for individual offspring, which would achieve maximal fitness when parents invest everything in a single offspring. Hence a "parent–offspring conflict" exists (Trivers, 1974; Alexander, 1974; Brockelman, 1975). The exact shape of the curve relating progeny fitness to parental expenditure in a real organism is influenced by a virtual plethora of environmental variables, including length of life, body size, survivorship of adults and juveniles, popula-tion density, and spatial and temporal patterns of resource availability. The competitive environment of immatures is likely to be of particular importance because larger, better endowed offspring should usually enjoy higher survivorship and generally be better competitors than smaller ones.

Juveniles and adults are often subjected to very different selective pressures. Reproductive effort should reflect environmental factors operating upon adults, whereas expenditure per progeny will be strongly influenced by juvenile environments. Since any two parties of the triumvirate determine the third, an optimal clutch or litter size is a direct consequence of an optimum current reproductive effort coupled with an optimum expenditure per progeny (indeed, clutch size is equal to reproductive effort divided by expenditure per progeny). Of course clutch size can be directly affected by natural selection as well. Recall the example of horned lizards, which are long-lived and relatively *K*-selected as adults, but which because of their tanklike body form have a very large reproductive effort and produce many tiny offspring which must suffer very high mortality (see pp. 92–93).

Patterns in Avian Clutch Sizes

A good deal of data has now been accumulated demonstrating op-timal clutch sizes in birds (see Lack, 1954, 1966, and 1968, for re-views). These elegant studies show that, compared with very small and very large clutches, clutches of intermediate size leave pro-portionately more offspring that survive to breed in the next generation. This is an excellent example of stabilizing selection. Young birds from large clutches leave the nest at a lighter weight (Figure 5.22) and have a substantially reduced postfledging survivorship. The optimal clutch apparently represents the number of young for which the parents can, on the average, provide just enough food. Perrins (1965) has good evi-dence for this in a population of great tits, *Parus major,* which varied their average clutch size from 8 to 12 over a 17-year period, apparently in response to the density of their major food, caterpillars (Figure 5.23). Wynne-Edwards (1962) interprets the optimal clutch as that which

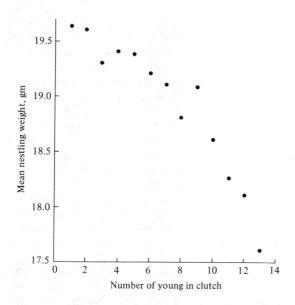

Figure 5.22   Average weight of nestling great tits plotted against clutch size, showing that individual young in larger clutches weigh considerably less than those in smaller clutches. [From data of Perrins (1965).]

produces a net number of young just replacing the parents during their lifetime of reproduction. As such, his explanation involves group selection (see Chapter 1) because individual birds do not necessarily raise as many young as possible, but rather produce only as many as are required to replace themselves. Clearly a "cheater" which produced more offspring would soon swamp the gene pool. We return to this point at the end of this chapter.

Even within the same widely ranging species, many birds and some mammals produce larger clutches (or litters) at higher latitudes than they do at lower latitudes (Figure 5.24). Such latitudinal increases in clutch size are widespread and have intrigued many population ecologists because of their general occurrence. The following hypotheses, which are not mutually exclusive, have been proposed to explain latitudinal gradients in avian clutch sizes.

THE DAYLENGTH HYPOTHESIS. As indicated in Chapter 2, during the late spring and summer, days are longer at higher latitudes than they are at lower latitudes. Diurnal birds therefore have more daylight hours in which to gather food, and thus they are able to feed larger

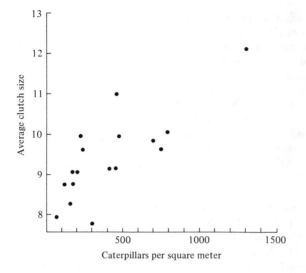

Figure 5.23   Average clutch size of great tit
populations over 17 years plotted against caterpillar
density. Clutches tend to be larger when caterpillars
are abundant than they are when this insect food is
sparse. [After Perrins (1965).]

numbers of young. However, clutch and litter sizes also increase with
latitude in nocturnal birds and mammals, which clearly have a shorter
period for foraging.

THE SPRING BLOOM OR COMPETITION HYPOTHESIS. Many
temperate-zone birds are migratory, while few tropical birds migrate.

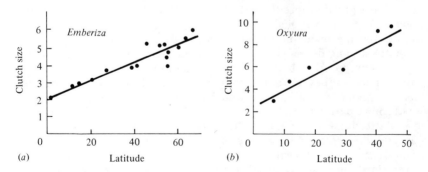

Figure 5.24   Graphs of clutch size against latitude for the avian genera
Emberiza and *Oxyura*. [After Cody (1966).]

During the spring months at temperate latitudes, there is a great surge of primary production and insects dependent upon these sources of matter and energy rapidly increase in numbers. Winter losses of both resident and migratory birds are often heavy so that spring populations may be relatively small. Hence returning individuals find themselves in a competitive vacuum with abundant food and relatively little competition for it. In the tropics wintering migrants ensure that competition is keen all year long, whereas in the temperate zones competition is distinctly reduced during the spring months. Thus, because birds at higher latitudes are able to gather more food per unit time, they are able to raise larger numbers of offspring to an age at which the young can fend for themselves.

THE PREDATOR HYPOTHESIS. There seem to be proportionately more predators, both individuals and species, in tropical than in temperate habitats (see also Chapter 8). Nest failure due to nest predation is extremely frequent in the tropics. Skutch (1949, 1967) proposed that many nest predators locate bird nests by watching and following the parents. Since the parents must make more trips to the nest if they have a large clutch, larger clutches should suffer heavier losses than smaller ones (this effect, however, has not yet been demonstrated empirically). A fact in support of the hypothesis is that hole-nesting birds, which are relatively free of nest predators, do not show as great an increase in clutch size with latitude as birds that do not nest in holes (Cody, 1966). Moreover, on tropical islands known to support fewer predators than adjacent mainland areas, birds tend to have larger clutches than mainland populations.

An observation that is somewhat difficult to reconcile with any of the three hypotheses is that clutch size often increases with altitude—as it does in the song sparrow on the Pacific coast (Johnston, 1954). Neither daylength, competition, nor predation need necessarily vary altitudinally. Cody (1966) suggested that climatic uncertainty, both instability and/or unpredictability, may well result in reduced competition at higher elevations.

Using the principle of allocation, Cody (1966) merged the above hypotheses and developed a more general theory of clutch size involving a compromise between the conflicting demands of predator avoidance, competitive ability, and clutch size. His model fits observed facts reasonably well, as one would expect of a more complex model with more parameters.

Latitudinal gradients in clutch or litter size could also be influenced by the tradeoff between expectation of future offspring and optimal current investment in reproduction. If hazards of migration or overwintering at high latitudes inevitably result in steeper mortality, expectation of life and residual reproductive value will both be reduced at higher latitudes. This, in turn, would favor an increased effort in current reproduction—and hence larger clutches.

Evolution of Death Rates and Old Age

Why do organisms become senile as they grow old? One might predict quite the opposite, since older organisms have had more experience and should therefore have learned how to avoid predators and have more antibodies, etc.; in general they should be wiser and better adapted, both behaviorally and immunologically. The physiological processes of aging have long been of interest and have received considerable attention, but only fairly recently has the evolutionary process been examined (Medawar, 1957; Williams, 1957; Hamilton, 1966; Emlen, 1970). Here again, Fisher (1930) foreshadowed thought on the subject and paved the way for its development with his concept of reproductive value.

Medawar nicely illustrated the evolution of senescence by setting up an inanimate model. A chemist's laboratory has a stock of 1000 test tubes and a monthly breakage rate of 10 percent. Every month 100 test tubes are broken completely at random and 100 new ones are added to replace them. (Although the rate of breakage is thus deterministic, the model could easily be rephrased in probabilistic or stochastic terms.) All new test tubes are marked with the date of acquisition so that their age (in months) can be determined at a later date. Every test tube has exactly the same probability of survival from any one month to the next: 900/1000 or 0.9. Thus initially older test tubes have the same mortality as younger ones and there is no senility. All test tubes are potentially immortal. The probability of surviving two months is the product of the probability of surviving each month separately, or 0.9 times 0.9 ($0.9^2 = 0.81$), while that of surviving three months is 0.9^3, and that of surviving x months is 0.9^x. After some years, the population of test tubes reaches a stable age distribution with 100 age 0 months, 90 age 1, 81 age 2, 73 age 3, . . . , 28 age 12, . . . , 8 age 24, . . . , 2.25 age 36, . . . , and less than one test tube in age groups over 48 months, . . . etc., totaling 1000 test tubes. (Of course these numbers are merely the *expected* numbers of tubes of a given age; random sampling and stochastic variations will result in some of the numbers in the various age groups being slightly above—and others slightly below—expected value.) Figure 5.25 shows part of the expected stable age distribution, with younger test tubes greatly outnumbering older ones. Virtually no test tubes are over five years old, even though individual test tubes are potentially immortal. With time, increased handling results in almost certain breakage.

Next, Medawar assigned to each of the 900 test tubes surviving every month an equal share of that month's "reproduction" (i.e., the 100 tubes added that month). Hence each surviving test tube reproduces $\frac{1}{9}$ of one tube a month. Fecundity does not change with age, but the proportions of tubes reproducing does. Younger age groups contribute much more to each month's reproduction than do older ones, simply because there are more of them; moreover, as a group their

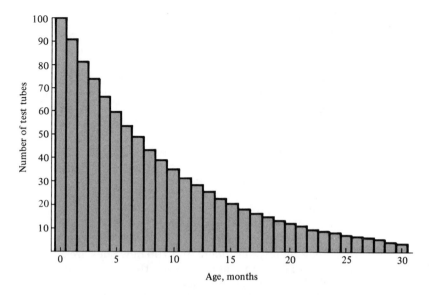

Figure 5.25 Stable age distribution of test tubes with a 10 percent breakage rate per month. Because very few test tubes survive longer than 30 months, the distribution is arbitrarily truncated at this age.

age-specific expectation of future offspring, or total reproductive value, is also higher (Figure 5.25), because as an aggregate their total expectation of future months of life is greater. In such an equally fecund and potentially immortal population, reproductive value of individuals, however, remains constant with age.

Now, pretending test tubes have "genes," consider the fate of a mutant whose phenotypic effect is to make its bearer slightly more brittle than an average test tube. This gene is clearly detrimental; it reduces the probability of survival and therefore the fitness of its carrier. This mutant is at a selective disadvantage and will eventually be eliminated from the population. Next consider the fate of another set of mutant alleles at a different locus which control the time of expression of the first gene for brittleness. Various alleles at this second locus alter the time of expression of the brittle gene differently, with some causing it to be expressed early and others late. Obviously a test tube with the brittle gene and a "late" modifier gene is at an advantage over a tube with the brittle gene and an "early" modifier because it will live longer on the average and therefore produce more offspring. Thus, even while the brittle gene is slowly being eliminated by selection, "late" modifiers accumulate at the expense of "early" modifiers. The later the time of expression of brittleness, the more nearly normal a test tube is in its contribution to future generations of test tubes. In the extreme, after reproductive value decreases to zero, natural selection, which operates only by differential reproductive success, can no longer postpone the expression of a detrimental trait and it is expressed as senescence. Thus traits that have been postponed into old age by selection of modifier genes have effectively been removed from the population gene pool. For this reason, old age has been referred to as a "genetic dustbin." The process of selection that postpones the expression of detrimental genetic traits is termed *recession of the overt effects of an allele*.

Exactly analogous arguments apply to changes in the time of expression of beneficial genetic traits, except that here natural selection works to move the time of expression of such characters to early ages, with the result that their bearers benefit maximally from possession of the allele. Such a sequence of selection is termed *precession of the beneficial effects of an allele*.

In the test tube case, reproduction begins immediately and reproductive value is constant with age. However, in most real populations organisms are not potentially immortal; furthermore, the onset of reproduction is usually delayed somewhat so that reproductive value first rises and then falls with increasing age (see Figure 5.5). Thus individuals at an intermediate age have the highest expectation of future offspring. In this situation, detrimental traits first expressed *after* the period of peak reproductive value can readily be postponed by selection of appropriate modifiers, but those first expressed *before* the period of maximal reproductive value can be quite another case, particularly if they prevent their bearers from reproducing at all. It is difficult to see how selection acting upon the bearer of such a trait could postpone the time of its expression.

In two laboratory studies on *Tribolium* beetles (Sokal, 1970; Mertz, 1975) artificial selection for early reproduction resulted in decreased longevity, demonstrating that senescence does in fact evolve. In Mertz's experiments, selected strains produced more eggs at early ages but total lifetime fecundity remained fairly constant as might be expected due to the principle of allocation.

The Joint Evolution of Rates of Reproduction and Mortality

A species with high mortality obviously must possess a correspondingly high fecundity to persist in the face of its inevitable mortality. Similarly, a very fecund organism must on the average suffer equivalently heavy mortality or else its population increases until some balance is reached. Likewise, organisms with low fecundities enjoy low rates of mortality while those with good survivorship have low fecundities. Figure 5.26 shows the inverse relationship between survivorship and fecundity in a variety of lizard populations. Rates of reproduc-

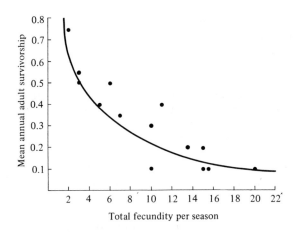

Figure 5.26   Total fecundity per reproductive season plotted against the probability of surviving to a subsequent reproductive year for 14 lizard populations. [After Tinkle (1969).] The curve represents populations with a net reproductive rate of one. Deviations from this curve can occur both because R_0 is greater or less than one and because of differences in juvenile survivorship.

tion and death rates evolve together and must stay in some kind of balance. Changes in either of necessity affect the other.

A convenient indicator of an organism's potential for increase in numbers is the maximal instantaneous rate of increase, r_{max}, which takes into account simultaneously processes of both births and deaths. Table 5.2 shows the great range of values r_{max} takes. Since actual instantaneous rate of increase averages zero over a sufficiently long time period, organisms with high r_{max} values like *Escherichia coli* are farther from realizing their maximal rate of increase than organisms with lower r_{max} values such as *Homo sapiens*. In fact, as Smith (1954) points out, r_{max} is a measure of the "harshness" of an organism's average natural environment. As such, it is one of the better indicators of an organism's position along the $r \rightarrow K$ selection continuum. Furthermore, organisms with high r_{max} values usually have much more variable actual rates of increase (and decrease) than organisms with low r_{max} values (see also Figure 5.14).

The hyperbolic inverse relationship between r_{max} and generation time T is illustrated in Figure 5.27. Small organisms tend to have much higher r_{max} values than larger ones, primarily because of their shorter generation times. Figure 5.28 shows the positive correlation between body size and generation time. The causal basis for this correlation is obvious; it takes time for an organism to attain a large body size. Moreover, delayed reproduction invariably reduces r_{max}.

Nevertheless, gains of increased body size must often outweigh losses of reduced r_{max}, or large organisms would never have evolved. The frequent tendency of phyletic lines to increase in body size during geological time as evidenced by the fossil record (Newell, 1949) has given rise to the notion of "phyletic size increase." Many advantages of large size are patently obvious, but so are some disadvantages. Certainly a larger organism is less likely to fall victim to predators and would therefore have fewer potential predators and better survivorship. Small organisms are at the mercy of their physical environment, and very slight changes in it can often be devastating; larger organisms, however, are comparatively better buffered and therefore better protected. Some disadvantages of large size are that (1) larger organisms require more matter and energy per individual per unit time

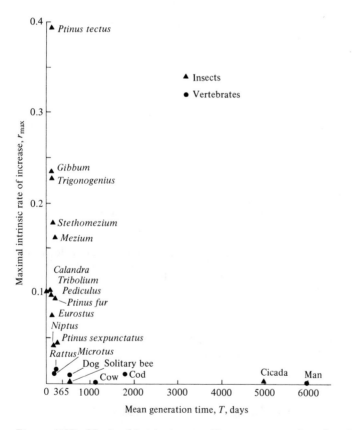

Figure 5.27   Maximal intrinsic rate of increase, r_{max}, plotted against mean generation time, T, in days, to show the strong inverse hyperbolic relationship between these two population parameters. [From Pianka (1970), partially adapted from Smith (1954).]

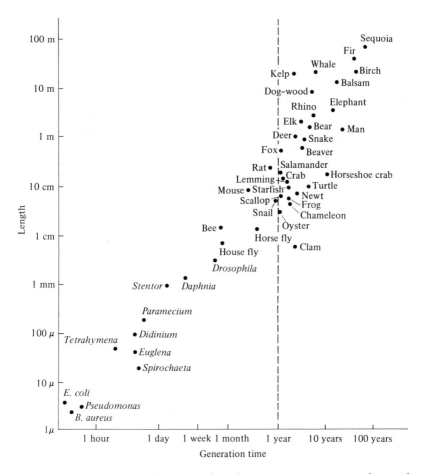

Figure 5.28   Log-log plot of organism length against generation time for a wide variety of organisms. [From John Tyler Bonner, *Size and Cycle: An Essay on the Structure of Biology* (Copyright © 1965 by Princeton University Press), Fig. 1, p. 17. Reprinted by permission of Princeton University Press.]

than smaller ones and (2) fewer refuges, hiding places, and safe sites exist for large animals than for small ones.

Use of Space: Home Range and Territoriality

Most habitats consist of a spatial–temporal mosaic of many different, often intergrading, elements, each with its own complement of organisms and other resources. Because of this extensive environmental heterogeneity, an individual's exact location is often a major determinant of its immediate fitness. Members of a prey species that are well protected from their predator(s) in one environmental patch type may be extremely vulnerable to the same predator(s) in another patch. Natural selection, by favoring those individuals that select better microhabitats, should produce a correlation between preference for a given patch type and fitness within it. The density of other individuals in a particular patch, of course, strongly influences the suitability of a given patch. Moreover, because most animals use several to many microhabitats for different purposes and/or at different times of day, the fitness relations of the use of space are usually quite complex. Some consequences of patchy environments are discussed here; others are examined in Chapters 6 and 7.

Organisms can be spaced about the landscape in two extreme ways: they may occur in groups (*clumped* or *contagious spatial distributions*) or individuals may be evenly spread out (*dispersed spatial distributions*). Intermediate between these extremes are *random spatial distributions*, in which organisms are spread randomly over the landscape. Statistical techniques have been developed to quantify the spatial relationships of individuals in a population; one such technique uses the ratio of variance to mean in the numbers of individuals per quadrat. When this ratio is unity, the distribution of organisms in the quadrats fits the Poisson distribution and the organisms are randomly distributed with respect to the quadrats. Ratios less than unity indicate dispersion while those greater than unity reflect clumped distributions. Dispersed spatial distributions are generally indicative of competition and *K* selection; however, random and clumped distributions, in themselves, indicate little about factors influencing the distribution of the organisms concerned.

Organisms vary widely in their degree of mobility (frequent ecological synonyms are *motility* and *vagility*). Some, such as terrestrial plants and sessile marine invertebrates like barnacles, spend their entire adult life at one spot, with their gametes (and/or larvae) being the dispersal stages. Others, such as earthworms and snails, although vagile, seldom move very far. Still others, such as the monarch butterfly and migratory birds, regularly move distances of many kilometers during their lifetimes.

Returns on many species of banded birds have shown that, even after migrating thousands of kilometers, individuals often return to the same general area where they were raised. Similarly, fruit flies (*Drosophila*) labeled with radioactive tracers do not usually move very far. Such restricted movements presumably allow individuals to become genetically adapted to local conditions. Two extreme types of populations are distinguished, although once again there are all degrees of intermediates: *viscous populations*, in which individuals do not usually move very far, and *fluid populations*, in which individuals cover great distances. In viscous populations there is little gene flow and great genetic variability can occur from place to place, whereas the opposite is true of fluid populations. [Note that kin selection (pp. 167–169) is much more likely to occur in viscous populations than in fluid ones.] At a local level, however, inbreeding in a viscous population may lead to reduced genetic variability.

Important differences occur between organisms in the way they use space. The distinction between two-dimensional and three-dimensional patterns of utilization is fundamental (one-dimensional use of space may also be approximated, as along the shore of a stream, lake, or ocean). Thus we find certain ecological similarities between organisms as diverse as plankton, pelagic fish, flying insects, many birds, and bats—all of which live in a three-dimensional world.

The area or volume over which an individual animal roams during the course of its usual daily wanderings and in which it spends most of its time is the animal's *home range*. Often home ranges of several individuals overlap; home ranges are not defended and are not used to the exclusion of other animals. In contrast, *territories* are defended and used exclusively by an individual, a pair, a family, or a small inbred group of individuals. Nonoverlapping territories normally give rise to dispersed spacing systems and are invariably indicative of competition for some resource in short supply.

Several different kinds of territories are classified by the functions they serve. Many sea birds, such as gulls, defend only their nest and the area immediately adjacent to it, that is, they have a *nesting territory*. Some male birds and mammals, such as grouse and sea lions, defend territories used solely for breeding, termed *mating territories*. By far the most widespread type of territory, however, is the *feeding territory*, which occurs in a few insects, some fish, numerous lizards, many mammals, and most birds.

For territoriality to evolve, some resource must be in short supply and that resource must also be defendable (Figure 5.29). Food items are generally not defendable, since most animals eat their prey as soon as they encounter it. However, the space in which prey occurs can often be defended with a reasonable amount of effort. Sometimes even space is not easily defended, especially when food items are very sparse or extremely mobile; under such circumstances feeding territories cannot be evolved because costs of defense exceed benefits that could be gained (Brown, 1964). An optimal territory size exists where the difference between benefit gained and costs of defense is greatest (Figure 5.30). Often birds that nest in colonies (i.e., seabirds and swallows) defend nesting territories, but because their prey are very mobile and thus indefensible, they have no feeding territories and feed in flocks. Territoriality is particularly prominent in insectivorous and carnivorous birds, probably largely as a result of the energetic economy of flight (Chapter 4) and their consequent great mobility, which makes territorial defense economically feasible. Typically, males of these birds set up territories during the early spring, often even before wintering

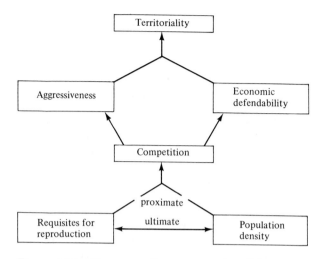

Figure 5.29 Diagrammatic representation of the various factors influencing evolution of territoriality. [From Brown (1964).]

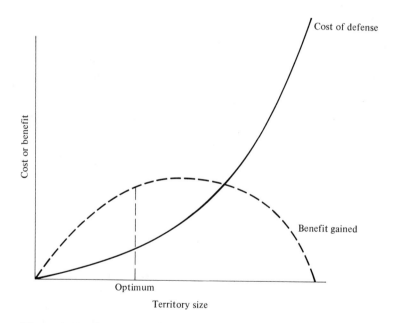

Figure 5.30 Cost of territorial defense increases monotonically with territory size, but profits gained decrease above a certain size. At the optimum territory size, the *difference* between benefit and cost is maximal.

females return. During this period many disputes over territories occur and there is much fighting among males. Once breeding has begun, however, disputes over boundaries are usually greatly reduced, with males advertising that they are still "on territory" only briefly during the morning and evening hours.

Male ovenbirds recognize individual neighbors by their territorial songs; Weedon and Falls (1959) played tape-recorded songs back in different places and at different rates and times and watched responses of various males to playbacks. When a tape recording of a non-neighbor's call is played from a neighboring territory (after the original neighbor has been removed), a male ovenbird reacts by responding vigorously and singing frequently and loudly. The male thus recognizes the substitution of calls, because the same male does not react strongly to the call of his original neighbor. Falls (1969) used this technique to study the function of the territorial song of the white-throated sparrow; a greater frequency of playback elicits a stronger response—which suggests that highly motivated birds sing more often than birds that are less likely to win a territorial encounter.

There are definite advantages to recognizing territorial calls of established neighbors and not reacting strongly to them; it would be a waste of time and energy to respond vigorously to such calls once territorial boundaries have been well established. Both individuals stand only to benefit from a "gentleman's agreement" over where such boundaries lie.

Possession of home ranges and territories also serves other important functions: by becoming familiar with a small local area, an animal can learn (1) when and where food is likely to be found, (2) the locations of safe retreats from predators, and (3) in some cases, when and where those predators are likely to be encountered. Thus resident individuals will normally have distinct advantages over nonresidents (so-called *vagrants*).

An enormous literature on territoriality exists; the interested reader can find an entry to it through the references at the end of this chapter.

Foraging Strategies

Another essential aspect of the use of space is foraging tactics, which involve the ways in which animals gather matter and energy. Matter and energy constitute the profits gained from foraging, in that they are used in growth, maintenance, and reproduction. But foraging has its costs as well; a foraging animal may often expose itself to potential predators, and much of the time spent in foraging is rendered unavailable for other activities, including reproduction. An optimal foraging tactic maximizes the difference between foraging profits and their costs. Presumably natural selection, acting as an efficiency expert, has often favored such optimal foraging behavior. Consider, for example, prey of different sizes and what might be termed "catchability." How great an effort should a foraging animal make to obtain a prey item with a given catchability and of a particular size (and, therefore, matter and energy content)? Clearly an optimal consumer should be willing to expend more energy to find and capture food items that return the most energy per unit of expenditure upon them. Moreover, an optimal forager should take advantage of natural feeding routes and should not waste time and energy looking for prey either in inappropriate places or at inappropriate times. What is optimal in one environment is seldom optimal in another, and an animal's particular anatomy strongly constrains its optimal foraging tactic. There is considerable evidence that animals actually do attempt to maximize their foraging efficiencies, and a substantial body of theory on optimal foraging tactics exists (see also pp. 260–262).

Many aspects of optimal foraging theory are concisely summarized in an excellent chapter, "The Economics of Consumer Choice," by MacArthur (1972). He makes several preliminary assumptions: (a) Environmental structure is repeatable, with some statistical expectation of finding a particular resource (such as a habitat, microhabitat, and/or prey item). (b) Food items can be arranged in a continuous and unimodal spectrum, such as size distributions of insects (Schoener and Janzen, 1968; Hespenhide, 1971). (This assumption is clearly violated by foods of some animals, such as monophagous insects or herbivores generally, because plant chemical defenses are typically discrete; see pp. 221–224.) (c) Similar animal phenotypes are usually closely equivalent in their harvesting abilities; an intermediate phenotype is thus best able to exploit foods intermediate between those which are optimal for two neighboring phenotypes (see also pp. 253–255). Conversely, similar foods are gathered with similar efficiencies; a lizard with a jaw length that adapts it to exploit best 5-mm-long insects is only slightly less efficient at eating 4-mm and 6-mm insects. (d) The principle of allocation applies, and no one phenotype can be maximally efficient on all prey types; improving harvesting efficiency on one food type necessitates reducing the efficiency of exploiting other kinds of items. (e) Finally, an individual's economic "goal" is to maximize its total intake of food resources. (Assumptions b, c, and d are not vital to the argument.)

MacArthur then breaks foraging down into four phases: (1) deciding where to search, (2) searching for palatable food items, (3) upon locating a potential food item, deciding whether or not to pursue it, and (4) pursuit itself, with *possible* capture and eating. Search and pursuit efficiencies for each food type in each habitat are entirely determined by the above assumptions about morphology (assumption c) and environmental repeatability (assumption a); moreover, these efficiencies dictate the probabilities associated with the searching and pursuing phases of foraging (2 and 4, respectively). Thus MacArthur considers only the two decisions—where to forage and what prey items to pursue (phases 1 and 3 of foraging). Clearly an optimal consumer should forage where its expectation of yield is greatest—an easy decision to make, given knowledge of the above efficiencies and the structure of its environment (in reality, animals have incomplete information, of course). The decision as to which prey items to pursue is also simple. Upon finding a potential prey item, a consumer has only two options: either pursue it or go on searching for a better item and pursue that one instead. Both decisions end in the forager's beginning a new search, so the best choice is clearly the one which returns the greatest yield per unit time. Thus an optimal consumer should opt to pursue an item only when it cannot expect to locate, catch, and eat a better item (i.e., one that returns more energy per unit of time) during the time required to capture and ingest the first prey item.

Many animals, such as foliage-gleaning insectivorous birds, spend much of their foraging time searching for prey, but expend relatively little time and energy pursuing, capturing, and eating small sedentary insects which are usually easy to catch and quickly swallowed. In such "searchers," mean search time per item eaten is large compared to average pursuit time per item; hence the optimal strategy is to eat essentially all palatable insects encountered. Other animals ("pursuers") which expend little energy in finding their prey but a great deal of effort in capturing it (such as, perhaps, a falcon or a lion) should select prey with small average pursuit times (and energetic costs). Hence pursuers should generally be more selective and more specialized than searchers. Moreover, because a food-dense environment offers a lower mean search time per item than does a food-sparse area, an optimal consumer should restrict its diet to only the better types of food items in the former habitat. To date, optimal foraging theory has been developed primarily in terms of the rate at which energy is gathered per unit time: limiting materials such as nutrients in short supply and the risks of predation have so far been largely neglected. (Optimal foraging strategies are considered further on pp. 260–262.)

Sex, Sex Ratio, Sexual Selection and Mating Systems

Sexual reproduction probably occurs in most organisms, although a number of plants and invertebrate animals employ it only at very infrequent intervals. The evolutionary origin and selective advantage(s) of sexual reproduction are still major unresolved problems in biology (Williams, 1971, 1975). Sexual processes allow the genes in a gene pool to be mixed up each generation and recombined in various new combinations; as such, genetic variability is produced by sexual reproduction. The potential rate of evolution of a sexual population is greater than that of a group of asexual organisms because a variety of beneficial mutations are readily combined into the same individual in a sexual species. But a rapid potential rate of evolution is of no advantage to an individual organism. Certainly sexual reproduction is very basic in diploid organisms and is doubtless an ancient and primitive trait. Considered from an individual's standpoint, however, sex is expensive because an individual's genes are mixed with those of another organism and hence each of its offspring carries only one-half of its genes (i.e., heritability is halved). A female reproducing asexually (including parthenogenesis), in contrast, duplicates only her own genome in each of her offspring. Even Fisher (1930) believed that sex could have evolved for the benefit of the group by way of group selection. Strangely enough, although many temporary losses of sexuality have been secondarily evolved, relatively few known organisms seem to have completely lost the capacity to exchange genes with other organisms for any geologically long period of time. Presumably, evolutionary benefits of genetic recombination and increased variability more than offset the disadvantage of one organism's perpetuating another's genes. One advantage to an individual might be that, by reproducing sexually, an organism can mix its genes with other very fit genes, thereby increasing the fitness of its progeny. Of course this can work both ways, for by mating with a less fit partner an organism could decrease its own fitness. (To the extent that heterozygosity in itself confers increased fitness, however, sexual reproduction can clearly be advantageous to individuals.)

Numerous varieties of sexual reproduction exist. Organisms may be *monoecious,* or hermaphroditic, ranging from *simultaneous hermaphrodites* (in which one individual has both male and female gonads at the same time—as in many invertebrates and many plants) to *sequential hermaphrodites* (an example is *protandry,* in which a given individual is a male when young and subsequently develops into a female). Perhaps most familiar are *dioecious* organisms, in which the sexes are separate (as in most vertebrates and some plants).

In populations of many dioecious organisms there are approximately equal numbers of males and females. The *sex ratio* is defined as the proportion of males in the population. To be more precise, we distinguish the sex ratio at conception, or the *primary sex ratio,* from that at the end of the period of parental care, the *secondary sex ratio.* The sex ratio of newly independent nonbreeding animals (as recently fledged birds) is the *tertiary sex ratio,* while that of the older breeding adult population is the *quarternary sex ratio.*

Why are various sex ratios often near equality (i.e., 50 : 50, or 0.5)? Darwin (1871) speculated that sex ratios of 1 : 1 might benefit groups by minimizing intrasexual fighting over mates. Other workers have reasoned that since one male can easily serve a number of females, it might "be better for the *species*" if the population sex ratio were biased in favor of females, since this would increase the total number of off-spring produced. Similarly, males are sometimes viewed as super-numerous and therefore "dispensable." Such interpretations invoke group selection, and it is preferable to look for an explanation of sex ratio in terms of selection at the level of the individual. Here again, Fisher (1930) first solved the problem. He noted that, in sexually re-producing diploid species, exactly half the genes (more precisely, half those on autosomal chromosomes) come from males and half from females each and every generation. This statement merely asserts that every individual organism has a mother and a father, but its implica-tions regarding the sex ratio are extensive. Fisher's early statement on the sex ratio is logically reasoned and concise.

> In organisms of all kinds the young are launched upon their careers en-dowed with a certain amount of biological capital derived from their par-ents. This varies enormously in amount in different species, but, in all, there has been, before the offspring is able to lead an independent exis-tence, a certain expenditure of nutriment in addition, almost universally, to some expenditure of time or activity, which the parents are induced by their instincts to make for the advantage of their young. Let us consider the reproductive value of these offspring at the moment when this parental expenditure on their behalf has just ceased. If we consider the aggregate of an entire generation of such offspring it is clear that the total reproductive value of the males in this group is exactly equal to the total value of all the females, because each sex must supply half the ancestry of all future gen-erations of the species. From this it follows that the sex ratio will so adjust itself, under the influence of Natural Selection, that the total parental ex-penditure incurred in respect of children of each sex, shall be equal; for if this were not so and the total expenditure incurred in producing males, for instance, were less than the total expenditure incurred in producing females, then since the total reproductive value of the males is equal to that of the females, it would follow that those parents, the innate tendencies of which caused them to produce males in excess, would, for the same ex-penditure, produce a greater amount of reproductive value; and in conse-quence would be the progenitors of a larger fraction of future generations than would parents having a congenital bias towards the production of females. Selection would thus raise the sex-ratio until the expenditure upon males became equal to that upon females. If, for example, as in man, the males suffered a heavier mortality during the period of parental ex-penditure, this would cause them to be more expensive to produce, for, for every hundred males successfully produced expenditure has been in-curred, not only for these during their whole period of dependence but for a certain number of others who have perished prematurely before incurring the full complement of expenditure. The average expenditure is therefore greater for each boy reared, but less for each boy born, than it is for girls at the corresponding stages, and we may therefore infer that the condition toward which Natural Selection will tend will be one in which boys are the more numerous at birth, but become less numerous, owing to their higher death-rate, before the end of the period of parental expenditure. The actual sex-ratio in man seems to fulfil these conditions somewhat closely, espe-cially if we make allowance for the large recent diminution in the deaths of

infants and children; and since this adjustment is brought about by a somewhat large inequality in the sex ratio at conception, for which no *a priori* reason can be given, it is difficult to avoid the conclusion that the sex-ratio has really been adjusted by these means.

The sex-ratio at the end of the period of expenditure thus depends upon differential mortality during that period, and if there are any such differences, upon the differential demands which the young of such species make during their period of dependency; it will not be influenced by differential mortality during a self-supporting period; the relative numbers of the sexes attaining maturity may thus be influenced without compensation, by differential mortality during the period intervening between the period of dependence and the attainment of maturity. Any great differential mortality in this period will, however, tend to be checked by Natural Selection, owing to the fact that the total reproductive value of either sex, being, during this period, equal to that of the other, whichever is the scarcer, will be the more valuable, and consequently a more intense selection will be exerted in favour of all modifications tending towards its preservation. The numbers attaining sexual maturity may thus become unequal if sexual differentiation in form or habits is for other reasons advantageous, but any great and persistent inequality between the sexes at maturity should be found to be accompanied by sexual differentiations, having a very decided bionomic value. [Fisher (1930) pp. 142–143.]

Fisher concludes that, at equilibrium, an optimal organism should allocate exactly half its reproductive effort to progeny of each sex; thus if each male offspring costs approximately as much to produce as each female offspring, the optimal family sex ratio is near 50 : 50, provided that the population is in or near equilibrium. Note that Fisher's argument does *not* depend upon competition for mates in any way, as it assumes that each male has the same probability of mating as every other male (likewise all females are assumed to be of equal fitness).

Figure 5.31 and Table 5.5 illustrate Fisher's principle in two hypothetical populations. In the first case, there is no sexual dimorphism in energy demands of progeny and the parental investment argument can be translated directly into numbers; thus this case leads to an optimal family sex ratio at equilibrium of 0.5. This is not true in the second case, where there is an investment differential such that individual offspring of one sex require twice as much parental expenditure as individuals of the other sex; this case leads to an optimal family sex ratio at equilibrium of either 0.33 or 0.67, depending on which sex is more expensive to produce. In both cases the optimal family sex ratio differs when the population sex ratio deviates from the optimal family sex ratio at equilibrium. In such a circumstance, families producing the sex in deficit (compared to the equilibrium sex ratio) have a selective advantage; this results in excess production of the underrepresented sex, which then forces the population sex ratio to the equilibrium sex ratio. Sex ratio is a special case of *frequency-dependent selection*, which arises whenever the selective value of a trait changes with its frequency of occurrence. Figure 5.32 shows the

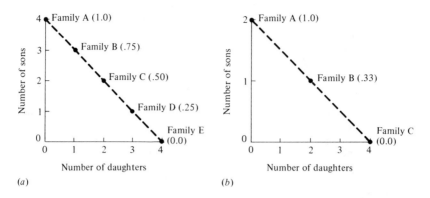

(a)   (b)

Figure 5.31   Two hypothetical situations illustrating possible family structures without a sexual dimorphism in costs to the parent (*a*) and with such a sexual differential (*b*). Parents are assumed to invest a given fixed amount in reproduction. In (*a*) sons cost parents the same amount as daughters and the optimal family sex ratio (provided the population is at equilibrium) is 0.50, at which sex ratio parental investment is equalized on offspring of the two sexes. In (*b*) male progeny cost their parents twice as much as female progeny, and expenditure on offspring of the two sexes is equalized at a family sex ratio of 0.33 (again, provided the population at large is near equilibrium). Table 5.5 and the text develop the reasons why parental investment should be divided equally between offspring of each sex when the population is at equilibrium.

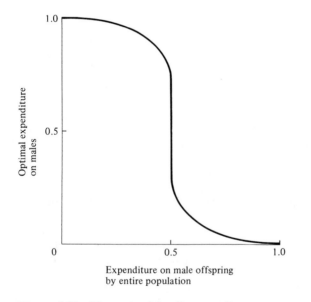

Figure 5.32   The optimal family expenditure on sons varies with expenditure on male offspring in the population at large.

Table 5.5 Comparison of the Contributions to Future Generations of Various Families in Case *a* and Case *b* of Figure 5.31 in Populations with Different Sex Ratios

CASE *a*:

	Number of Males	Number of Females
Initial Population	100	100
Family A	4	0
Family C	2	2
Subsequent Population (sum)	106	102

$C_A = 4/106 = 0.03773$
$C_C = 2/106 + 2/102 = 0.03846$ (family C has a higher reproductive success)

CASE *b*:

	Number of Males	Number of Females
Initial Population	100	100
Family A	2	0
Family B	1	2
Subsequent Population	103	102

$C_A = 2/103 = 0.01942$
$C_B = 1/103 + 2/102 = 0.02932$ (family B is more successful)

	Number of Males	Number of Females
Initial Population	100	100
Family B	1	2
Family C	0	4
Subsequent Population	101	106

$C_B = 1/101 + 2/106 = 0.02877$
$C_C = 4/106 = 0.03773$ (family C is more successful than family B)

Natural selection will favor families with an excess of females until the population reaches its equilibrium sex ratio (below)

	Number of Males	Number of Females
Initial Population	100	200
Family B	1	2
Family C	0	4
Subsequent Population	101	206

$C_B = 1/101 + 2/206 = 0.01971$
$C_C = 4/206 = 0.01942$ (family B now has the advantage)

Note: The contribution of family x is abbreviated C_x.

way in which optimal family sex ratio varies with population sex ratio when the two sexes are equally expensive to produce. This plot is more general in that it shows how optimal investment of a family varies as average expenditure of the population at large changes. Note that when average expenditure on offspring of the two sexes by the entire population is near equilibrium (1 : 1), a relatively broad range of family strategies are near optimal; as overall population expenditure deviates more from this equilibrium, the optimal family tactic rapidly converges on producing families containing only the more valuable underrepresented sex.

Thus the only factor that can influence the primary and secondary sex ratios is a sexual difference in costs of progeny to the parents. A special case is differential mortality of the sexes *during the period of parental care* (Figure 5.33). Because differential energetic requirements and differential mortality after this period do not interact with parental expenditure, they cannot directly alter either the primary or the secondary sex ratios unless their effects are also manifest during the period of parental care. Sexual dimorphisms, or physiological, morphological, and/or behavioral differences between the sexes, are of paramount importance in any discussion of sex ratio. We next discuss the great variety of ecological factors that can influence sexual dimorphism, especially sexual selection and mating systems, which are complexly intertwined.

Fisher (1958a) quotes an unnamed "modern" biologist as having asked the question, "Of what advantage could it be *to any species* for the males to struggle for the females and for the females to struggle for the males?" (italics mine). As Fisher points out, this is really a pseudo question because the fundamental units of natural selection are individuals rather than species. The question thus reflects the attitude of a group selectionist. In the next few pages we see exactly why this is not a biologically meaningful question.

Given that an organism is to mix its genes with those of another individual (i.e., that it is to reproduce sexually), just which other individual those genes are mixed with can make a substantial difference.

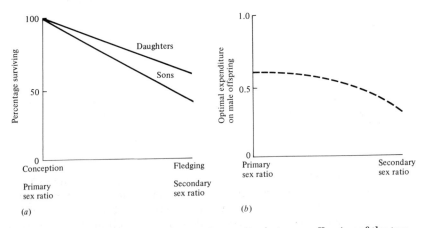

Figure 5.33   When there is differential mortality between offspring of the two sexes during the period of parental care, parental expenditure on the two sexes can only be equalized through skewed primary and secondary sex ratios. For example, if sons die off more rapidly than daughters (*a*), the optimal primary sex ratio will be baised toward sons while the optimal secondary sex ratio is biased in favor of daughters (*b*). The line curves because the proportion of males to females decreases more and more rapidly as surviving daughters accumulate.

An organism mating with a very fit partner by virtue of associating its genes with "good" genes passes its own genes on to future generations more effectively than another genetically identical individual (twin) mating with a less fit partner. Thus those members of any population that make the best matings leave a statistically greater contribution to future generations. As a result, *within* each sex there is competition for the best mates of the opposite sex; this leads to the *intrasexual* component of *sexual selection*. Intrasexual selection usually generates antagonistic and aggressive interactions between members of a sex, with those individuals best able to dominate other individuals of their own sex being at a relative advantage. Often direct physical battle is unnecessary and mere gestures (and/or various other signals of "strength") are enough to determine which individual "wins" an encounter. This makes some selective sense, for if the outcome of a fight is relatively certain, little if anything can be gained from actually fighting; in fact there is some disadvantage due to the finite risk of injury to both contestants. Similar considerations apply in the defense of territories.

Maynard Smith (1956) convincingly demonstrated mating preferences in laboratory populations of the fruit fly *Drosophila subobscura*. Females of these flies usually mate only once during their lifetime and store sperm in the seminal receptacle. Males breed repeatedly. Maynard Smith mated genetically similar females to two different strains to males, one inbred (homozygous) and one outbred (heterozygous), and collected all eggs laid by these females during their entire lifetimes. He found that similar total number of eggs were laid by both groups of females, but the percentage of eggs which hatched differed markedly. Females mated to inbred males laid an average of only 264 fertile eggs each, whereas those bred to outbred males produced an average of 1134 fertile eggs per female (and hence produced over four times as many viable offspring). Maynard Smith reasoned that there should therefore be strong selection for females to mate with outbred rather than inbred males. When virgin females were placed in a bottle with outbred males, mating occurred with an hour in 90 percent of the cases; however, when similar virgins were offered inbred males, only 50 percent mated during the first hour. Both kinds of males courted females vigorously and repeatedly attempted to mount females, but outbred males were much more successful than inbred ones. By carefully watching the elaborate courtship behavior of these little flies, Maynard Smith discovered that inbred males responded more slowly than outbred males to the rapid side-step dance of females. Presumably as a result of this lagging, females often rejected advances of inbred males and flew away before being inseminated. These observations clearly show that females exert a preference as to which males they will accept. Similar mating preferences almost certainly exist in most natural populations, although they are usually difficult to demonstrate. Over evolutionary time, natural selection operates to produce a correlation between male fitness and female preference because those females preferring the fittest males associate their own genes with the best male genes and therefore produce the fittest sons (female fitness is correlated with male preference for the same reason).

As a result of such mating preferences, populations have *breeding structures*. At one extreme is inbreeding, in which genetically similar organisms mate with one another *(homogamy)*; at the other extreme is outbreeding, in which unlikes mate with each other *(heterogamy)*. Outbreeding leads to association of unlike genes and thus generates genetic variation. Inbreeding produces genetic uniformity at a local level, although variability may persist over a broader geographic region. Both extremes represent nonrandom breeding structures; randomly mating *panmictic* populations described by the Hardy-Weinberg equation of population genetics lie midway between them. It is very doubtful, however, that any natural population is truly panmictic.

Animal populations also have *mating systems*. Most insectivorous birds and carnivorous birds and mammals are *monogamous*, with a pair bond between one male and one female. In such a case, both parents typically care for the young. *Polygamy* refers to mating systems in which one individual maintains simultaneous pair bonds with more than one member of the opposite sex. There are two kinds of polygamy, depending upon which sex maintains multiple pair bonds. In some birds, such as marsh wrens and yellowheaded blackbirds, one male may have pair bonds with two or more females at the same time *(polygyny)*. Much less common is *polyandry*, in which one female has simultaneous pair bonds with more than one male; polyandry is, however, thought to occur in a few bird species, such as some jacanas, rails, and tinamous. In *polybrachygyny* (Selander, 1972) one male has several short pair bonds with different females in sequence; typically each such pair bonds lasts only long enough for completion of copulation and insemination. Polybrachygyny occurs in a variety of birds (including some grackles, hummingbirds, and grouse) and mammals (many pinnipeds and some ungulates). Finally, an idealized mating system (perhaps, more appropriately, a lack of a mating system) is *promiscuity*, in which each organism has an equal probability of mating with every other organism. True promiscuity is extremely unlikely and probably nonexistent; it would result in a panmictic population. It may be approached in some invertebrates like certain polychaete worms and crinoid echinoderms, which shed their gametes into the sea, or in terrestrial plants that release pollen to the wind, where they are mixed by currents of water and air. However, various forms of chemical discrimination of gametes—and therefore mating preferences—probably occur even in such sessile organisms.

The intersexual component of sexual selection (that occurring between the sexes) is termed *epigamic selection*. It is often defined as "the reproductive advantage accruing to those genotypes which provide the stronger heterosexual stimuli," but it is also aptly described as "the battle of the sexes." Epigamic selection operates by mating preferences. Of prime importance is the fact that what maximizes an individual male's fitness is not necessarily coincident with what is best for an individual female, and vice versa. As an example, in most vertebrates individual males can usually leave more genes under a polygynous or polybrachygynous mating system, whereas an individual female is more likely to maximize her reproductive success under a monogamous or polyandrous system. Sperm are small, energetically inexpensive to produce, and are produced in large numbers. As a result, vertebrate males have relatively little invested in each act of reproduction and can and do mate frequently and rather indiscriminately (i.e., males tend toward promiscuity). Vertebrate females, on the other hand, often or usually have much more invested in each act of reproduction because eggs and/or offspring are usually energetically expensive. Because these females have so much more at stake in each act of reproduction, they tend to exert much stronger mating preferences than males and to be more selective as to acceptable mates. By refusing to breed with promiscuous and polygynous males, vertebrate females can sometimes "force" males to become monogamous and to contribute their share toward raising the offspring. In effect, polygyny is the outcome of the battle of the sexes when the males win out (patriarchy), whereas polyandry is the outcome when females win out (matriarchy). Monogamy is a compromise between these two extremes. Under a monogamous mating system, a male must be certain that the offspring are his own; otherwise he might expend energy raising offspring of another male (note that females do not have this problem). No wonder monogamous males jealously guard their females against stolen copulations! On the other hand, females mated monogamously are vulnerable to desertion once reproduction is underway.

Let us now examine ecological determinants of mating systems. Some assert that sex ratios "drive" mating systems; under such an interpretation polygyny arises when males are in short supply and polyandry occurs when there are not enough females to go around. According to this explanation, many species are monogamous simply *because* sex ratios are often near equality. In fact quite the reverse is true, with sexual selection and mating systems indirectly and directly determining sexual dimorphisms and hence various sex ratios. In many birds and some mammals, floating populations of nonbreeding males exist. These can be demonstrated by simply removing breeding individuals; typically they are quickly replaced with younger and less experienced animals (Stewart and Aldrich, 1951; Hensley and Cope, 1951; Orians, 1969b).

Only 14 of the 291 species (5 percent) of North American passerine birds are regularly polygynous (Verner and Willson, 1966). Some 11 of these 14 (nearly 80 percent) breed in prairies, marshes, and savanna habitats. Verner and Willson suggested that, in these extremely productive habitats, insects are continually emerging and thus food supply is rapidly renewed; as a result several females can successfully exploit the same feeding territory. However, a similar review of the nesting habitats used by polygynous passerines in Europe (which also constitute about 5 percent of the total number of species) showed no such prevalence toward grassland or marshy habitats (Haartman, 1969). Indeed, for elusive reasons, Haartman suggested that closed-in, safe nests were a more important determinant of polygynous mating systems than were breeding habitats. Crook (1962, 1963, 1964, 1965) has suggested that among African weaverbirds monogamy is evolved when food is scarce and both parents are necessary to raise young, whereas polygyny evolves in productive habitats with abundant food where male assistance is less essential. This argument, of course, ignores entirely "the battle of the sexes" (epigamic selection).

One of the best field studies of polygyny to date is that of Verner (1964), who studied long-billed marsh wrens in Washington state. Male wrens build dummy nests (not used by females) scattered around their territories; during courtship females are escorted around a male's entire territory and shown each dummy nest (this presumably allows females to assess the quality of a male's territory). Some males possessed two females (one male had three), whereas males on adjacent territories had only one female or none at all. Territories of bigamous and trigamous males were not only larger than those of bachelors and monogamous males, but they contained more emergent vegetation (where female wrens forage). Verner reasoned that a female must be able to raise more young by pairing with a mated male on a superior territory than by pairing with a bachelor on an inferior territory even though she obtains less help from her mate. This has since been demonstrated in redwinged blackbirds (Figure 5.34). Verner noted that evolution of polygyny depends on males being able to defend territories containing enough food to support more than one female and her offspring; this condition for the evolution of polygyny requires fairly productive habitats. Female wrens are antagonistic toward each other and as a result males cannot make a second mating until their first female is incubating; this produces a temporal staggering of the females (Verner, 1965). Building on Verner's work and studies on blackbirds, Verner and Willson (1966) defined the *polygyny threshold* as the minimum difference in habitat quality of territories held by males in the same general region that is sufficient to favor bigamous matings by females (Figure 5.35).

Polygyny is much more prevalent in mammals than in birds, presumably because in most mammals females nurse their young and, at least among herbivorous species, males can do relatively little* to assist females in raising young (such species typically have a pronounced sexual dimorphism). A notable exception is carnivorous mammals that

---

* Why male mammals do not lactate is unresolved.

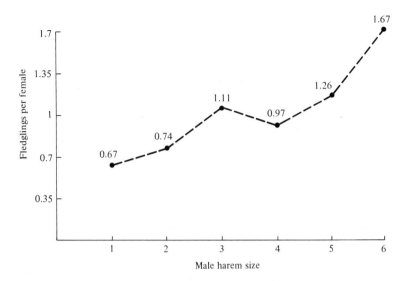

Figure 5.34   Reproductive success of female redwinged blackbirds mated with males having different harem sizes. [After Alcock (1975). *Animal Behavior*. From data of Holm (1973). Copyright by the Ecological Society of America.]

are often monogamous during the breeding season, with males participating in feeding both the female and the young (typically sexual dimorphisms are slight in such species). Similarly, most carnivorous and insectivorous birds are monogamous, and males can and do gather food for nestlings. Often sexual dimorphism in such species is slight, and those that are dimorphic are usually migratory (Hamilton, 1961, suggested that sexual dimorphism promotes rapid pairing as well as species recognition). Birds whose young are well developed at hatching (*precocial* as opposed to *atricial* birds) typically have little male parental care and are frequently polygynous or polybrachygynous, with pronounced sexual dimorphisms.

Given a population sex ratio near equality and a monogamous mating system, every individual (both male and female) has a substantial opportunity to breed and to pass on its genes; however, with an equal population sex ratio and polygyny, a select group of the fittest males make a disproportionate number of matings. Dominant battle-scarred males of the northern sea lion, *Eumetopias jubata*, which have "won" the rocky islets where most copulations occur, often have harems of 10 to 20 females. Under such circumstances, these males father most progeny and their genes constitute half the gene pool of the subsequent generation. Because those heritable characteristics making them good fighters and dominant animals are passed on to their sons, contests

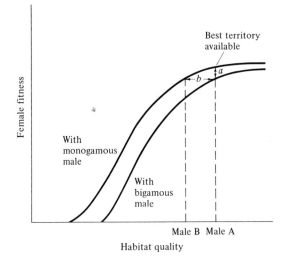

Figure 5.35 Graphic model of the conditions necessary for the evolution of polygyny. Reproductive success of females is correlated with environmental quality and females select mates which give them the highest individual fitness. Distance *a* is the difference in fitness between a female mated monogamously and a female mated bigamously in the same environment. Distance *b* is the *polygyny threshold*, or the minimum difference in territory quality held by males in the same region which is sufficient to favor bigamous matings by females. [After Orians (1969b).]

over the breeding grounds may be intensified in the next generation. It is essentially an "all or none" proposition in that only the winning males are able to perpetuate their genes. As a result of this intense competition between males for the breeding grounds, intrasexual selection has favored a striking sexual dimorphism in size. Whereas adult females usually weigh less than 500 kg, adult males may weigh as much as 1000 kg. Sexual dimorphism in size is even more pronounced in the California sea lion, *Zalophus californianus*, where females attain weights of only about 100 kg, while males reach nearly 500 kg. Presumably the upper limit on such a differential size escalation is set by various other ecological determinants of body size, such as predation pressures, foraging efficiency, and food availability (see also Chapters 4 and 7).

Somewhat analogous situations occur among various polygynous and polybrachygynous birds. Many species of grouse are polybrachygynous, with males displaying their sexual prowess in groups termed "leks." Dominant, usually older, males occupy the cen-

tral portion of the communal breeding grounds and make a disproportionate percentage of the matings. Receptive females rush past periph-
eral males to get to central males for copulation. Strong sexual dimorphisms in size, plumage, color, and behavior exist in many grouse.
In addition to intrasexual selection, epigamic selection operating
through female choice can also produce and maintain sexual dimorphisms; usually both types of sexual selection occur simultaneously, and it is often difficult to separate their effects. Indeed, by
choosing to mate with gaudy and conspicuous males, females have
presumably forced the evolution of some bizarre male sexual
adaptations, such as the long tails of some male birds of paradise.
Bower birds have avoided becoming overly gaudy (and hence danger-
ously conspicuous) by evolving a unique behavioral adaptation; males

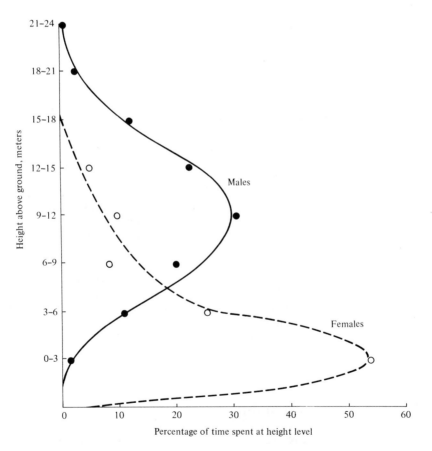

Figure 5.36   Male red-eyed vireos tend to forage higher than females. [Adapted
from P. Williamson (1971). Copyright by the Ecological Society of America.]

build highly ornamented bowers that are used to attract and to court females and that signal the male's intersexual attractiveness. Frequently, if not usually, the same sexual characteristics (such as size, color, plumage, song, behavior) advertise both intrasexual prowess and intersexual attractiveness. This makes evolutionary sense because an individual's overall fitness is determined by its success at coping with both types of sexual selection, which should usually be positively correlated; moreover, economy of energy expenditure is also obtained by consolidation of sexual signals.

Sexual dimorphisms sometimes serve still another ecological function, by reducing niche overlap and competition between the sexes (Figure 5.36). In certain island lizards (Schoerner, 1967, 1968a) and some birds (Selander, 1966), strong sexual dimorphisms in the feeding apparatus (jaws and beaks, respectively) are correlated with differential utilization of food resources (see also Chapter 7).

Fitness and the Individual's Status in the Population

An organism's fitness is determined by the interaction between its phenotype and the totality of its environment. In K-selected organisms, fitness is determined largely by the biotic environment, especially the individual's status within its own population; fitness of r-selected organisms, however, may often depend less upon the biotic environment and be more strongly influenced by the physical environment.

Compared with breeding individuals, members of a nonbreeding floating population have very low immediate fitnesses (although their fitness rises if they are able to breed later). Even within a breeding population, various individuals may often differ substantially in fitness; for instance, among long-billed marsh wrens, males vary greatly in their individual fitness, with bigamous and trigamous males being more fit than monogamous ones. Predators tend to crop mainly the "excess" members of bobwhite quail and muskrat populations (Errington, 1956, 1963); presumably resident prey individuals know their own territories and home ranges—making them more difficult to capture than vagrant individuals with a less stable status in the population.

Maintenance of Variability

The fundamental source of variation between individuals is sexual reproduction; reassortment and recombination of genes each generation assure that new genotypes will arise regularly in any population with genetic variability. In most higher organisms no two individuals are genetically identical (except identical twins and progeny produced asexually). Population biologists are interested in understanding factors that create and maintain genetic variability in natural populations (Selander and Johnson, 1972). Numerous genetic mechanisms produce genetic variability, both within and between populations, including linkage, chromosomal inversions, translocations, and heterosis (see also Chapter 1).

At the outset we must distinguish phenotypic from genotypic variation. The phenotypic component of variability is the total observable variability; the genotypic component is that with a genetic basis. It is usually difficult to distinguish genetically induced variation from environmentally induced variation. However, by growing clones of genetically identical individuals (i.e., with the same genotype) under differing environmental conditions, biologists have been able to determine how much interindividual variation is due to the developmental plasticity of a particular genotype in different environments. Pedigree studies show that approximately half the phenotypic variation in height observed in human populations has a genetic basis and the remaining variation is environmentally induced. Since natural selection can act only on heritable traits, many phenotypic variants may have little direct selective value. The degree of developmental flexibility of a given phenotypic trait strongly influences an organism's fitness; such a trait is said to be *canalized* when the same phenotypic character is produced in a wide range of genetic and environmental backgrounds. Presumably some genes are rather strongly canalized, such as those that produce "wild-type" individuals, whereas others are less determinant, allowing individuals to adapt and regulate via developmental plasticity. Such environmentally induced phenotypic varieties are common in plants; they are less common among animals, probably because mobile organisms can easily select an appropriate environment. It may be selectively advantageous for certain genetically induced traits to be under tight control while others increase individual fitness by allowing some flexibility of response to differing environmental influences.

Genotypic and phenotypic variation between individuals, in itself, is probably seldom selected for directly. But it may often arise and be maintained in a number of more or less indirect ways. Especially important are changing environments; in a temporally varying environment, selective pressures vary from time to time and the phenotype of highest fitness is always changing. There is inevitably some lag in response to selection, and organisms adapted to tolerate a wide range of conditions are frequently at an advantage. (Heterozygotes may often be better able to perform under a wider range of conditions than homozygotes.) Indeed, in unpredictably changing environments, reproductive success may usually be maximized by production of offspring with a broad spectrum of phenotypes (this may well be the major advantage of sexual reproduction).

Similar considerations apply to spatially varying environments because phenotypes best able to exploit various "patches" usually differ (see also Chapter 7). On a broader geographic level, differences from one habitat to the next presumably often result in different selective milieus and therefore in different gene pools adapted to local conditions. Gene flow between and among such divergent populations can result in substantial amounts of genetic variability, even at a single spot (this has been termed the "gene flow-variation hypothesis"—pp. 319–320).

Competition among members of a population for preferred resources (Chapter 6) may often confer a relative advantage on variant individuals that are better able to exploit marginal resources; thus competition within a population can directly favor an increase in its variability. By virtue of such variation between individuals, the population exploits a broader spectrum of resources more effectively and has a larger populational "niche breadth"; the "between phenotype" component of niche breadth is great (Roughgarden, 1972; see also pp. 253–256 and 317–319). Because such increased phenotypic variability between individuals promotes a broader populational niche, this has been termed the "niche-variation hypothesis" (Soulé and Stewart, 1970). Similarly, environments with low availability of resources usually require that individuals exploiting them make use of a wide variety of available resources; in this case, however, because each individual must possess a broad niche, variation between individuals is not great (i.e., the "between-phenotype" component of niche breadth is slight while the "within-phenotype" component is great).

One further way in which variability can be advantageous involves the interactions between individuals belonging to different species (Chapter 6), especially interspecific competition and predation. Fisher (1958b) likened such interspecific interactions and coevolution to a giant evolutionary game in which moves alternate with countermoves. He suggested that it may well be more difficult to evolve against an unpredictable and variable polymorphic species than against a better standardized and more predictable monomorphic species. A possible example may be foraging birds developing a "search image" for prey items commonly encountered, often bypassing other less abundant kinds of suitable prey.

Social Behavior and Kin Selection

A wide variety of ecological phenomena have been interpreted as having been evolved for the benefit of the population rather than for the benefit of individuals. Clutch size, sex ratio, and sexual selection have already been discussed; predator alarm calls and so-called "prudent predation" are discussed in Chapter 6, and "selection at the level of the ecosystem" is considered in Chapter 8. Another broad area latent with opportunities for interpretation of events and phenomena as having arisen for the advantage of a group is social behavior. Why, for example, does a worker honey bee sacrifice her own reproduction for the good of the colony? As is well known, a worker bee will even give up her own life in defense of the hive. To find the probable answer to the above question, we must first develop some basic considerations and definitions.

True altruism occurs only when an individual behaves so that it suffers a net loss while a neighbor (or neighbors) somehow gain from its loss (Figure 5.37). Obviously selfish behavior will always be of selective advantage; the problem is to explain the occurrence of apparently altruistic behavior such as that of the worker honey bee.

Hamilton (1964) points out that the very existence of social behavior implies that *individuals* living in cooperating groups are in fact leaving more genes in the population gene pool than hermits. Thus social behavior may be expected to evolve when distinct advantages are inherent in group participation. To develop his thesis, Hamilton defines *kin selection* as selection operating between closely related individuals to produce cooperation. As an extreme example, an individual should theoretically sacrifice its own life if it can thereby save the lives of more than two siblings, each of which shares half its genes. Such behavior furthers its own genotype even more effectively than

	Neighbor(s) gain	Neighbor(s) lose
Individual gains	Pseudo–altruistic behavior (kin-selection)	Selfish behavior (selected)
Individual loses	True altruistic behavior (counter–selected)	Mutually disadvantageous behavior (counter–selected)

Figure 5.37 The four possible situations involving an individual's behavior and its influence upon a neighbor. Kin selection occurs when an individual actually gains more than it loses (due to the advantage the individual obtains in perpetuating its genes among its neighbors who are relatives); thus it is pseudo-altruistic behavior. True altruistic behavior, in which an individual actually loses while its neighbor gains, is virtually unknown (except, perhaps, in humans); moreover, group selection is necessary for its evolution.

living to reproduce itself. This is not true altruistic behavior because the individual making the "sacrifice" actually gains more than it loses. Kin selection operates at a much more subtle level: closely related relatives are much more likely to benefit from such pseudo-altruistic behavior than distantly related ones; for the latter to occur, loss to the benefactor should be very small and/or number of distant relatives benefited must be large. Thus the condition necessary for evolution of such pseudo-altruistic behavior is that total gain(s) to relative(s) must exceed the loss to the pseudo-altruist. Parental care is, of course, a special case of kin selection.

Here once again, ideas were foreshadowed by Fisher (1930), in his consideration of the evolution of distastefulness and warning coloration in insects. Many distasteful or even poisonous insects, especially the larvae of some moths and butterflies, are brightly colored; vertebrate predators, especially birds, quickly learn not to eat such warningly colored insects. Fisher noted the difficulty of explaining the origin of this gaudy coloration since new ultraconspicuous mutants would be expected to suffer the first attacks of inexperienced predators and therefore be at a selective disadvantage to less gaudily colored genotypes. He suggested that benefits to siblings on the same branch accruing under a gregarious family system could favor evolution of warning coloration; hence such ultraconspicuous mutants would be "pseudo-altruists." Fisher even made his argument quantitative by noting that although each sibling shares only half the individual's genes, their numbers ensure that the total number of shared genes benefited greatly exceeds the number destroyed in the individual's own genome.

Members of the insect order Hymenoptera, which includes ants, bees, wasps, and hornets, often form colonies and exhibit apparent altruistic social behavior. Hymenoptera have a peculiar haploidiploid genetic system: males are produced asexually and are haploid. Thus all sperm produced by any given male are genetically identical (barring somatic mutation) and carry that male's entire genome. (Males have no father but do have a grandfather.) Females are normal diploids, with each ovum carrying only one-half their genomes. In many hymenopteran species, a queen mates just once during her entire lifetime and stores one male's sperm in a spermatotheca; thus all her progeny have the same father. A result of this strange genetic system is that sisters are more closely related to one another than mothers are to their own daughters; the former share three-fourths of their genes, the latter only one-half. Hence one would predict that worker bees should help raise their sisters to reproductive maturity in preference to mating themselves, which in fact they often do. Interestingly, workers can and sometimes do lay haploid male-producing eggs. Males share fewer genes with both siblings and progeny and never work for the benefit of the colony.

Termites (order Isoptera) also form highly organized colonies, complete with queens and "kings" and both male and female workers with distinct castes. But because both sexes of termites have a normal diploid genetic system, the kin selection argument is inadequate to explain the evolution of sociality in termites. To date the best "explanation" for this apparent evolutionary enigma involves the fact that termites are highly dependent upon one another for continual replenishment of their intestinal protozoa. (These endosymbionts, which produce the cellulases that enable termites to digest wood, are lost with each molt and must be replenished continually during the lifetime of an individual.) Presumably a pair of termites (the king and queen) maximize their own reproductive success by producing many nonreproductive progeny (workers) which in turn allow production of many successful reproductive offspring (new kings and queens). Natural selection among workers, however, should operate to release them from such parental "control." Doubtlessly, termites also enjoy other advantages from group cooperation, such as protection from predators and the elements.

Another form of pseudo-altruism, termed "reciprocal altruism" (Trivers, 1971), does not require genetic affinity or kin selection to operate. In reciprocal altruism, some behavioral act incurs a relatively minor loss to a donor but provides a recipient with a large gain; thus two entirely unrelated animals can both benefit from mutual assistance. An example perhaps explained by reciprocal altruism is the posting of sentinels. A sentinel crow spends a brief time period sitting in a tree watching for predators while the rest of the flock forages; in turn it receives continual sentinel protection from the remainder of the flock during the much longer period of time in which it forages. (Such crows may not be sentinels at all, but may merely have had their fill of food to eat; their squawk could be a signal intended to inform the predator that it has been detected—see also pp. 219–220.) Reciprocity is, of course, absolutely essential for evolution and maintenance of this sort of altruistic behavior unless kin selection is also operating to promote it.

Kin selection and reciprocal altruism are appealing concepts that facilitate explanation of the evolution of social behavior by natural selection at the level of individuals. However, neither mechanism is readily verified by observation. Future empirical work in these areas, while difficult, will be of some interest.

Selected References

Introduction

Cole (1954b, 1958); Gadgil and Bossert (1970); Harper (1967); MacArthur and Connell (1966); Mettler and Gregg (1969); Slobodkin (1962); Wilson and Bossert (1971).

Life Tables and Tables of Reproduction

Bogue (1969); Botkin and Miller (1974); Caughley (1966); Cole (1965); Deevey (1947); Fisher (1930); Harper and White (1974); Lotka (1925, 1956); Mertz (1970); Pearl (1928); Slobodkin (1962); Spinage (1972); Zweifel and Lowe (1966).

Net Reproductive Rate and Reproductive Value

Emlen (1970); Fisher (1930, 1958a); Hamilton (1966); Mertz (1970, 1971a, 1971b); Slobodkin (1962); Turner *et al.* (1970); Vandermeer (1968); Wilson and Bossert (1971).

Stable Age Distribution

Emlen (1973); Keyfitz (1968); Krebs (1972); Leslie (1945, 1948); Lotka (1922, 1925, 1956); Mertz (1970); Vandermeer (1968); Williamson (1967); Wilson and Bossert (1971).

Intrinsic Rate of Natural Increase

Andrewartha and Birch (1954); Birch (1948, 1953); Cole (1954b, 1958); Evans and Smith (1952); Fenchel (1974); Fisher (1930); Gill (1972); Goodman (1971); Leslie and Park (1949); May (1976b); Mertz (1970); F. E. Smith (1954, 1963a).

Population Growth and Regulation

Allee *et al.* (1949); Andrewartha and Birch (1954); Ayala (1968); Bartlett (1960); Beverton and Holt (1957); Chitty (1960, 1967a, 1967b); Christian and Davis (1964); Clark *et al.* (1967); Cole (1965); Ehrlich and Birch (1967); Errington (1946, 1956); Fretwell (1972); Gadgil and Bossert (1970); Gibb (1960); Green (1969); Grice and Hart (1962); Hairston, Smith, and Slobodkin (1960); Horn (1968a); Krebs (1972); Lack (1954, 1966); McLaren (1971); Murdoch (1966a, 1966b, 1970); Nicholson (1933, 1954, 1957); Pearl (1927, 1930); Pimentel (1968); Slobodkin (1962); F. E. Smith (1952, 1954, 1963a); Solomon (1949, 1972); Southwood (1966); Williamson (1971).

Density Dependence and Density Independence

Andrewartha (1961, 1963); Andrewartha and Birch (1954); Brockelman and Fagen (1972); Davidson and Andrewartha (1948); Ehrlich *et al.* (1972); Gunter (1941); Horn (1968a); Lack (1954, 1966); McLaren (1971); Nicholson (1957); Orians (1962); Pianka (1972); F. E. Smith (1961, 1963b); Solomon (1972); St. Amant (1970).

Opportunistic versus Equilibrium Species

Anderson (1971); Charlesworth (1971); Clarke (1972); Dobzhansky (1950); Force (1972); Gadgil and Bossert (1970); Gadgil and Solbrig (1972); Grassle and Grassle (1974); King and Anderson (1971); Lewontin (1965); MacArthur

(1962); MacArthur and Wilson (1967); Menge (1974); Pianka (1970, 1972); Roughgarden (1971); Wilson and Bossert (1971).

Population "Cycles": Cause and Effect

Chitty (1960, 1967a); Christian and Davis (1964); Cole (1951, 1954a); Elton (1942); Freeland (1974); Gilpin (1973); Keith (1963, 1974); Krebs (1964, 1966, 1970); Krebs and DeLong (1965); Krebs, Keller, and Myers (1971); Krebs, Keller, and Tamarin (1969); Pitelka (1964); Platt (1964); Schaeffer and Tamarin (1973); Schultz (1964, 1969); Tamarin and Krebs (1969); Wellington (1960).

Evolution of Reproductive Tactics

Ashmole (1963); Baker (1938); Brockelman (1975); Charnov and Krebs (1973); Chitty (1967b); Cody (1966, 1971); Cole (1954b); Gadgil and Bossert (1970); Goodman (1974); Harper and Ogden (1970); Hirshfield and Tinkle (1975); Istock (1967); Johnson and Cook (1968); Johnston (1954); Klomp (1970); Lack (1954, 1966, 1968, 1971); Mertz (1970, 1971a); Millar (1973); Murphy (1968); Perrins (1964, 1965); Pianka (1976b); Pianka and Parker (1975a); Royama (1969); Salisbury (1942); Schaffer (1974); Skutch (1949, 1967); Snell and Burch (1975); Snell and King (1977); Stearns (1976); Taylor *et al.* (1974); Tinkle (1969); Tinkle, Wilbur, and Tilley (1970); Wilbur (1977); Wilbur, Tinkle, and Collins (1974); Williams (1966a, 1966b); Willson (1972b, 1973); Wynne-Edwards (1955, 1962).

Evolution of Death Rates and Old Age

Emlen (1970); Fisher (1930); Haldane (1941); Hamilton (1966); Medawar (1957); Mertz (1975); Pearl (1922, 1928); Snell and King (1977); Sokal (1970); Williams (1957); Willson (1971).

The Joint Evolution of Rates of Reproduction and Mortality

Bonner (1965); Cole (1954b); Frank (1968); Gadgil and Bossert (1970); Lack (1954, 1966); Newell (1949); F. E. Smith (1954); Tinkle (1969); Williams (1966a, 1966b).

Use of Space: Home Range and Territoriality

Ardrey (1966); Brown (1964, 1969); Brown and Orians (1970); Carpenter (1958); Falls (1969); Howard (1920); Hutchinson (1953); Kohn (1968); McNab (1963); Menge (1972); Morse (1971); Orians and Horn (1969); Orians and Willson (1964); Pielou (1969); C. C. Smith (1968); Tinbergen (1957); Weedon and Falls (1959).

Foraging Strategies

Charnov (1976a, 1976b); Charnov *et al.* (1976); Emlen (1966, 1968a); MacArthur (1959, 1972); MacArthur and Pianka (1966); Rapport (1971); Royama (1970); Schoener (1969a, 1969b, 1971); Werner and Hall (1974).

Sex, Sex Ratio, Sexual Selection and Mating Systems

Crook (1962, 1963, 1964, 1965, 1972); Darwin (1871); Dawkins and Carlisle (1976); Downhower and Armitage (1971); Emlen (1968b); Fisher (1930, 1958a); Haartman (1969); Hamilton (1961); Hensley and Cope (1951); Holm (1973); Howard (1974); Kolman (1960); Lack (1968); Maynard Smith (1956, 1958, 1971); Orians (1969b, 1972); Schoener (1967, 1968a); Selander (1965, 1966, 1972); Smouse (1971); Stewart and Aldrich (1951); Trivers (1972); Trivers and Willard (1973); Verner (1964, 1965); Verner and Engelsen (1970); Verner and Willson (1966); Wiley (1974); Williams (1971, 1975); Willson and Pianka (1963); Wittenberger (1976).

Fitness and the Individual's Status in the Population

Errington (1946, 1956, 1963); Fretwell (1972); C. C. Smith (1968); Verner (1964, 1965); Verner and Engelsen (1970); Wellington (1957, 1960).

Maintenance of Variability

Ehrlich and Raven (1969); Fisher (1958b); Mettler and Gregg (1969); Selander and Johnson (1972); Somero (1969); Soulé (1971); Soulé and Stewart (1970); Van Valen (1965); Wilson and Bossert (1971).

Social Behavior and Kin Selection

Alexander (1974); Brown (1966, 1975); Crook (1965); Dawkins (1976); Eberhard (1975); Fisher (1930, 1958a); Hamilton (1964, 1967, 1970, 1971, 1972); Horn (1968b); Maynard Smith (1964); Price and Maynard Smith (1973); N. Smith (1968); Trivers (1971, 1974); Wallace (1973); Wiens (1966); E. O. Wilson (1971, 1975); Wynne-Edwards (1962).

Interactions
Between Populations 6

Introduction

Two populations may or may not affect each other; if they do, the influence may be beneficial or adverse. By designating a detrimental effect with a minus, no effect with a zero, and a beneficial effect with a plus, all possible population interactions can be conveniently classified. Thus, when neither of two populations affects the other, the interaction is designated as (0, 0). Similarly, a mutually beneficial relationship is $(+, +)$ and a mutually detrimental one is $(-, -)$. Other possible interactions are $(+, -)$, $(-, 0)$, and $(+, 0)$, making a total of six fundamentally different ways in which populations can interact (Table 6.1).

Competition $(-, -)$ takes place when each of two populations affects the other adversely. Typically both require the same resource(s) which is (are) in short supply; the presence of each population inhibits the other. *Neutralism* (0, 0) occurs when the two populations do not interact and neither affects the other in any way whatsoever; it is thus of little ecological interest. True neutralism is likely to be very rare or even nonexistent in nature because there are probably indirect interactions between all the populations in any given ecosystem. Interactions that benefit both populations $(+, +)$ are classified as *mutualism* if the association is obligatory (neither population can exist without the other) and as *protocooperation* when the interaction is not an essential condition for survival of either population (Müllerian mimicry, p. 221, falls under this heading). *Predation* $(+, -)$ occurs when one population affects another adversely but benefits itself from the interaction. Usually a predator kills its prey and consumes part or all of the prey organism. (Exceptions include lizards losing their tails to predators and plants losing their leaves to herbivores.) *Parasitism* $(+, -)$ is essentially identical to predation, except that the host (a member of the population being adversely affected) is usually not killed outright but is exploited over some period of time. Thus parasitism might be best considered as a "weak" form of predation; Batesian mimicry (pp. 220–

173

Table 6.1 Summary of the Various Sorts of Interactions that May Occur Between Two Populations

Type of Interaction	Species A	Species B	Nature of the Interaction
Competition	−	−	Each population inhibits the other
Neutralism	0	0	Neither population affects the other
Mutualism	+	+	Interaction is favorable to both but is obligatory
Protocooperation, Müllerian mimicry	+	+	Interaction is favorable to both but is not obligatory
Predation	+	−	Population A, the predator, kills and consumes members of population B, the prey
Parasitism, Batesian mimicry	+	−	Population A, the parasite, exploits members of population B, the host, which is affected adversely
Commensalism	+	0	Population A, the commensal, benefits whereas B, the host, is not affected
Amensalism	−	0	Population A is inhibited, but B is unaffected

Source: Adapted from Odum (1959) after Haskell (1947).

221) and herbivory could be placed here. When one population benefits while the other is unaffected, the relationship is termed *commensalism* (+, 0). *Amensalism* (−, 0) is said to occur when one population is affected adversely by another but the second is unaffected. Two of the six population interactions, competition and predation, are of overwhelming importance; most of this chapter is devoted to them.

Competition

Competition occurs when two or more organismic units use the same resources and when those resources are in short supply. It may also occur through direct interference such as in the production of toxins. Interaction between the two organismic units reduces the fitness and/or equilibrium population size of each. This can occur in several ways. By requiring that an organismic unit expend some of its time and/or matter and/or energy on competition or the avoidance of competition, a competitor may effectively reduce the amounts left for maintenance and reproduction. By using up or occupying some of a scarce resource, competitors directly reduce the amount available to other organismic units. Competition by way of direct interaction, such as production of toxins or aggressive encounters between competitors, is termed *interference competition;* more indirect inhibitory effects, such as those arising from reduced availability of a common resource, are known as *exploitation competition*. Clearly mechanisms of interference competition, such as interspecific territoriality, will be favored by natural selection only when there is a potential for overlap in the use of limited resources at the outset (i.e., when exploitation competition is potentially possible).

Intra specific competition, considered in Chapter 5, is competition between individuals belonging to the same species, and usually to the same population. *Inter* specific competition occurs between individuals belonging to different species and is of chief interest in the present chapter. Because it is always advantageous, *when possible,* for either party in a competitive relation to avoid the interaction, competition has been an important evolutionary force that has led to niche separation, specialization, and diversification. If, however, avoidance of a competitive interaction is impossible, natural selection may sometimes favor convergence.

Competition is not an on–off process; rather, its level presumably varies continuously as the ratio of demand over supply changes. Thus there is little, if any, competition in an ecological vacuum while competition is keen in a fully saturated environment. All degrees of intermediates exist.

Lotka-Volterra Equations and Competition Theory

Competition was placed on a fairly firm, if greatly oversimplified, theoretical basis nearly 50 years ago by Lotka (1925) and Volterra (1926, 1931). Their equations describing competition have strongly influenced the development of modern ecological theory and nicely illustrate a mathematical model of an important ecological phenomenon. They also help to develop a number of very useful concepts, such as competition coefficients, the community matrix, and diffuse competition, that are conceptually independent of the equations.

The Lotka-Volterra competition equations are a modification of the Verhulst-Pearl logistic equation (Chapter 5), and they share its assumptions. Consider two competing species, N_1 and N_2, with carrying capacities K_1 and K_2 in the absence of one another. Each species also has its own maximal instantaneous rate of increase per head, r_1 and r_2. The simultaneous growth of the two competing species occurring together is described by a pair of differential logistic equations:

$$\frac{dN_1}{dt} = r_1 N_1 \left(\frac{K_1 - N_1 - \alpha_{12} N_2}{K_1} \right) \tag{1}$$

$$\frac{dN_2}{dt} = r_2 N_2 \left(\frac{K_2 - N_2 - \alpha_{21} N_1}{K_2} \right) \tag{2}$$

where α_{12} and α_{21} are competition coefficients; α_{12} is a characteristic of species 2 which measures its competitive inhibition (per individual) on the species 1 population; and α_{21} is a similar characteristic of species 1 that measures its inhibitory effects on species 2. Competition coefficients are subscripted to show at a glance which population is affected and which is having the effect; thus α_{12} measures the inhibitory effect of one N_2 individual on the growth of the N_1 population, while α_{21} represents the effect of one N_1 individual on the N_2 population. In the absence of any interspecific competition [α_{12} or N_2 equal zero in Equation (1); α_{21} or N_1 equal zero in Equation (2)], both populations grow sigmoidally according to the Verhulst-Pearl logistic equation and reach an equilibrium population size at their carrying capacity.

By definition, the inhibitory effect of each individual in the N_1 population on its own population's growth is $1/K_1$ (see also Chapter 5); similarly the inhibition of each N_2 individual on the N_2 population is $1/K_2$. Likewise, from inspection of (1) and (2), the inhibitory effect of each N_2 individual upon the N_1 population is α_{12}/K_1 and the inhibitory effect of each N_1 individual on the N_2 population is α_{21}/K_2. Competition coefficients are normally, though not always (see below), numbers less than 1. The outcome of competition depends upon the relative values of K_1, K_2, α_{12}, and α_{21}. Four possible cases of competitive interaction correspond to different combinations of values for these constants (Table 6.2).

To see this, we ask at what density of N_1 individuals is the N_2 population held exactly at zero, and vice versa? In other words, what density of each species will always prevent the other from increasing? By inspection, note that at an N_2 of K_1/α_{12}, N_1 can never increase, and that when N_1 reaches K_2/α_{21}, N_2 can never increase.

Therefore, in the absence of the other species, populations of both species increase at any density below their own carrying capacity and decrease at any value above it. As developed above, in the presence of K_1/α_{12} individuals in the N_2 population, N_1 decreases at all densities; and, in the presence of K_2/α_{21} individuals in the N_1 population, N_2 decreases at every density.

Table 6.2   Summary of the Four Possible Cases of Competition Under the Lotka-Volterra Competition Equations

	Species one can contain species two ($K_2/\alpha_{21} < K_1$)	Species one cannot contain species two ($K_2/\alpha_{21} > K_1$)
Species two can contain species one ($K_1/\alpha_{12} < K_2$)	Either species can win (Case 3)	Species two always wins (Case 2)
Species two cannot contain species one ($K_1/\alpha_{12} > K_2$)	Species one always wins (Case 1)	Neither species can contain the other: stable coexistence (Case 4)

Recall that, under the Verhulst-Pearl logistic equation, r_a decreases linearly with increasing N, reaching a value of zero at density K (see Figure 5.9). Exactly the same relationships hold for the Lotka-Volterra competition equations, except that here a family of straight lines relate to r_1 to K_1 and r_2 to K_2; each line corresponds to a different population density of the competing species (Figure 6.1a and b).

The r axis is omitted in Figure 6.2 and N_1 is simply plotted against N_2. Points in this N_1–N_2 plane thus correspond to different proportions of the two species and to different population densities of each. Setting dN_1/dt and dN_2/dt equal to zero in equations (1) and (2), and solving, gives us equations for the boundary conditions between increase and decrease for each population:

$$\frac{K_1 - N_1 - \alpha_{12}N_2}{K_1} = 0 \quad or \quad N_1 = K_1 - \alpha_{12}N_2 \tag{3}$$

$$\frac{K_2 - N_2 - \alpha_{21}N_1}{K_2} = 0 \quad or \quad N_2 = K_2 - \alpha_{21}N_1 \tag{4}$$

These two linear equations are plotted in Figure 6.3, giving dN/dt isoclines for each species; below these isoclines populations of each species increase, above them they decrease. Thus the lines represent equilibrium population densities or saturation values; neither species can increase when the combined densities of the two lie above its isocline.

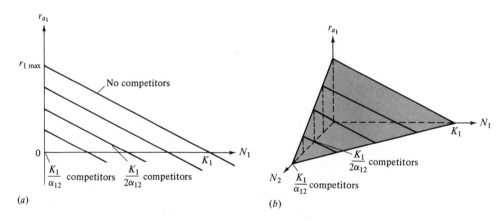

(a)   (b)

Figure 6.1 Two plots showing how the actual instantaneous rate of increase per individual (r_{a_1}) varies with the densities of a population (N_1) and its competitor (N_2) under the Lotka-Volterra competition equations. (a) Two-dimensional graph with four lines, each of which represents a given density of competitors (compare with Figure 5.9). (b) Three-dimensional graph with an N_2 axis showing the plane on which each of the four lines in (a) lie. At densities of N_1 and N_2 above K_1 and K_1/α_{12}, respectively, the plane continues with r_{a_1} becoming negative [compare with (a)].

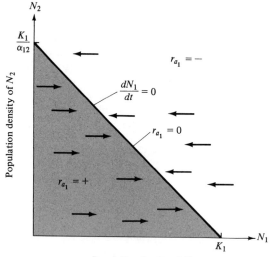

Figure 6.2   A plot identical to that of Figure 6.1*b*, but only the $N_1 - N_2$ plane is shown at r_{a_1} equal to zero. The line represents all equilibrium conditions ($dN_1/dt = 0$) at which the N_1 population just maintains itself ($r_{a_1} = 0$). In the shaded area below the line r_{a_1} is positive and the N_1 population increases (arrows); above the isocline r_{a_1} is always negative and the N_1 population decreases (arrows). An exactly equivalent plot could be drawn on the same axes for the competitor, N_2, except that the intercepts of the N_2 isocline ($dN_2/dt = 0$) are K_2 and K_2/α_{21} and arrows parallel the N_2 axis rather than the N_1 axis. See also Figure 6.3.

The four cases of competitive interaction are summarized in Table 6.2 and are shown graphically in Figure 6.3. Only one combination of values (Case 4) leads to a stable equilibrium of the two species; this case arises when neither species is able to reach densities high enough to eliminate the other—that is, when both $K_1 < K_2/\alpha_{21}$ and $K_2 < K_1/\alpha_{12}$. Implicit in these inequalities are the conditions necessary for coexistence: each population must inhibit its *own* growth more than that of the other species. For this to occur, if K_1 equals K_2, α_{12} and α_{21} must be numbers less than 1. When carrying capacities are not equal, α_{12} or α_{21} can take on a value greater than 1 and coexistence may still be possible as long as the product of the two competition coefficients is less than one ($\alpha_{12}\alpha_{21} < 1$) and the ratio of K_1/K_2 falls between α_{12} and $1/\alpha_{21}$ (see MacArthur, 1972, p. 35). Population sizes at the joint equilibrium (N_{e_1} and N_{e_2}) of the two species in Figure 6.3*d* are below their respective carrying capacities K_1 and K_2; thus in competition neither population reaches densities as high as it does without

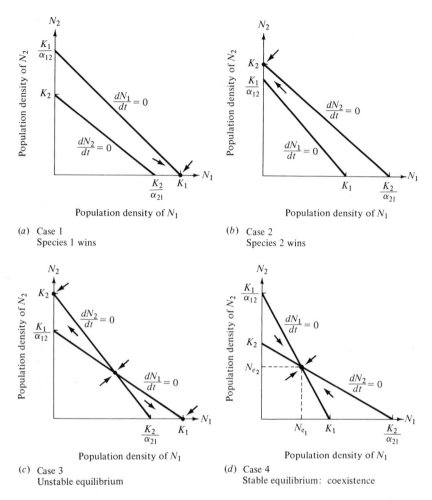

(a) Case 1
Species 1 wins

(b) Case 2
Species 2 wins

(c) Case 3
Unstable equilibrium

(d) Case 4
Stable equilibrium: coexistence

Figure 6.3   Plots like that of Figure 6.2, but with isoclines of both species superimposed upon one another. (*a*) Case 1: The N_1 isocline lies above the N_2 isocline and species 1 always wins in competition. The only stable equilibrium (dot) is at $N_1 = K_1$ and $N_2 = 0$. (*b*) Case 2: The reverse, in which species 2 is the superior competitor and always excludes species 1. Here the only stable equilibrium (dot) is at $N_2 = K_2$ and $N_1 = 0$. (*c*) Case 3: Each species is able to contain the other (Table 6.2); that is, each inhibits the other population's growth more than its own. Three possible equilibria exist (dots), but the joint equilibrium of both species (where the two isoclines cross) is unstable. Alternate stable equilibria are $N_2 = K_2$ and $N_1 = 0$ or $N_1 = K_1$ and $N_2 = 0$. Depending on initial proportions of the two species, either can win. (*d*) Case 4: Neither species can contain the other, but both inhibit their own population growth more than that of the other species. Only one equilibrium exists at N_{e_1} and N_{e_2}; both species thus *coexist* at densities *below* their respective carrying capacities.

competition. None of the other three cases leads to stable coexistence of both populations and hence they are of less interest. However, in Case 3 an unstable equilibrium exists with each species inhibiting the *other's* growth rate more than its own; in this case the outcome of competition depends entirely upon the initial proportions of the two species.

It is sometimes useful to rearrange equations (1) and (2) by multiplying through the bracketed term by $r_1 N_1$ or $r_2 N_2$, respectively:

$$\frac{dN_1}{dt} = r_1 N_1 - \frac{r_1 N_1^2}{K_1} - \frac{r_1 N_1 \alpha_{12} N_2}{K_1} \tag{5}$$

$$\frac{dN_2}{dt} = r_2 N_2 - \frac{r_2 N_2^2}{K_2} - \frac{r_2 N_2 \alpha_{21} N_1}{K_2} \tag{6}$$

These equations can be written in simpler form by consolidating constants:

$$\frac{dN_1}{dt} = r_1 N_1 - z_1 N_1^2 \quad \beta_{12} N_1 N_2 \tag{7}$$

$$\frac{dN_2}{dt} = r_2 N_2 - z_2 N_2^2 - \beta_{21} N_1 N_2 \tag{8}$$

where z_1 and z_2 are equal to r_1/K_1 and r_2/K_2, respectively; similarly β_{12} is $z_1 \alpha_{12}$ and β_{21} is $z_2 \alpha_{21}$. In equations (7) and (8), the first term to the right of the equals sign is the density-independent rate of population increase, while the second and third terms measure intraspecific self-damping and interspecific competitive inhibition of this rate of increase, respectively.

The Lotka-Volterra equations can also be written in a more general form for a community composed of n different species:

$$\frac{dN_i}{dt} = r_i N_i \left\{ \frac{K_i - N_i - \left(\sum\limits_{j \neq i}^{n} \alpha_{ij} N_j \right)}{K_i} \right\} \tag{9}$$

where the is and js subscript species ranging from 1 to n. At steady state, dN_i/dt must equal zero for all i, and equilibrium population densities are given by an equation analogous to (3) and (4):

$$N_{ei} = K_i - \sum\limits_{j \neq i}^{n} \alpha_{ij} N_j \tag{10}$$

Notice that the more competitors a given species has, the larger the term $\left(\sum\limits_{j \neq i}^{n} \alpha_{ij} N_j \right)$ becomes, and the farther that species' equilibrium population size is from its K value; this accords well with biological intuition. The total competitive effect of the remainder of the community on a particular population is known as *diffuse competition* (see also pp. 248–249).

Implicit in the Lotka-Volterra competition equations are a number of assumptions; some can be relaxed, although the mathematics rapidly become unmanageable. Maximal rates of increase, competition coefficients, and carrying capacities are all assumed to be *constant* and immutable; they do not vary with population densities, community composition, or anything else. As a result, all inhibitory relationships within and between populations are strictly linear, and every N_1 individual is identical as is every N_2 individual. Similar results can, however, be reached without assuming linearity even by a purely graphical argument (Figure 6.4). Response to changes in density is instantaneous. In addition, the two species are not allowed to diverge; thus the environment is assumed to be completely homogeneous. Competition in a patchy environment has also been modeled mathematically (Skellam, 1947; Levins and Culver, 1971; Horn and MacArthur, 1972; Slatkin, 1974; Levin, 1974).

In real populations, rates of increase, competitive abilities, and carrying capacities *do* vary from individual to individual, with population density, community composition, and in space and time. Indeed, temporal variation in the environment may often allow coexistence by continually altering the competitive abilities of populations inhabiting it. Time lags are doubtless of some importance in real populations. Finally, a heterogeneous environment may allow real competitors to evolve divergent resource utilization patterns and to reduce interspecific competitive inhibition.

Almost all theory built on the Lotka-Volterra competition equations

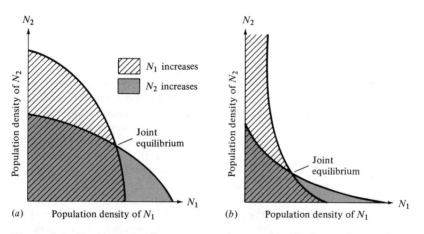

Figure 6.4   Nonlinear isoclines representing graphically the conditions for stable coexistence of two competitors. Concave upward isoclines have been observed in bottle experiments with *Drosophila* flies (Ayala, Gilpin, and Ehrenfeld, 1973) and have also emerged from some innovative new models of competition such as those of Schoener (1973, 1977b).

deals only with equilibrium conditions. But real ecological systems (and portions thereof) may often be partially unsaturated, which in itself could allow coexistence of otherwise competitively intolerant populations. For instance, either predation or density-independent rarefaction might hold down population levels and thereby decrease competition. However, increasing rarefaction in the Lotka-Volterra equations does not reduce the intensity of competitive inhibition, except by reduction of population densities. A more realistic model might incorporate the ratio of instantaneous demand to supply; one such possibility might be to make competition coefficients variables and functions of the combined densities of both populations. At saturation values (K_1, K_2, or $N_{e_1} + N_{e_2}$ in Figure 6.3d), demand/supply is unity and the αs would take on maximal values, but as a perfect competitive vacuum is approached these same αs would become vanishingly small.

The Lotka-Volterra competition equations play an extremely prominent role in modern ecological theory (Levins, 1968; MacArthur, 1968, 1972; Vandermeer, 1970, 1972; May, 1976); however, their numerous biologically unrealistic assumptions reveal the inadequacy of existing competition theory. Although these equations have been overworked and overextended by enthusiastic theorists, they have nevertheless contributed substantially to the development of many important ecological concepts, such as the actual rate of increase, r and K selection, intraspecific and interspecific competitive abilities, diffuse competition, competitive communities, and the community matrix (Chapter 8), which in fact are *independent* of the equations. Thus these equations help to provide a useful conceptual framework.

Competitive Exclusion

How does one population drive another to extinction? In Cases 1, 2, and 3 of Figure 6.3, one species ultimately eliminates the other entirely when the two come into competition and the system is allowed to go to saturation; we say then that competitive exclusion has occurred. Consider an ecological vacuum inoculated with small numbers of each species. At first both populations grow nearly exponentially at rates determined by their respective instantaneous maximal rates of increase. As the ecological vacuum is filled, the actual rates of increase become progressively smaller and smaller. Both populations are infinitely unlikely to have exactly the same rates of increase, competitive abilities, and carrying capacities. Hence, as the ecological vacuum is filled, a time must come when one population's actual rate of increase drops to zero while the other's rate of increase is still positive. This situation represents a turning point in competition, for the second population now increases still further and its competitive inhibition of the first is further intensified, reducing the actual rate of increase of the first population to a negative value. The first population is now declining while the second is still increasing; barring changes in competitive parameters, competitive exclusion (extinction of the first population) is only a matter of time. This process has been demonstrated experimentally (Figure 6.5).

Some rather strong statements concerning competitive exclusion have been made. Among them is a hypothesis called the competitive exclusion "principle": two species with identical ecologies cannot live together in the same place at the same time. Ultimately one must edge out the other; complete ecological overlap is impossible. The corollary is that, if two species coexist, there must be ecological differences between them. Since any two organismic units are infinitely unlikely to be *exactly* identical, mere observation of ecological differences between species does not constitute "verification" of the hypothesis. Untestable hypotheses like this one are of little scientific utility and are gradually forgotten by the scientific community.

The competitive exclusion "principle" has, however, served a useful purpose by emphasizing that some ecological difference may be necessary for coexistence in competitive communities in *saturated* environments. Ecologists are now asking more penetrating and

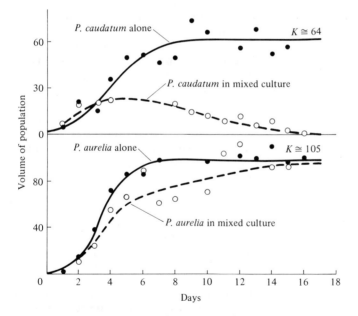

Figure 6.5 Competitive exclusion in a laboratory experiment with two protozoans, *Paramecium caudatum* and *P. aurelia*. [From Gause (1934), reprint ed., © 1964 The Williams & Wilkins Co., Baltimore, Md.] Other experiments with *P. bursaria* and *P. caudatum* resulted in coexistence of two species at densities below their carrying capacities.

dynamic questions: How much ecological overlap can two species tolerate and still coexist? How does this maximal tolerable overlap vary as the ratio of demand to supply changes? How high must migration rates be for competitively inferior fugitive species to persist in spatially and temporally varying patchy habitats? Can temporally-changing competitive abilities lead to coexistence?

The Balance Between Intraspecific and Interspecific Competition

Intraspecific and interspecific competition probably often have opposite effects on a population's tolerance, as depicted in Figure 4.2, as well as on its use of resources and its phenotypic variability. To see this, consider an idealized case with no interspecific competition and look at the effects of intrapopulational competition on the population's use of a resource or habitat. We will consider the former a resource continuum and the latter a habitat gradient, although an exactly parallel argument is easily developed for discrete resources and habitats (Figure 6.6). Individuals should spread themselves out more or less evenly along such a continuum or gradient. If all resources along the continuum are not being utilized to approximately the same degree, those individuals using relatively unused portions should encounter less intense intraspecific competition and, as a result, presumably will often have higher individual fitnesses. Hence we would predict that, in using such a continuous resource in its entirety, individuals should behave so as to equalize the ratio of demand to supply along the continuum, which in turn equalizes the level of intraspecific competition. (MacArthur, 1972, calls this the "principle of equal opportunity.")

Using the same argument, consider now the manner in which an *expanding* population makes use of a resource continuum or a habitat gradient (Figure 6.7). (Again the argument applies equally well to resource categories that are not continuous but discrete—see Figure 6.6.) The first individuals will no doubt select those resources and/or habitats that are optimal in the absence of competition. However, as the density of individuals increases, competition among them reduces benefits to be gained from these optimal resources and/or habitats and favors deviant individuals that use less "optimal," but also less hotly contested, resources and/or habitats. By these means, intraspecific competition can often act to *increase* the variety of resources and habitats utilized by a population.

Interspecific competition, on the other hand, generally tends to *restrict* the range of habitats and resources a population uses, because different species normally have differing abilities at exploiting habitat types and harvesting resources. Individuals using marginal habitats presumably cannot compete as effectively against members of another population as can individuals exploiting more "optimal" habitats. Thus, in most communities, any given population is "boxed in" by

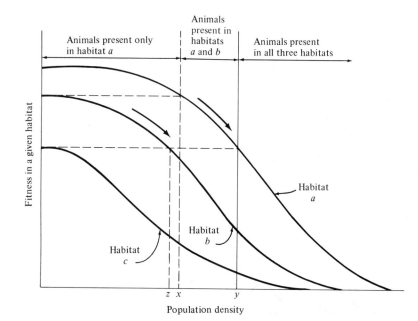

Figure 6.6 Fitnesses of individuals plotted against population density in three hypothetical habitats, *a, b,* and *c.* Fitness decreases as habitats are filled in all three habitats, but at any given density fitness is highest in *a,* intermediate in *b,* and lowest in *c.* Habitat *a* is preferred until densities reach level *x,* at which point it pays to invade habitat *b* in low densities. Habitat *c,* however, remains vacant. As habitats *a* and *b* continue to fill (arrows), it eventually pays to colonize the least desirable habitat *c* (when densities in *a* and *b* reach densities of *y* and *z,* respectively). For fitness to remain constant as the habitats are filled, population density must always be lowest in habitat *c,* intermediate in habitat *b,* and highest in the best habitat *a.* [Modified after Fretwell and Lucas (1969).]

other populations that are superior at exploiting adjacent habitats (see also Chapter 7). Indeed, since these two forces oppose each other, one could even hypothesize that, at equilibrium, total intraspecific competition should be exactly balanced by total interspecific competition. In fact this is not quite true because inherent genetic and physiological limitations must also restrict the range of habitats and resources used by an organism.)

Evolutionary Consequences of Competition

Many of the long-term consequences of competition have been mentioned in Chapters 1 and 5; others are treated in Chapters 7, 8, and 9. For instance, natural selection in saturated environments (*K* selection) favors competitive ability. A raft of presumed populational prod-

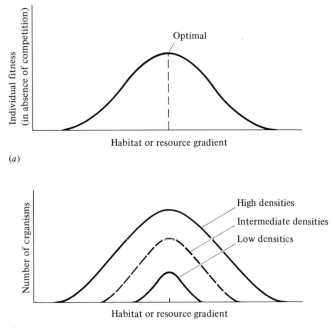

(a)

(b)

Figure 6.7   Diagrammatic representation of how an expanding population might use a gradient of habitats or resources. *(a)* Individual fitness as a function of the habitat or resource gradient in the *absence* of any intraspecific competition. *(b)* At low densities, most individuals select near-optimal environmental conditions, but as density increases, intraspecific competition for more optimal habitats (or resources) increases, favoring individuals that exploit less optimal, but also less hotly contested, habitats or resources. Variety of habitats or resources exploited thus *increases* with increasing population density.

ucts of intraspecific competition over evolutionary time include: rectangular survivorship, delayed reproduction, decreased clutch size, increased size of offspring, parental care, mating systems, dispersed spacing systems, and territoriality (see also Chapter 5). Perhaps the most far-reaching evolutionary effect of interspecific competition is ecological diversification, also termed niche separation. This in turn has made possible, and has led to, the development of complex biological communities. Another presumed result of both intraspecific and interspecific competition is increased efficiency of utilization of resources in short supply.

Although the concept of competition is a central theme in much of modern ecological theory, it has proven to be surprisingly difficult to study in the field and is still poorly understood as an actual phenomenon. Some ecologists consider competition among the most important of ecological generalizations, yet others maintain that it is of little utility in understanding nature. At least three possible reasons, not necessarily mutually exclusive, for this difference in viewpoint have been suggested: (1) competition in nature is often elusive and very difficult to study and to quantify, (2) ecologists with an evolutionary approach might often consider competition more important than workers concerned with explaining more immediate events (Orians, 1962),* and (3) there may be a natural dichotomy of relatively r-selected and relatively K-selected organisms, especially in terrestrial communities (Pianka, 1970).

Laboratory Experiments

Competition is often fairly easily studied by direct experiment, and many such studies have been made. Gause (1934) was one of the first to investigate competition in the laboratory; his classic early experiments on protozoa verified competitive exclusion. He grew cultures of two species of *Paramecium* in isolation and in mixed cultures, under carefully controlled conditions and nearly constant food supply (Figure 6.5). "Carrying capacities" and respective rates of population growth of each species when grown separately and in competition were then calculated (competition coefficients are also readily computed from such data). Interestingly enough, the protozoan with the highest maximal instantaneous rate of increase per individual *(P. caudatum)* was the inferior competitor, as expected from considerations of r and K selection (Chapter 5).

An effect of environment on the outcome of competition was demonstrated by Park (1948, 1954, 1962) and his many colleagues. Working with two species of flour beetles *(Tribolium)*, these investigators showed that, depending upon conditions of temperature and humidity, either species could eliminate the other (Table 6.3). In early experiments they could not always predict the outcome of competition under particular environmental conditions; this set of environments was termed the "indeterminate zone." More recently, however, this zone has been substantially reduced by taking into account genotypes of beetles (Lerner and Ho, 1961; Park, Leslie, and Mertz, 1964). Under some environmental conditions, cultures begun with a numerical preponderance of a given species always resulted in the extermination of the other species (Figure 6.8), thereby constituting an empirical verification of Case 3 of the Lotka-Volterra equations (Figure 6.3c).

In a laboratory study of competitive interactions among ciliate protozoans, Vandermeer (1969) cultured each of four species separately

* See pages 15–16 for a discussion of the difference between the "proximate" and the "ultimate" approaches to biological phenomena.

Table 6.3   Outcome of Competition Between Two Species of Flour Beetles, *Tribolium confusum* and *T. castaneum*, in Many Replicates of Laboratory Experiments at Different Temperatures and Humidities

Temperature (°C)	Relative Humidity (%)	Climate	Single Species Numbers	Mixed Species (% wins)	
				confusum	*castaneum*
34	70	Hot–moist	*confusum = castaneum*	0	100
34	30	Hot–dry	*confusum > castaneum*	90	10
29	70	Temperate–moist	*confusum < castaneum*	14	86
29	30	Temperate–dry	*confusum > castaneum*	87	13
24	70	Cold–moist	*confusum < castaneum*	71	29
24	30	Cold–dry	*confusum > castaneum*	100	0

Source: From Krebs (1972) after Park.

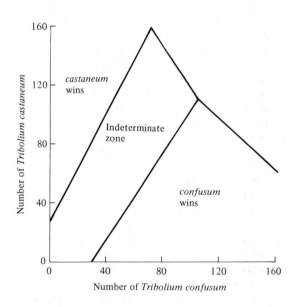

Figure 6.8   Outcome of competition between laboratory strains of two beetles, *Tribolium confusum* and *T. castaneum*, is a function of initial densities of the species; an initial preponderance of either species increases the likelihood of its winning. The "indeterminate zone" represents initial conditions under which either species can win. Note the similarity with Case 3 in Figure 6.3c. [From Krebs (1972) after Neyman, Park, and Scott. Originally published by the University of California Press; reprinted by permission of The Regents of the University of California.]

and in all possible pairs. These experiments allowed estimation of rs, Ks, and αs. Observed pair-wise interactive effects were similar to those actually observed when all four species were grown together in mixed culture—an indication that higher-order interactions among these species were slight. However, several similar studies (Hairston *et al.*, 1968; Wilbur, 1972; Neill, 1974) suggest strong interactive effects among species, with the competitive effects between any two species depending strongly upon the presence or absence of a third.

Competition and competitive exclusion have now been demonstrated in laboratory experiments on a wide variety of plants and animals. Potential flaws in many of these investigations are that, for practical reasons, they are carried out in constant and simple environments, almost invariably on small, often relatively r-selected, organisms which may not encounter high levels of competition regularly under natural circumstances.

Evidence from Nature

Competition is notoriously difficult to demonstrate in natural communities, but a variety of observations and studies suggest that it does indeed occur regularly in nature and that it has been important in molding the ecologies of many species of plants and animals. Even if competition did not occur on a day to day basis, it could nevertheless still be a significant force; active avoidance of interspecific competition in itself implies that competition has occurred sometime in the past and that the species concerned have adapted to one another's presence. Also it might be difficult to find competition actually occurring in nature because inefficient competitors should be eliminated by competitive exclusion and therefore might not normally be observable. We might not expect to find abundant evidence of competition in small, short-lived organisms, such as insects and annual plants, but would look for it in larger, longer-lived organisms such as vertebrates and perennial plants.

Ecologists have several different sorts of evidence, much of which is circumstantial, suggesting that competition either has occurred or is occurring in natural populations. These include: (1) studies on the ecologies of closely related species living in the same area; (2) character displacement; (3) studies on "incomplete" floras and faunas and associated changes in niches, or "niche shifts"; and (4) taxonomic composition of communities.

Closely related species, especially those in the same genus, or "congeneric" species, are often quite similar morphologically, physiologically, behaviorally, and ecologically. As a result, competition is intense between pairs of such species that live in the same area, known as *sympatric* congeners; selection may be strong to render their ecologies more different or to lead to ecological separation. Many groups of closely related sympatric species have been studied; almost without exception, detailed investigations on relatively K-selected organisms have revealed subtle but important ecological differences between such species. Usually the differences are of one or more of three basic types: the species exploit different habitats or microhabitats (differential spatial utilization of the environment), they eat different foods, or they are active at different times (differential patterns of temporal activity). Such ecological differences are known as "niche dimensions" because they are important in defining a species' role in its community and its interactions with other species (see Chapter 7).

Many examples of clear-cut habitat and microhabitat differences could be cited. MacArthur (1958) studied spatial utilization patterns in five species of sympatric warblers (genus *Dendroica*) by noting the time spent in precise locations by foraging individuals of each species. Each species has its own unique pattern of exploiting the forest (Figure 6.9).

Competition for space between two species of barnacles, *Balanus balanoides* and *Chthamalus stellatus,* was studied by Connell (1961a, 1961b) along a rocky Scottish seacoast. The two crustaceans occupy sharply defined horizontal zones, as do most sessile organisms in the marine intertidal, with *Chthamalus* occupying an upper, and *Balanus,* a lower zone (Figure 6.10). Larvae of both species settle and attach over a wider vertical range than the zone occupied by adults. By manipulation of wire cages to exclude a snail predator *(Thais)* and by periodic removal of barnacles, Connell elucidated the various forces causing the zonation. He demonstrated that, although *Chthamalus* loses in competition with *Balanus*, adults persist in a narrow band because *Chthamalus* are more tolerant of physical desiccation than *Balanus*. Connell suggests that the lower limit of distribution of intertidal organisms is usually determined mainly by biotic factors, such as competition with other species and predation, whereas the upper limit is more often set by physical factors, such as dry conditions prevailing during low tides.

Overdispersion in general, and territoriality in particular (Chapter 5), are indicative of competition in that they reduce its intensity. Overdispersion is widespread in plants, as is territoriality in vertebrate populations (Chapter 5). Indeed, *inter*specific territoriality has been documented in many species of birds (Orians and Willson, 1964) and probably occurs in other taxa as well; thus both intraspecific and interspecific competition have led to territorial behavior.

Differences in time of activity among ecologically similar animals can effectively reduce competition, *provided that resources differ at different times.* This is true in situations where resources are rapidly renewed, since the resources available at any one instant are relatively unaffected by what has happened at previous times. Perhaps the most obvious type of temporal separation is that between day and night; animals active during the daytime are "diurnal," those active at night are "nocturnal." Examples of pairs apparently separated by such tem-

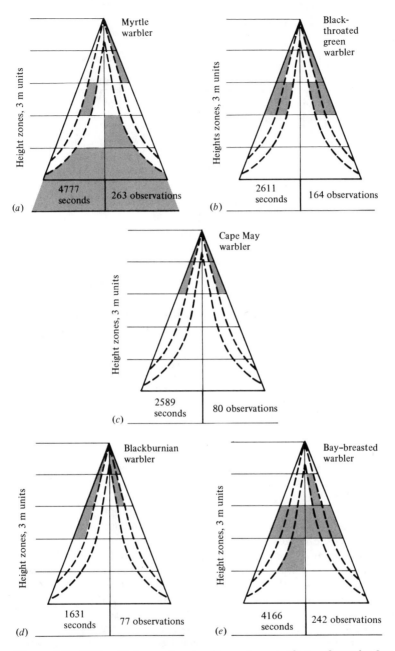

Figure 6.9 Differential use of parts of trees in a coniferous forest by five sympatric species of congeneric warblers (genus *Dendroica*). Shading indicates parts of trees in which foraging activities of each species are most concentrated. The right side of each schematic tree represents use based on the total number of birds observed (sample size given below each "tree"); the left side represents use based on total number of seconds of observation (again, sample size is given at the bottom of each "tree"). [After MacArthur (1958). By permission of Duke University Press.]

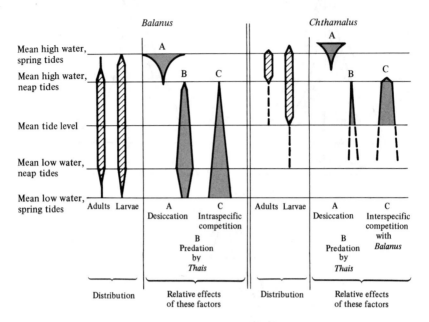

Figure 6.10 Vertical distributions of larvae and adults of two barnacle species, *Balanus balanoides* and *Chthamalus stellatus*, in the rocky intertidal of Scotland. Relative intensities of various limiting factors are represented diagrammatically by widths of shaded areas. [After Connell (1961b). By permission of Duke University Press.]

poral differences are hawks and owls, swallows and bats, or grasshoppers and crickets. Patterns of activity within the course of the day alone also differ, with some species being active early in the morning, others at midday, etc. Seasonal separation of activity also occurs among some animals, such as certain lizards. In many animals, daily time of activity changes seasonally (Figure 6.11).

Dietary separation among closely related animal species has been shown repeatedly. For example, Table 6.4 shows that several sympatric congeneric species of the marine snail genus *Conus* (commonly called "cone shells") eat distinctly different foods (Kohn, 1959). Similarly, three species of stoneflies eat prey of different sizes (Figure 6.12). Among desert lizards, diets of several sympatric species are composed predominantly of ants, termites, other lizards, and plants (Pianka, 1966b). Similar cases of food differences among related sympatric species are known in many birds and mammals.

Simultaneous differences in the use of space, time, and food have also been documented for some sympatric congeneric species. In lizards of the genus *Ctenotus* (Pianka, 1969), for instance, seven sympatric species forage at different times, in different microhabitats,

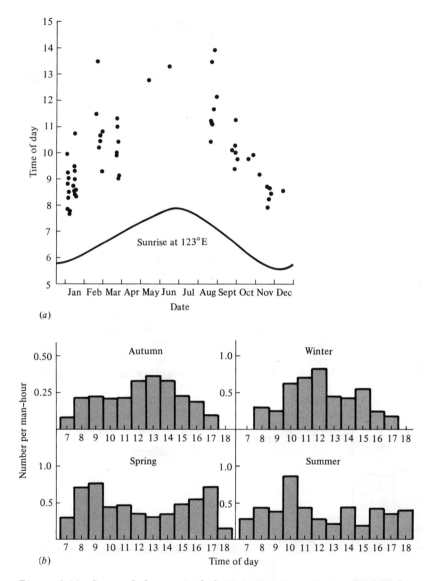

(a)

(b)

Figure 6.11 Seasonal changes in daily activity patterns in two species of Australian desert lizards. *(a)* A small blue-tailed skink, *Ctenotus calurus;* each dot represents one lizard. [From Pianka (1969). By permission of Duke University Press.] *(b)* The "military dragon," an agamid, *Amphibolurus isolepis;* numbers observed per man-hour of observation at different times of day in each season. [After Pianka (1971b).]

Table 6.4 Major Foods (Percentages) of Eight Species of Cone Shells, *Conus,* on Subtidal Reefs in Hawaii

Species	Gastro-pods	Entero-pneusts	Nereids	Eunicea	Tere-bellids	Other polychaetes
flavidus		4			64	32
lividus		61		12	14	13
pennaceus	100					
abbreviatus				100		
ebraeus			15	82		3
sponsalis			46	50		4
rattus			23	77		
imperialis				27		73

Source: From data of Kohn (1959). By permission of Duke University Press.

and/or on different foods. Frequently in such cases pairs of species with high overlap in one niche dimension have low overlap along another, presumably reducing competition between them (see also Figure 7.8).

The phenomenon of "character displacement," which refers to increased differences between species where they occur together, is also evidence that competition occurs in nature. Sometimes two widely ranging species are ecologically more similar in the parts of their ranges where each occurs alone without its competitor (i.e., in *allopatry*) than they are where both occur together (in *sympatry*). This

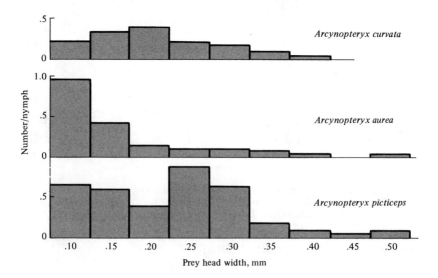

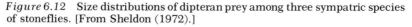

Figure 6.12 Size distributions of dipteran prey among three sympatric species of stoneflies. [From Sheldon (1972).]

sort of ecological divergence can take the form of morphological, behavioral, and/or physiological differences. One way in which character displacement occurs is in the size of the food-gathering or "trophic" apparatus, such as mouthparts, beaks, or jaws. Prey size is usually strongly correlated with the size of an animal's beak or jaw as well as with its structure (Figure 6.13). Although the evidence is circumstantial (see Grant, 1972, for a review), character displacement in either body size or the size of the trophic apparatus is thought to have occurred in some lizards, snails (Figure 6.14), birds, mammals, and insects, presumably separating food niches. Such niche shifts in the presence of a potential competitor suggest that each population has adapted to the other by evolving a means to reduce interspecific competition.

Morphological character displacement in the size of mouthparts need not evolve if the populations concerned have diverged in other ways; hence it is expected only in situations where both competitors occur side by side, exploiting identical microhabitats (i.e., true *syntopy*). Animals that forage in different microhabitats, such as the warblers of Figure 6.9, have adapted to one another primarily by means of behavioral, rather than morphological, character displacement.

There may be a definite limit on how similar two competitors can be and still avoid competitive exclusion; character displacement in average mouthpart sizes is typically about 1 to 1.3, and the ratio of 1.3 may thus be a crude estimate of just how different two species must be to coexist syntopically (Hutchinson, 1959; Schoener, 1965). Clearly the preceding argument applies only in competitive communities, and ecological overlap, or a greater similarity between syntopic species, can

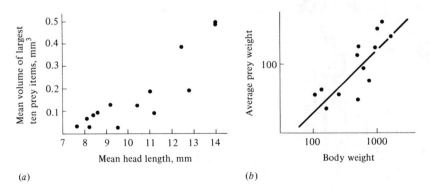

(a)   (b)

Figure 6.13 Two plots of prey size versus predator size. (a) Plot of the average volume of ten largest prey items (in cubic millimeters) against mean head length in 14 species of lizards (genus *Ctenotus*). [After Pianka (1969). By permission of Duke University Press.] (b) Average prey weight plotted against mean body weight among 13 species of hawks (log-log plot). [After Schoener (1968). By permission of Duke University Press.]

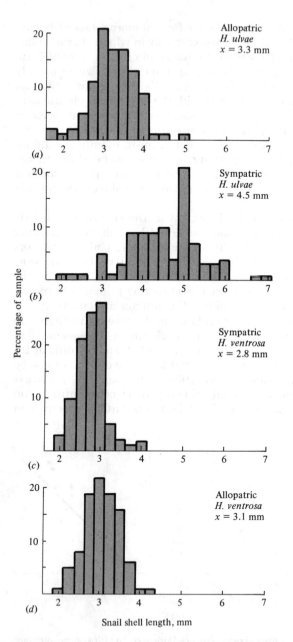

Figure 6.14 Frequency distributions of shell lengths of two species of mud snails in Denmark. The two species, *Hydrobia ulvae* and *H. ventrosa*, are similar in size where each occurs alone in allopatry (top and bottom panels); however, their shell lengths are markedly divergent in the area of sympatry (center two panels). [After Fenchel (1975).]

presumably be greater in unsaturated habitats where competition is reduced.

Another type of evidence for competition comes from studies on so-called "incomplete" biotas, such as islands, where all the usual species are not present (see also Chapter 9). Those species which invade such areas often expand their niches and exploit new habitats and resources that are normally exploited by other species on areas with more complete faunas. On the island of Bermuda, for example, considerably fewer species of land birds occur than on the mainland, with the three most abundant being the cardinal, catbird, and white-eyed vireo. Crowell (1962) found that, compared with the mainland, these three species are much more abundant on Bermuda and that they occur in a wider range of habitats. In addition, all three have somewhat different feeding habits on the island, and one species at least (the vireo) employs a greater variety of foraging techniques (see also pp. 246–247).

Mountain tops represent islands on the terrestrial landscape just as surely as Bermuda is an island in the ocean, and they often show similar phenomena. For example, two congeneric species of salamanders, *Plethodon jordani* and *P. glutinosus* occur in sympatry on mountains in the eastern United States. In sympatry, the two species are altitudinally separated, with *glutinosus* occurring at lower elevations than *jordani;* vertical overlap between the two species never exceeded 70 meters (Hairston, 1951). On mountain tops where *jordani* occurs, *glutinosus* is restricted to lower elevations, whereas on adjacent mountains that lack *jordani, glutinosus* is found at higher elevations, often right up to the peak.

Niche expansion under reduced interspecific competition has been termed "ecological release." Further evidence of competition stems from a corollary of ecological release: when mainland forms are introduced onto islands, native species are frequently driven to extinction, presumably via competitive exclusion. Thus many birds that once occurred only on Hawaii (called *endemic* Hawaiian species) became extinct shortly after the introduction of mainland birds such as the English sparrow and the starling. Similar extinctions have apparently occurred in the Australian marsupial fauna (e.g., the Tasmanian wolf) with the introduction of placental mammal species (e.g., the dingo dog and the European fox). Of course fossil history is replete with cases of natural invasions and subsequent extinctions. The simplest and most plausible explanation for many of these observations is that surviving species were superior competitors and that niche overlap was too great for coexistence. Before natural selection could produce character displacement and niche separation, one species had become extinct. Elton (1958) discusses many other examples of ecological invasions among both plants and animals.

A final observation, involving the taxonomic composition of communities, can be used to assess whether or not competition has been an important force in nature. Because closely related species should be strong competitors, one might predict that fewer pairs of congeneric species will occur within any given natural community than would be found in a completely random sample from the various species and genera occurring over a broader geographic area. Such a paucity of sympatric congeners, if observed, would suggest that competitive exclusion occurs more often among congeneric species than it does in more distantly related ones. This test was applied to many communities by Elton (1946), who was well aware of the problem of defining a "community." Frequently, where there are two abutting communities, each supports its own member of a congeneric pair and such pairs must be excluded wherever possible. Despite this potential bias toward an increased proportion of congeneric species pairs, Elton found fewer congeners than expected on a strictly random basis. Elton's analysis has since been shown to be incorrect by Williams (1964), who gave a corrected statistical approach to the problem. Using this correct technique, Terborgh and Weske (1969) calculated the expected number of congeneric species pairs of Peruvian birds in seven habitats (Figure 6.15). They found that the four habitats richest in total num-

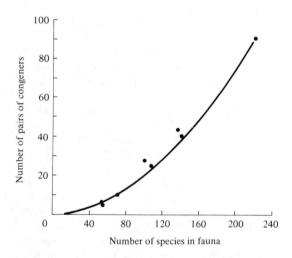

Figure 6.15   Observed (points) and expected (curve) numbers of pairs of congeneric bird species in seven Peruvian habitats. The combined avifauna of all seven areas contained 92 different pairs of congeners among a total of 221 species (plotted in the uppermost right-hand corner). Random subsamples of this total fauna would contain the numbers of congeneric pairs indicated by the curve. If competitive exclusion occurred more frequently among congeners, the observed number of such pairs (points) would fall below the expected curve. [From Terborgh and Weske (1969). By permission of Duke University Press.]

ber of species had greater than the expected number of congeneric pairs—thus refuting any increased incidence of competitive exclusion among congeners in this particular avifauna. I also failed to find any consistent impoverishment in the number of congeneric species pairs in a number of lizard communities (Pianka, 1973). More analyses of this sort on a broad range of taxa in different communities might be worthwhile.

Future Prospects

Although competition is presumably central to numerous ecological processes and phenomena, current understanding of competitive interactions remains inadequate from both theoretical and empirical points of view. Possibilities abound for significant future work. Clearly the great temporal and spatial heterogeneity of the real world demands a dynamic approach to competitive interactions.

Innovative new models that depart from the notion of competitive communities at equilibrium with their resources will be of interest. The notion of competition coefficients itself may be somewhat illusory and may often obscure the real mechanisms and dynamics of competition. Even so, theory could be improved substantially simply by treating alphas as *variables* in both ecological and evolutionary time. For example, the actual shapes of resource utilization curves might be allowed to change (subject to an appropriate constraint such as holding the area under the curves constant), either in ecological time by behavioral release or in evolutionary time via directional selection favoring deviant phenotypes. Because the competitive effects between any given pair of species are often sensitive to the presence or absence of a third species, theory is needed on *interactive* competition coefficients (see also pp. 188–189 and 275–276).

An attractive alternative to competition coefficients is to measure the intensity of interactions between species by the sensitivity of each species' own density to changes in the density of the other. As such, these interactions are represented mathematically by $\partial N_i/\partial N_j$ and $\partial N_j/\partial N_i$ terms (partial derivatives). Note that this approach involves population dynamics and that it can be applied to prey–predator and symbiotic interactions as well as to competitive ones; if the presence of species j is detrimental to species i, $\partial N_i/\partial N_j$ is negative, whereas a beneficial interaction has a positive sign.

Great potential also exists for further development of theory on diffuse competition. Consider two communities with similar numbers of species but different guild structures. In the first, several distinct clusters of competing species have strong competitive interactions among themselves but weak interactions with members of other guilds. In the second community, all members interact more or less equally and diffuse competition is more intense. (Differences in the degree of niche dimensionality would produce such a difference between communities—see also pp. 247–251.) What effects will such differences in degree of competitive "connectedness" have upon various community-level properties like stability? Will maximal tolerable overlap be less in the second community? To what extent do the resources available force guild structure? Ecologists are currently seeking answers to such questions.

Prospects for future empirical work are even brighter, although certainly more difficult and challenging. Well-designed and executed removal and addition experiments or perturbations of equilibrium densities will certainly allow quantification of competitive effects, but, as Schoener (1974) points out, in themselves they probably will not provide much insight into the actual mechanisms of competition. Clever empirical work will probably provide greater insights into competitive mechanisms than further theoretical explorations. Unfortunately, however, the precise nature of these crucial observations is difficult to anticipate.

Predation

Predation is readily observed and easily studied, and neither its existence nor its importance in nature is doubted. It is directional in the sense that one member of the pair (the predator) benefits from the association while the other (the prey) is affected adversely. In contrast, competition is a symmetric process in that both species are affected adversely and, where possible, each tends to evolve mechanisms whereby the relationship with the other is avoided.

Individual predators that are better able to capture prey should have more resources at their disposal and should therefore normally be more fit than those which are less proficient at capturing prey. Hence natural selection acting on the predator population tends to increase the predator's efficiency at finding, capturing, and eating its prey. However, members of the prey population that are better able to escape predators should normally be at a selective advantage within the prey population. Thus selection on the prey population favors new adaptations that allow prey individuals to avoid being found, caught, and eaten. Obviously these two selective forces oppose one another; as the prey become more adept at escaping from their predators, the predators in turn evolve more efficient mechanisms for capturing them. Hence, in the evolution of a prey–predator relationship, the prey evolves so as to dissociate itself from the interaction while the predator continually maintains the relationship. Long-term evolutionary escalations of this sort have resulted in some rather intricate and often exceedingly complex adaptations. Consider, for instance, the complex social hunting behavior of lions and wolves, the long sticky tongues and accurate aim of some fish, toads, and certain lizards, the folding fangs and venom-injection apparatus of viperine snakes, spiders and their webs, the deep-sea angler fish, and snakes such as boas that suffocate their prey by constriction. Other examples include the rapid and very accurate strikes of predators as diverse as praying mantids, dragonflies, fish, lizards, snakes, mammals, and birds. Prey have equally elaborate predator escape mechanisms, such as the posting of sentinels, predator alarm calls, background color matching, and thorns (see also pp. 215–222). Many prey organisms recognize their predators at some distance and employ appropriate avoidance tactics well before the predator gets close enough to make a kill; this behavior, in turn, has forced many predators to hunt by ambush.

One of my favorite examples of the joint adaptations of a predator and its prey is provided by the starling and the peregrine falcon. The peregrine is a magnificent bird hawk whose hunting behavior must be seen to be fully appreciated. These falcons take other birds as large as themselves; nearly all prey is captured on the wing and in the air. Peregrines have exceedingly keen vision; foraging individuals climb high up into the sky and move across country. When a potential prey is sighted flying along below, the peregrine closes its wings and dives or "stoops." To make its ambush most effective, the falcon often "comes out of the sun" at its prey. Diving peregrines have been estimated to reach speeds of over 300 kilometers/hour (nearly 100 meters/second!). Most prey are killed instantly by the sudden jolt of the peregrine's talons. Large prey are allowed to fall to the ground and eaten there, but smaller items may be carried away in the air. (Little wonder that falconers and their dogs find it extremely difficult to get small birds to fly when a peregrine is "waiting on" overhead! On occasion game birds allow themselves to be overtaken on the ground by dogs in preference to taking to the air and risking the falcon's deadly stoop.) Starlings normally fly in loose flocks, but when a peregrine is sighted, often at a considerable distance, they quickly assume a very tight formation. Tight flocking is a response specific to the peregrine and is not employed with other hawks. Falcons are much less likely to attack a tight flock than a single bird; indeed, "stragglers" slightly out of the starling flock formation are often taken by peregrines. Presumably the falcon itself could be injured if it were to stoop into a tight flock. Thus even a predator as effective as the peregrine falcon has had its hunting efficiency impaired by appropriate behavioral responses of its prey.

There is an important difference between predation on animals and predation on plants. In most animals predation is an all-or-none proposition in that the predator kills the prey outright and consumes most or all of it; however, when plants are eaten, usually only a portion of the plant is consumed by its predator. Hence predation on plants (herbivory) is more like parasitism among animals. But even partial predation must often reduce a prey individual's ability to survive and/or to reproduce. Because of this fundamental difference, though, selective pressures on animals to avoid being eaten may be stronger than they are on plants. Nevertheless, plants have evolved elaborate antipredator devices (see pp. 221–224).

Theories: Predator–Prey Oscillations

The theory of predation has, in many ways, lagged behind that for competition; perhaps its asymmetry makes it more difficult to model. Lotka (1925) and Volterra (1926, 1931) wrote a simple pair of predation equations:

$$\frac{dN_1}{dt} = r_1 N_1 - p_1 N_1 N_2 \tag{11}$$

$$\frac{dN_2}{dt} = p_2 N_1 N_2 - d_2 N_2 \tag{12}$$

where N_1 is prey population density, N_2 is population density of the predator, r_1 is the instantaneous rate of increase of the prey population (per head), d_2 is the death rate of the predator population (per head), and p_1 and p_2 are predation constants. Each population is limited by the other and there are no self-limiting density effects (that is, no second order N_1^2 or N_2^2 terms). Thus, in the absence of the predator, the prey population expands exponentially and the rate of increase of the prey population is potentially unlimited. The product of the densities of the two species, $N_1 N_2$, reflects the number of contacts between them; after multiplication by the constant, p_2, this term becomes the maximal rate of *increase* of the predator population $(p_2 N_1 N_2)$. The same term multiplied by the constant, p_1, appears with negative sign in the prey equation and acts to *decrease* the rate of growth of the prey population.

The equations are solved by setting dN/dt equal to zero, factoring out the appropriate N to get the actual rate of increase, and setting this r_a equal to zero. These algebraic manipulations show that the prey reaches an equilibrium population density when the predator's density is r_1/p_1; similarly, the predator is at equilibrium when the prey's density is d_2/p_2 (Figure 6.16). Thus each species' isocline corresponds to a particular (constant) density of the other species, and again there is no self-damping term such as the $-zN^2$ term in the competition equations. Below some threshold prey density, predators always decrease, while above that threshold they increase; similarly, prey increase below a particular predator density but decrease above it (Figure 6.16). A joint equilibrium exists where the two isoclines cross, but prey and predator densities do not converge on this point. Rather, any given initial pair of densities results in oscillations of a certain magnitude. Initial densities near the joint equilibrium point result in repeating oscillations of low amplitude; initial densities farther from the joint equilibrium point generate oscillations of greater amplitude. Thus this pair of differential equations has a periodic solution, with the population densities of both prey and predator changing cyclically and out of phase over time. The

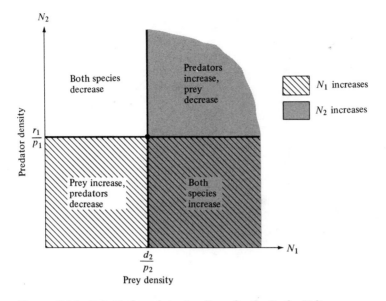

Figure 6.16 Prey and predator isoclines for the Lotka-Volterra prey–predator equations (see text).

amplitude of the fluctuations depends on initial conditions. Mathematically, a system of such repeating and undamped oscillations is termed neutrally stable. Neutral stability probably does not exist in the biological world since most individuals and populations encounter either self-regulation or density-dependent feedback.

Addition of a simple self-damping term ($-zN_1^2$) to the prey equation results either in a rapid approach to equilibrium or in damped oscillations, both of which lead eventually to the joint equilibrium (Figure 6.17). However, a self-damping term for the predator should include the prey's density as a determinant of the predator's carrying capacity. Perhaps a more realistic (although mathematically less tractable) pair of simple equations for modeling the prey–predator relationship is

$$\frac{dN_1}{dt} = r_1 N_1 - z_1 N_1^2 - \beta_{12} N_1 N_2 \tag{13}$$

$$\frac{dN_2}{dt} = \gamma_{21} N_1 N_2 - \beta_2 \frac{N_2^2}{N_1} \tag{14}$$

The prey equation is the simple Lotka-Volterra competition equation (7), but the predator equation has a new twist in that competitive inhibition of the predator population is now a function of the relative densities of predator and prey. Thus inhibition of the predator popula-

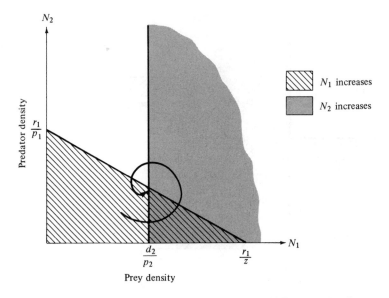

Figure 6.17   Prey and predator isoclines with self-damping in the prey population. Population densities converge on the stable joint equilibrium.

tion increases both with increased predator density and with decreased prey density. Notice also that the predator population cannot increase unless there are some prey. However, even though this pair of equations overcome some of the faults of previous pairs, they are still unrealistic in at least one important way. Imagine a situation in which there are more prey than the predator population can possibly exploit; in such a case, growth rate of the predator cannot be simply proportional to the product of the two densities as in (14), but some sort of threshold effects must be taken into account.

    Equations like the above omit entirely many important subtleties from the prey–predator interaction. For instance, Solomon (1949) distinguished two separate components of the way in which predators respond to changes in prey density. First, individual predators capture and eat more prey per unit time as prey density increases until some satiation threshold is reached, above which the number of prey taken per predator is more or less constant (Figures 6.18 and 6.19*a*); second, increased prey density raises the predator's population size and a greater number of predators eat an increased number of prey (Figure 6.19*b*). Solomon termed the former the functional response and the latter the numerical response of the predator. Three types of functional responses are recognized, representing pure forms among a continuum of possibilities (Figure 6.18). [Equations (12) and (14) model a Type 1 linear functional response without a ceiling.] Note that a predator's functional response can allow regulation of prey density without an increase in predator numbers (no numerical response). Using the "systems" approach (see also Chapter 8) that relies on continued feedback between observation and model, Holling (1959a, 1959b, 1966) developed elaborate models of predation incorporating both the functional and the numerical responses as well as other parameters including various time lags and hunger level. These models are more realistic and descriptive than the others (above), but they are also more complex and restricted. Obviously a realistic model of prey–predator relationships *must* be quite complex!

A simple graphical model of the prey–predator interaction was developed by Rosenzweig and MacArthur (1963), who reasoned somewhat as follows. In the absence of predators, the maximum equilibrium population density of the prey is K_1, the prey's carrying capacity. Similarly, some lower limit on prey density is likely to exist, below which contacts between individuals are too rare to ensure reproduction and the prey population thus decreases to extinction. Likewise, at any given density of prey, there must be some maximal predator density that can just be supported without either an increase or a decrease in the prey population. Using these arguments, a prey isocline ($dN_1/dt = 0$) can be drawn in the $N_1 - N_2$ plane (Figure 6.20) similar to those drawn earlier in Figures 6.3 and 6.4. As long as the prey isocline has but a single peak, the exact shape of the curve is not important to the conclusions that can be derived from the model. Above this line, prey populations decrease; below it they increase. Next consider the shape of the predator isocline ($dN_2/dt = 0$). Below some threshold prey density, individual predators cannot gather food enough to replace themselves and the predator population must decrease; above this threshold prey density, predators will increase. For simplicity, first assume (this assumption is relaxed below) that there is little interaction or competition between predators, as would occur when predators are limited by some factor other than availability of prey. Given this assumption the predator isocline should look somewhat like that shown in Figure 6.21a. If there is competition between predators, higher predator den-

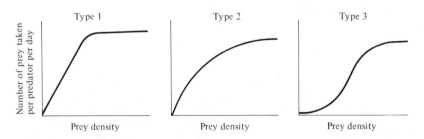

Figure 6.18   Three types of functional responses. [After Holling (1959a).]

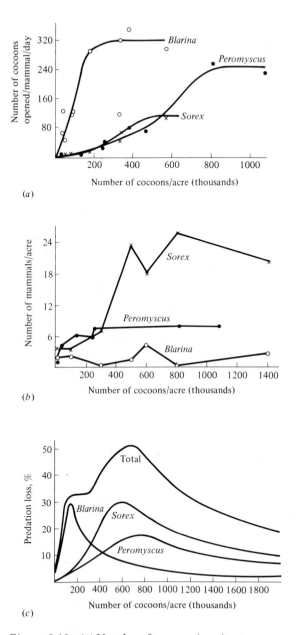

(a)

(b)

(c)

Figure 6.19 (a) Number of cocoons (prey) eaten per mammal per day by three small mammals plotted against the density of their prey (the so-called functional response). (b) Density of each of the three mammal predators plotted against prey density (the so-called numerical response of the predators). (c) Combined functional plus numerical responses of each predator species and the total, which represents the overall intensity of predation on the prey population, as a function of prey density. [From Holling (1959a).]

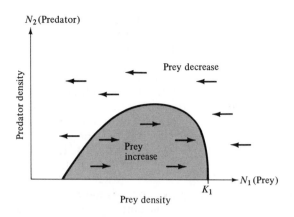

Figure 6.20 Hypothetical form of the isocline of a
prey species ($dN/dt = 0$) plotted against densities of
prey and predator. Prey populations increase within
the shaded region and decrease above the line
enclosing it. Prey at intermediate densities have a
higher turnover rate and will support a higher
density of predators without decreasing.

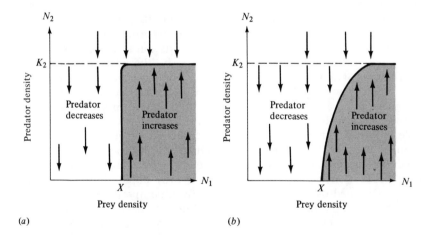

Figure 6.21 Two hypothetical predator isoclines. (*a*) Below some threshold
prey density, X, individual predators cannot capture enough prey per unit time
to replace themselves. To the left of this threshold prey density, predator
populations decrease; to the right of it they increase provided the predators are
below their own carrying capacity, K_2 (that is, within the shaded area). So long
as predators do not interfere with one another's efficiency of prey capture, the
predator isocline rises vertically to the predator's carrying capacity, as shown in
(*a*). (*b*) Should competition between predators reduce their foraging efficiency at
higher predator densities, the predator isocline might slope somewhat like the
curve shown. More rapid learning of predator escape tactics by prey through
increased numbers of encounters with predators would have a similar effect.

sities will require denser prey populations for maintenance and the predator isocline will slope somewhat as in Figure 6.21*b*. In both examples, the carrying capacity of the predator is assumed to be set by something other than prey density. Only one point in the $N_1 - N_2$ plane represents a stable equilibrium for both species—the point of intersection of the two isoclines (where dN_1/dt and dN_2/dt are both zero). Consider now the behavior of the two populations in each of the four quadrants marked A, B, C, and D in Figure 6.22. In quadrant A, both species are increasing; in B, the predator increases and the prey decreases; in C, both species decrease; and in D, the prey increases while the predator decreases. Arrows or vectors in Figure 6.22 depict the above changes in population densities.

Relative magnitudes of the changes in the population densities of prey and predator determine another important property of this model—that is, whether or not a stable equilibrium exists. There are three cases, corresponding to vectors that (1) spiral inward, (2) spiral outward, or (3) form a closed circle (Figure 6.22*a, b, c*). These three cases correspond to damped oscillations, oscillations of neutral stability, and oscillations increasing in amplitude until a limit cycle is reached, respectively (Figure 6.22*a, b, c*). Such oscillations of prey and predator could perhaps produce population "cycles" like those of lemmings and their predators (pp. 123–125). Given time, the case with damped oscillations will reach its equilibrium value at which neither prey nor predator population densities change; this case corresponds to a predator which is relatively *inefficient* at gathering prey (the predator cannot even begin to exploit the prey population until prey are fairly near their own carrying capacity). Similarly, the case that produces oscillations of increasing amplitude corresponds to a very *efficient* predator, which can exploit the prey population nearly down to its limiting rareness. Such an overly efficient predator should rapidly exterminate its prey (and thus go extinct itself unless alternative prey are available); little wonder that increasing oscillations are never observed in nature! Because of predator escape tactics of prey, many (or most) real predators are probably relatively inefficient, tending to crop only those prey present in excess of a substantial prey population (Errington, 1946). Damped oscillations also result from competition among predators producing a sloping predator isocline (Figure 6.21*b*). Hence the case with damped oscillation is probably the most realistic reflection of nature.

Individual predators that reproduce successfully at low prey densities will normally outcompete and replace less efficient individuals that require higher prey densities; hence natural selection acting on the predator moves its isocline to the left and *reduces* the stability of the interacting system. However, selection operating in favor of those prey individuals best able to escape predators opposes the action of selection on the predators and forces the predator isocline to the right (presumably it also raises the prey isocline); thus natural selection on the prey

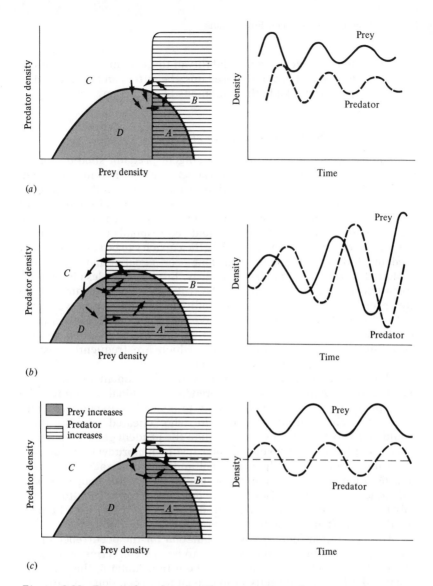

Figure 6.22 Prey and predator isoclines superimposed upon one another to show stability relationships. (*a*) An inefficient predator that cannot successfully exploit its prey until the prey population is near its carrying capacity. Vectors spiral inward, prey–predator population oscillations are damped, and the system moves to its joint stable equilibrium point (where the two isoclines cross). (*b*) An extremely efficient predator that can exploit very sparse prey populations near their limiting rareness. Vectors now spiral outward and the amplitude of population oscillations increases steadily until a limit cycle is reached, often leading to the extinction of either the predator or both the prey and the predator. (*c*) A moderately efficient predator that can begin to exploit its prey at some intermediate density. Vectors here form a closed circle, and populations of prey and predator oscillate in time with neutral stability, as in Figure 6.16. [After MacArthur and Connell (1966).]

population tends to *increase* the stability of the system. Indeed, unless the prey is one step ahead of the predator, the latter can be expected to overeat its prey and take both populations to extinction.

"Prudent" Predation and Optimal Yield

Some have suggested that an intelligent predator should crop its prey so as to maximize the prey's turnover rate and therefore the predator's yield. Such a "prudent" predator would maintain the prey population at that density which gives the maximum rate of production of new prey biomass. In terms of the prey isoclines depicted in Figure 6.22, this prey density of "optimal yield" corresponds to the density at the peak of the prey isocline. Man has the capacity to be such a prudent predator; indeed, optimal yield has long been a goal in management of exploited populations in fisheries biology. But do other, less intelligent predators also maximize their yield? A truly "prudent" predator should prey preferentially upon those prey individuals with low growth rates and low reproductive values, and leave those with rapid growth rates and higher reproductive values alone. In fact predators often do take aging and decrepit prey individuals, which are frequently easy to catch, while younger, more vigorous ones escape.

However, there is a potential flaw in the prudent predation interpretation, provided that several predator individuals encounter the same prey items. If, say, young juicy prey are less experienced and easier to catch, an individual predator who "cheated" and ate them would be likely to leave more genes than the prudent genotypes which did not exploit this food supply; as a result, nonprudence would become incorporated into the gene pool and spread. Exactly the same considerations apply to a competing species that is able to use the prey individuals in question. Hence we would expect prudence to evolve only in a situation where a single predator has exclusive use of a prey population; perhaps some feeding territories are examples.

Another, much more likely, explanation for the occurrence of apparent "prudent" predation in nature can be made in terms of the prey organisms themselves. As was pointed out in Chapter 5, the intensity of natural selection is directly proportional to expectation of future offspring (reproductive value); thus one might predict that individual prey with high reproductive values would have more to gain from escaping predators than would those with lower reproductive values. After one's expectation of future offspring has dropped to zero, nothing further can be gained from being able to escape a predator and predator avoidance cannot be evolved. Thus many cases of apparent "prudent" predation may well be simply part and parcel of old age; the evolution of senescence has wide significance! Viewed in this way, the susceptibility of prey to predation should be inversely related to their reproductive value.

Selected Experiments and Observations

Predation is readily studied in the laboratory and, under certain favorable circumstances, in the field. Gause (1934) studied a simple prey–predator system in the laboratory using two microscopic protozoans, *Paramecium caudatum* and *Didinium nasutum*. *Didinium* are voracious predators on *Paramecium*. When both species were placed together in a test tube containing clear medium (which supports a culture of bacteria, the food for *Paramecium*), *Didinium* overate its food supply, exterminated it, and then starved to death itself (Figure 6.23*a*). When some sediment was added to the medium (making it "heterogeneous" rather than "homogeneous"), providing a refuge or "safe site" for the prey, *Didinium* went extinct but the *Paramecium* population recovered (Figure 6.23*b*). In a third experiment (Figure 6.23*c*), Gause introduced new individuals of each species at regular time intervals; such "immigrations" resulted in two complete cycles of prey and predator. In other experiments, using *Paramecium aurelia* as the predator and a yeast, *Saccharomyces exiguus*, as prey, Gause (1935) obtained nearly three complete cycles.

In an experiment like Gause's with *Paramecium* and *Didinium* (but with *P. aurelia* as prey), Luckinbill (1973) showed that an unstable prey–predator interaction could be stabilized simply by adding methyl cellulose to the medium: this slowed down movements of both prey and predator and reduced the rate of contact between them.

Huffaker (1958) performed similar laboratory experiments on two species of mites, using oranges as the plant food for the system. One mite was herbivorous, eating the oranges; the second mite was a predator on the herbivorous one. In simple systems with oranges close together and evenly spaced the predator simply overate its prey and both species became extinct. Increasing distances between oranges only lengthened the time required for extinction but did not allow coexistence. However, by introducing barriers to dispersal and making the system still more complex, Huffaker obtained three complete prey–predator cycles (Figure 6.24). Thus environmental heterogeneity increased the stability of the system of a predator and its prey. In addition, these experiments illustrate the existence of prey–predator oscillations predicted by theory.

In a very revealing (although more complex) laboratory investigation, Utida (1957) examined both competition and predation simultaneously. His system was composed of three species: a beetle *(Callosobruchus chinensis)* as prey, and two species of predatory wasps as competitors *(Neocatolaccus mamezophagus* and *Heterospilus prosopidis)*. The prey beetle was provided with a continually renewed food supply. Both species of wasps, which have similar life histories, were dependent upon the beetle population as a common food source. Population densities of the three species fluctuated widely and erratically (Figure 6.25), but after four years, some 70 generations later, all three

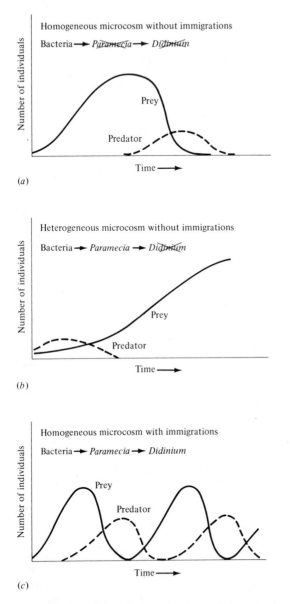

(a)

(b)

(c)

Figure 6.23 Three laboratory prey–predator experiments with protozoans. (*a*) In a simple homogeneous microcosm without immigration of new prey or predators, the predator quickly overeats and exterminates its prey and then all predators themselves starve to death. (*b*) In a more heterogeneous system, the predator goes extinct and the prey population recovers to expand to its carrying capacity. (*c*) Even in a homogeneous microcosm, immigration of prey and predators results in both populations oscillating in time. [From Gause (1934), reprint ed., © 1964 The Williams & Wilkins Co., Baltimore, Md.]

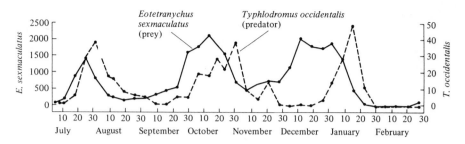

Figure 6.24 Fluctuations in the population size of a predatory mite, *Typhlodromus occidentalis,* and its prey (another mite), *Eotetranychus sexmaculatus,* in a spatially heterogeneous environment. [After Huffaker (1958).]

were still coexisting. Populations of the two predatory wasps tended to fluctuate out of phase with one another. Analysis showed that *Heterospilus* was more efficient at finding and exploiting the beetle population when it was at low densities but *Neocatolaccus* was more efficient at high prey densities (Figure 6.26); thus the competitive advantage shifted between the two wasps as the density of beetle prey changed in time. Utida thought that fluctuations of the beetle population were caused both by the effects of the two wasp predators and by density-dependent changes in the rate of reproduction of the beetle itself. Thus this system provides a neat example of coexistence of two competitors on a single resource due to changing abundance of that resource. The stability of the system is apparently a. result of biotic interactions.

A potent technique for studying the effects of predation in nature is to exclude predators from an area and then compare this experimental

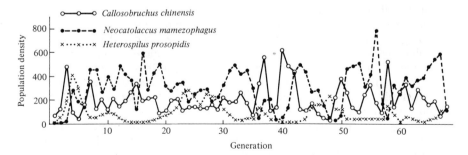

Figure 6.25 Changes in the population sizes of a beetle host *(Callosobruchus)* and two parasitic wasps over a four-year period. Although fluctuations appear erratic, all three species survived the entire period of some 70 generations. [After Utida (1957).]

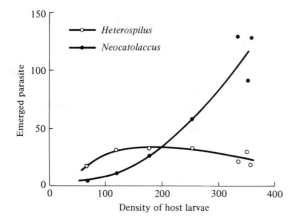

Figure 6.26 Number of parasitic wasps emerging from larvae of the host beetle plotted against the density of these larvae. *Neocatolaccus* is a more successful parasite at high host densities, whereas *Heterospilus* has the advantage at lower densities of host larvae. [After Utida (1957).]

site with an adjacent "control" area, which is similar but unaltered, with normal access for predators. Hopefully, resulting differences between the two areas can best be ascribed to predation. Paine (1966) performed just such a predator removal experiment along the rocky intertidal seacoast of the Olympic peninsula in Washington State. When the major top predator, the sea star *Pisaster ochraceus*, was removed, the number of species remaining was drastically reduced. Control areas with *Pisaster* supported some 15 species of marine invertebrates, but the area without the starfish had only 8 species. The rocky intertidal is a space-limited system. In the absence of predation, more efficient occupiers of space, especially the bivalve *Mytilus californianus* (a mussel), dominated the area. Presumably by preying on *Mytilus, Pisaster* continually open up new spaces that are rapidly colonized by fugitive species which are less efficient competitors for space. Thus, by reducing the level of competition at lower trophic levels, this starfish predator allows the coexistence of otherwise competitively incompatible species. Such *keystone* predators often have a powerful impact on community structure.

Islands typically support fewer species than are found in similar habitats on the adjacent mainland, with higher trophic levels being disproportionately "impoverished" (see pp. 312–317). As a result, it is often convenient to view islands as "natural" experiments in predator removal.

Evolutionary Consequences of Predation: Predator Escape Tactics

Generalized predators feeding on a variety of prey obviously must be adapted to cope with a wider variety of predator escape tactics than more specialized predators that need to deal with fewer types of prey. By diverging in their "strategies" of predator avoidance over evolutionary time, two (or more) prey species can make it increasingly difficult for a single predator to capture and exploit both prey types efficiently; thus prey evolution can "force" a predator to restrict the range of foods it eats.

Antipredator devices are extremely varied; some mechanisms for predator escape are quite simple and straightforward, but others may be exceedingly intricate and subtle. As an example of the former, many lizards dig special escape tunnels in their burrow systems that come up near the surface and allow the lizard to break out should a predator corner it underground.

Behavior and anatomy often make animals difficult to detect and/or to follow; such cryptic adaptations can involve sound, smell, color, pattern, form, posture, and/or movement. Concealing or cryptic coloration is widespread and often depends upon appropriate behavior; to hide itself, an animal must select the proper background and orient itself correctly. Some moths normally align themselves with the dark markings on their wings parallel to cracks and crevices of the tree bark substratum. Nearly all diurnal animals and some nocturnal ones are countershaded, with their dorsal (upper) parts darker than their ventral (lower) parts. Lighting from above casts shadows below; in a countershaded animal these balance dorsoventrally, reducing contrast and producing a neutral density—the net effect, of course, is to make the animal more difficult to see. Countershading occurs in most insects, fish, amphibians, lizards, snakes, birds, and many mammals. A counterexample which proves the point is provided by a few animals that are normally upside down in nature, such as the "upside down" catfish *Synodontis nigriventris* and certain moth larvae; these animals are darker ventrally than they are dorsally! Since a successful predator must also be inconspicuous in order to catch its prey, crypticity is equally as important to predators as it is to prey.

Many insects resemble parts of the plants on which they live, especially leaves, twigs, thorns, or bark; leaf butterflies and walking sticks are familiar examples. Both a green and a brown color phase often occur in such cryptically shaped animals. For example, females of two southern grasshoppers, *Syrbula admirabilis* and *Chortophaga viridifasciata*, have green and brown color phases (strangely, males are almost always brown!). Green females predominate in wetter, greener habitats; but in immediately adjacent drier and browner areas the brown form is most prevalent (Otte and Williams, 1972). Determination of a female's color is not under strict genetic control but is developmentally flexible in response to local conditions.

Actual demonstrations that coloration differences and background color matching have selective value are unfortunately rather scarce. The best documented example is that of the moth, *Biston betularia*, in England. This moth, along with several hundred other species, has evolved rapidly during the last century in response to human modification of its habitat. In the 1800s, *Biston* were pale colored, spending their daytime hours on pale, lichen-covered tree trunks. However, with the buildup of industry and concomitant air pollution, the lichens have died and tree trunks in some areas have been covered with a layer of soot and grime, becoming quite dark. In early collections, black moths (melanics) were very rare, but they have become increasingly more common until now these melanistic varieties comprise the vast majority of moth populations in polluted areas. This phenomenon of directional selection, termed *industrial melanism*, has also taken place in the United States and in Europe. In an elegant series of experiments, Kettlewell (1956) made reciprocal transfers of pale moths from a nonpolluted woods with melanic moths from a polluted area (Table 6.5). These moths, along with resident moths occurring at each locality, were marked with a tiny inconspicuous paint spot beneath their wings, and attempts were made to recapture them on later days. As expected, pale moths had lower survivorship in the polluted woods and melanic moths had lower survivorship in clean, lichen-covered forests. Moreover, Kettlewell actually observed foraging birds catching mismatched moths!

A black lava flow on white desert sands in New Mexico provides a "natural" experiment which strongly suggests that background color matching has evolved and is adaptive (Benson, 1933). This lava flow is completely surrounded by white sandy areas and has presumably been stocked mostly with animals derived from those that live on the white sands. Two closely related pocket mice live in the area: *Perognathus*

Table 6.5   Numbers of Typical and Melanic Marked Moths *(Biston betularia)* Released and Recaptured in a Polluted Woods Near Birmingham and an Unpolluted Woods Near Dorset[a]

	Polluted Woods	Unpolluted Woods
Numbers of Marked Moths Released		
Typical	64	496
Melanic	154	473
Number Recaptured		
Typical	16 (25%)	62 (12.5%)
Melanic	82 (53%)	30 (6.34%)

Source: From data of Kettlewell (1956).

[a] The wild population in the polluted woods was 87 percent melanic.

intermedius ater is nearly pitch black and occurs only on the lava; *P. apache gypsi* is pale white and lives only on the white sands.

The European land snail, *Cepaea nemoralis,* is polymorphic for shell color, with three phases: brown, pink, and yellow. In parts of England, a major predator on these snails is the song thrush *Turdus ericetorum.* These birds break open snail shells on stones ("thrush anvils"). Proportions of shells accumulated at these anvils usually differ from those in the population at large, thereby reflecting the relative intensity of thrush predation on the various snail morphs (Figure 6.27). During April, available backgrounds in the woods are largely brown; conditions become progressively greener during May. The percentage of yellow shells (with the animal inside, yellow shells take on a greenish hue) found at thrush anvils is higher than in the population at large in April, but lower than in the population in late May (Figure 6.27), indicating differential predation by thrushes. Thrush predation may well help to maintain the polymorphism in shell color since the brown and pink morphs have a fitness advantage early in the season while yellow snails are at an advantage later.

Some animals are covered with blotches of different colors, which tend to break up their shape and to make them more difficult to see. Good examples are rattlesnakes and boa constrictors; often these snakes look so much like a pile of dead leaves that they may go unde-

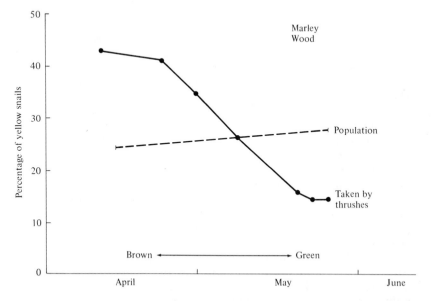

Figure 6.27   Percentages of yellow snails taken by thrushes versus those in the population at large, showing differential predation in time. [From Sheppard (1951).]

tected. This type of coloration, especially prevalent in larger animals that have difficulty in finding hiding places such as leopards, tigers, and giraffe, has been termed *disruptive coloration*.

A form of adaptive coloration which is not necessarily cryptic is called *flash coloration*. Many inconspicuous insects, including some butterflies and grasshoppers, are extremely cryptic when at rest; but when disturbed they fly away and reveal brightly colored underwings (often red, yellow, or orange). These insects thus suddenly become extremely visible and conspicuous, catching the predator's eye. When they land, they close their wings and quickly move away from the spot at which they landed. As anyone who has chased grasshoppers knows, it is very difficult to keep track of the position of such an animal. (Squid and octopi employ a remotely similar strategy when they squirt out their "ink," leaving a dense cloud in the water: typically the animal immediately changes both color and course, becoming pale and swimming at right angles to its original direction of flight, thereby evading the predator.) A rather derived kind of flash coloration occurs in some butterflies and moths which have large owl-like eyes on their underwings. These eyes are normally hidden by the upper pair of wings. When a small bird approaches, the upper wings are suddenly twitched aside, revealing an owl-like face beneath. Some small birds are apparently so startled that they fly away, leaving the insect alone! Similarly, the yellow "eyes" on some large green caterpillars may make them resemble green tree snakes.

Another, smaller, type of eyespot actually invites the attack of a predator. Many predators instinctively go for the eyes of their prey since eyes are usually one of the most vulnerable parts of an animal and their loss readily incapacitates them (lions and wolves have found another "Achilles heel" on large ungulates—they simply hamstring the animal). Many species of butterflies possess small "fake" eyespots along the periphery of their wings that may actually invite attack. Eyespots painted on such butterflies in places without eyespots are damaged when the animals are released and recaptured, apparently being pecked at by birds (Sheppard, 1959). Thus the butterfly obtains a second chance at escape by luring the predator's attack away from its own eyes. Behavioral adaptations may sometimes serve similar functions; certain snakes raise their blunt tails and wave them around in a very headlike manner, occasionally actually making short menacing lunges with their tails at the threatening predator! Should the unwary predator grab the snake by this "head," the serpent still has its real head free to bite back.

Many birds and mammals have various sorts of alarm signals, which, when given, warn other animals that a predator is in the immediate vicinity. Beaver warn one another of danger by slapping their tails loudly against the surface of the water. Similarly, rabbits in a warren (a colony of related individuals using the same burrow system) often "thump" with their hind feet to signal the approach of a predator.

The white underside of the tail of some rabbits and ungulates (such as the white-tailed deer) is raised when the animal flees from a predator, perhaps serving as a warning signal to nearby animals. Prairie dogs, many primates, and many foraging bird flocks, such as crows, frequently post sentinels (see also p. 169) that watch for predators from a good vantage point and warn the group should one appear in the distance. Among birds, alarm calls in response to the presence of bird-eating hawks are especially prevalent. Typically these faint shrill whistles are extremely difficult for a vertebrate predator to locate; often they are similar in widely different bird species, presumably having converged over evolutionary time.

Because different species may benefit from the call of one individual, hawk alarm calls appear to be somewhat altruistic. However, warning calls are normally used only during the nesting season and are therefore best interpreted as having arisen through kin selection (see Chapter 5). The ventriloquial nature of the call, coupled with the fact that the caller has already seen the predator and therefore knows exactly where the danger lies, ensures that the risk to the bird giving the alarm is slight. Likewise the fact that banded birds often return to breed in the same area where they themselves were raised (Chapter 5) ensures that birds in any given area will be related and share many genes, which in turn ensures that the total gain to the many relatives of the pseudo-altruist will frequently be quite large. If hawk alarm calls have evolved via kin selection, one would predict that the frequency of use of such calls should be inversely related to the distance the caller is from its own nest and immediate relatives (no one has yet demonstrated this empirically). The fact that these calls work across species lines is easily explained because individuals that recognize warning signals of other sympatric species as alarms should be at a relative advantage within their own population. The convergence of hawk alarm calls can be explained by either or both of two mechanisms. The first concerns the ventriloquial properties of the call. The number of ways in which an alarm call can be loud enough to function as a warning, and yet still be difficult to locate, are decidedly limited. Thus convergence may be a simple result of limitation inherent in the system. A second, equally likely, possibility is that natural selection has favored convergence because it facilitates interspecific recognition of alarm calls. Thus individuals tending to produce call variants more like those of another species would benefit because they would also be more likely to recognize the calls of the other species (as such, call convergence would be very similar to Müllerian mimicry, discussed below).

There are several alternative hypotheses for alarm calls. The calls may not be intended as "alarms" at all, but as signals intended to inform the predator that it has been detected and that it not need try to capture the caller (p. 169). Charnov and Krebs (1975) suggest that the caller could decrease its own chance of being captured by alerting other individuals because the predator's attention would be drawn away from the caller and toward the large numbers scurrying for cover. Still another argument for a selfish caller is based on the fact that many predators tend to stay nearby and return to an area where they have successfully captured prey in the past: to the extent that alarm calls prevent a predator from catching any prey at all, they may serve to keep the predator on the move and hence out of the immediate area of the caller.

Yet another evolutionary consequence of predation is *warning coloration;* unpalatable or poisonous animals have often evolved bright colors that advertise their distastefulness. Such animals are usually colored with the same conspicuous colors we use for signs along highways: reds, yellows, and blacks and whites. Examples of animals with warning coloration are bees and wasps, monarch butterflies, coral snakes, skunks, and certain brightly colored poisonous frogs and salamanders. Signals that warn potential predators may also involve pattern, posture, smell, or sound. A rattlesnake's buzzing presumably serves to warn other larger animals such as the American bison not to come too near (unfortunately for the rattler, however, this warning only attracts human attention, which usually results in the snake's demise). Experiments have shown that avian and lizard predators learn to avoid distasteful prey. In the case of poisonous prey, this learning may actually be incorporated into the gene pool and manifested as an "instinctual" avoidance.

Mimicry is an interesting sidelight of warning coloration that nicely demonstrates the power of natural selection. An organism that commonly occurs in a community along with a poisonous or distasteful species can benefit from a resemblance to the warningly colored species, even though the "mimic" itself is nonpoisonous and/or quite palatable. False warning coloration is termed Batesian mimicry after its discoverer. Many species of harmless snakes mimic poisonous snakes; in Central America some harmless snakes are so similar to poisonous coral snakes that only an expert can distinguish the mimic from the "model." Similarly, harmless flies and clearwing moths often mimic bees and wasps, and the palatable viceroy butterfly mimics the distasteful monarch. Batesian mimicry is disadvantageous to the model because some predators will encounter palatable or harmless mimics and thereby take longer to learn to avoid the model. The greater the proportion of mimics to models, the longer is the time required for predator learning and the greater the number of model casualties. In fact, if mimics became more abundant than models, predators might not learn to avoid the prey item at all but might actively search out model and mimic alike. For this reason Batesian mimics are usually much less abundant than their models; also these mimics are frequently polymorphic and mimic several different model species.

A different kind of mimicry occurs when two species, both distasteful or dangerous, mimic each other; this phenomenon is termed

Müllerian mimicry.* Both bees and wasps, for example, are usually banded with yellows and blacks. Since potential predators encounter several species of mimics more frequently than a single species, they learn faster to avoid them, and the relationship is actually beneficial to both prey species (Benson, 1972). The resemblance need not be as close as it must be under Batesian mimicry because neither species actually deceives the predator; rather each only reminds the predator of its dangerous or distasteful properties. Müllerian mimicry is beneficial to all parties including the predator; mimics can be equally common and are rarely polymorphic.

Plants, being sessile, cannot use many of the escape techniques of animals and are obviously and decidedly more limited in the ways they can deter potential predators. A plant with a patchy or spotty distribution in time and/or space may escape some predation simply by virtue of its unpredictable availability; thus an annual that is here today but gone tomorrow may be more difficult for herbivores to find and use than an evergreen perennial that is always relatively available (see also below). Some plants, especially perennials, have evolved morphological adaptations, such as hairs, spines, and hooks (Gilbert, 1971), that discourage many herbivores quite directly. By far the most widespread predator deterrent of plants is what might be termed chemical warfare. A great variety of secondary chemical substances occur in plants that are not known to serve any direct physiological function for their possessors. Many are evidently not breakdown products of larger molecules due to metabolic processes and wastes, but are secondary substances produced by active synthesis from smaller molecular precursors. Such secondary chemical substances often contain nitrogen and other elements that are available to the plant only in limited supply; moreover, it takes energy to produce these chemicals. Clearly there are definite costs to the plant in production of herbivore repellents. Nearly a century ago the German botanist Stahl (1888) suggested that these secondary substances might reduce the plant's palatability to herbivores. Stahl's prediction has now been amply verified for many different plant–herbivore systems.

Agriculturalists and plant breeders have produced many strains that are highly resistant to normal herbivores. Genetic varieties, or morphs, of plants in nature have been shown to be differentially palatable to herbivores; thus Jones (1962, 1966) found that a number of herbivores ranging from insects and snails to *Microtus* (a mammal) preferred a "noncyanogenic" morph of the plant *Lotus corniculatus* over a "cyanogenic" one. Tannins have been implicated as the agent that repels some herbivores; oak leaf tannin significantly reduces the growth rate of larvae of the moth *Operophtera brumata* (Feeny, 1968). Frequently herbivores eat only relatively new growth and do not utilize

---

* Actually the dichotomy between Batesian and Müllerian mimicry is somewhat artificial in that these two sorts of mimicry grade into one another.

older parts of plants, presumably because the latter contain tannins and other repellent chemicals (Feeny, 1970). Other chemical substances believed to protect plants from animals and fungi include essential oils and resins, alkaloids, terpenes, and terpenoids. The latter two classes of compounds have especially penetrating odors and tastes; sesquiterpenes are fatal to sheep. Wild herbivores that have evolved alongside poisonous plants would be unlikely to eat them, whereas domestic sheep and cattle will eat many poisonous fodders.

The argument presented at the beginning of this section suggests that sympatric plant species may usually gain by evolving qualitatively different antiherbivore secondary chemical defenses. As a result, plants in general are not continuous resources but constitute a spectrum of qualitatively distinct, discrete food types. Such divergent antiherbivore chemistries force evolution of herbivore specialists by reducing the efficiency of generalized herbivores. Because related plants have similar secondary chemistries, phylogenies of certain herbivorous insects (particularly butterflies) closely parallel the phyletic relationships of their host plants (Ehrlich and Raven, 1964; Benson, Brown, and Gilbert, 1975).

Coevolution

In its broadest sense, *coevolution* refers to the joint evolution of two (or more) taxa that have close ecological relationships but do not exchange genes, and in which reciprocal selective pressures operate to make the evolution of either taxon partially dependent upon the evolution of the other (Ehrlich and Raven, 1964). Thus coevolution includes most of the various forms of population interaction, from competition and predation to mutualism and protocooperation. The term coevolution is also often used in a more restricted sense to refer primarily to the interdependent evolutionary interactions between plants and animals, especially their herbivores and pollinators. A plant may evolve a secondary chemical substance that deters the vast majority of predators, but if a particular herbivore can in turn evolve a physiological means of coping with the chemical deterrent, it can thereby obtain an uncontested food supply. Through this kind of coevolution many herbivores have become strongly specialized on a single species or a few closely related species of plants. Thus *Drosophila pachea* is the only species of fruit fly that can exploit the "senita" group of cacti; these plants produce an alkaloid that is fatal to the larvae of all other fruit flies, but *D. pachea* has apparently evolved a means of detoxifying this chemical (Kircher *et al.,* 1976).

In many cases such specialized herbivores even use a plant's toxic chemicals as cues in locating and/or selecting their host plants. Some herbivores, such as the monarch butterfly, actually sequester plant poisons (cardiac glycosides in this case) which in turn make the herbivore unpalatable or even poisonous to its own potential predators.

Attempts have been made to generalize about the coevolution of herbivores and plant antiherbivore tactics (Orians, 1974). Feeny (1975) argues that rare or ephemeral plant species are hard for herbivores to find and hence are protected by escape in time and space; moreover, he asserts, such plant species should evolve a diversity of qualitatively different, chemically inexpensive, defenses that should constitute effective evolutionary barriers to herbivory by nonadapted generalized herbivores which are most likely to find such "cryptic" plants. These same secondary chemicals, however, will be only minimal ecological barriers to adapted specialized herbivores, against which the plant's primary antiherbivore tactic is escape in time and space (not being found). In contrast, Feeny reasons that abundant and/or persistent plant species cannot prevent herbivores from finding them either in ecological or evolutionary time. Such "apparent" plant species appear to have evolved more expensive quantitative defenses, including tough leaves of low nutrient or water content containing large amounts of relatively nonspecific chemicals such as tannins (Table 6.6). Feeny points out that such plant defenses should pose a significant ecological barrier to herbivores, although perhaps only a weak evolutionary barrier unless supplemented with qualitative chemical defenses (some plants have both).

Cates and Orians (1975) develop somewhat different but related predictions for early versus late successional plant species. Because early successional plants escape from herbivores in space and time, Cates and Orians reason that such plants should allocate fewer resources to chemical antiherbivore defenses than the more apparent plants of later stages in succession. Thus early successional plant species should make better foods for generalized herbivores than later

Table 6.6 Some of the Suggested Correlates of Plant Apparency

Apparent Plants	Unapparent Plants
Common and/or conspicuous	Rare and/or Ephemeral
Woody perennials	Herbaceous annuals
Slow growing, competitive species	Faster growing, often fugitive species
Late stages of succession, climax	Early stages of succession, second growth
Bound to be found by herbivores (cannot escape in time and space)	Protected from herbivores by escape in time and space (but still encountered by wide-ranging generalized herbivores)
Produce more expensive quantitative (broad-based) antiherbivore defenses (tough leaves, tannins)	Produce inexpensive qualitative chemical defenses (poisons and/or toxins) to discourage generalized herbivores.
Quantitative defenses constitute effective ecological barriers to herbivory, although perhaps only a weak evolutionary barrier unless supplemented with qualitative defenses	Qualitative defenses may be broken down over evolutionary time by coevolution of appropriate detoxification mechanisms in herbivores (host plant specific herbivore species result)

successional and climax plant species. Indeed, experimental studies on slug feeding indicate that early successional annuals were significantly more palatable than later successional species (Cates and Orians, 1975). However, the opposite result was obtained by Otte (1975) in similar experiments with grasshoppers: these generalized herbivores accepted more later successional plant species than early ones. Otte suggests that this difference may arise from the difference in mobility between slugs and grasshoppers. In a survey of lepidopteran feeding habits, Futuyma (1976) found greater degrees of specialization to host plant species among insects feeding on herbaceous plants than among those that feed on leaves of shrubs and trees (this pattern neatly fits Feeny's plant apparency dichotomy). Futuyma suggests that plant defense systems are more diverse in floristically rich plant communities than they are in less diverse communities.

Discrepancies among these studies indicate that generalizations are difficult to make and that they will have to allow for exceptions (for further discussion of this interesting area, see Rhoades and Cates, 1976, and/or Gilbert, 1977).

Populations of wild ginger *Asarum caudatum* in western Washington are polymorphic for growth rate, seed production, and palatability to a native slug, *Ariolimax columbianus* (Cates, 1975). In habitats where slugs were low in abundance, Cates found that populations of wild ginger were dominated by individuals allocating more energy to growth and seed production and less to production of antiherbivore chemicals. Presumably less palatable plants have a fitness advantage in habitats with more slugs: even though they grow more slowly, they lose less photosynthetic tissue to slug herbivory.

In some cases plants have actually formed cooperative relationships with animals that result in their protection from certain herbivore species. Janzen (1966) showed that some species of *Acacia* deprived of their normal epiphytic ant fauna are highly palatable to herbivorous insects, whereas species that do not normally have ants for protection from herbivores are less palatable. The acacias benefiting from ant protection actually produce nectaries and swollen thorns that attract and in turn benefit the ants! Thus these plants put matter and energy into attracting ants that defend their leaves, rather than into more direct chemical warfare. This antiherbivore ploy is broad based, since the ants ferociously attack a wide range of herbivores.

Because most land plants cannot move, they often exploit animals both for pollination and for seed dispersal (some rely on wind as well). Seeds of many fruits pass unharmed through the intestines of herbivores and germinate to grow a new plant from the droppings of the animal dispersing agent. Colorful flowers with nectar and brightly colored fruits can only be interpreted as having been evolved to attract appropriate animals. Here, as in plant–herbivore interactions, a high degree of plant–animal specificity has arisen. Animals that pollinate a particular plant are referred to as *pollinating vectors*. As an example, in Central America different species of male euglossine bees are highly specific to particular species of orchids; male bees travel long distances between orchids. Different bee species are attracted by different orchid fragrances (Dressler, 1968), as can be shown by putting out "baits" of artificially synthesized orchid "perfumes." These bees are probably necessary for, and may have allowed the evolution of, the great diversity of tropical orchids, many of which are evidently quite rare and far apart. Such specificity of pollinating vectors assures that the plant's pollen is transmitted to the ovules of its own species; the advantages of this specificity to male euglossine bees have not yet been determined. Whereas female euglossine bees are not as specific to the plant species they pollinate as males, individual females travel distances up to 23 km (Janzen, 1971a) and regularly move long distances between sparsely distributed plants in gathering nectar and pollen; thus they probably promote outcrossing among tropical plants at low densities. Indeed, Janzen suggests that such "traplining" by female bees may actually permit the very existence of plant species forced to very low densities by factors such as competition and predation on their seeds and seedlings.

Some pollinators, such as *Heliconius* butterflies (Gilbert, 1972), obtain amino acids from the pollen of plants they pollinate. Because production of nectar and pollen (and fruit) requires matter and energy, attracting animal pollinators (and seed dispersers) has its costs to the plant. Nectar and fruits are usually rich in sugars and other carbohydrates but contain relatively little protein; in contrast, pollen and seeds contain considerably greater amounts of nitrogen and other limiting materials. Due to the frequent scarcity of such vital nutrients, carbohydrates are presumably cheaper for a plant to produce than amino acids and proteins. Thus pollen-eating pollinators presumably cost a plant considerably more than strict nectar feeders. Returns from visiting a flower (or eating a fruit) must be great enough to the animal pollinator or seed disperser to make it worthwhile, yet small enough that the animal will travel the distances necessary to disperse the pollen or seeds. This intricate energetic interplay between plants and their pollinators is reviewed by Heinrich and Raven (1972) and Heinrich (1975).

Many plants protect their seeds either by enclosing them in a toxic matrix and/or by means of a hard shell. Some seeds are poisonous. Nevertheless the high nutrient content of seeds has resulted in the evolution of effective seed predators. Predation on seeds may often be heaviest where they occur in greatest concentrations (such as acorns underneath a parent oak) because seed predator populations will generally be largest where the most food is available (Janzen, 1971b). As a result, the probability of an individual seed's surviving to establish itself as a plant may often vary inversely with seed density. In many trees the majority of seeds fall to the ground near the parent tree, with a continually decreasing number of seeds ending up at distances farther from the parent tree (Figure 6.28). As a result of these opposing processes, Janzen suggests that recruitment is maximized at some distance from the parent tree (Figure 6.28). Janzen's model of seed predation and recruitment may help to explain the high species diversity of tropical trees, which suffer heavy seed losses to specialized seed predators that eat the seeds of particular tree species. (It neglects the question of why there are so many specialized seed predators in the tropics).

Intricacies of coevolutionary relationships between pine squirrels (*Tamiasciurus*) and their coniferous food trees were studied in the Pacific northwest by C. Smith (1970). Conifer seeds constitute the staple food supply of these squirrels; they can effectively strip a tree of most of its cones. Trees reduce the effectiveness of squirrel predation in many different ways: (1) by producing cones that are difficult for the squirrels to reach, open, and/or carry; (2) by putting fewer seeds into each cone (squirrels eat only the seeds themselves and must "husk" cones to get them); (3) by increasing the thickness of seed coats, requiring that the squirrels spend more time and energy extracting each seed; (4) or, similarly, by putting less energy into each seed (a drawback is that seedlings from smaller seeds have fewer resources at their disposal and are presumably poorer competitors than seedlings from larger seeds); (5) by shedding seeds from cones early, before the young squirrels of the year begin foraging; and (6) by periodic cone "failures" that decimate the squirrel population, thereby reducing the intensity of predation during the next year. Thus squirrel predation has had pro-

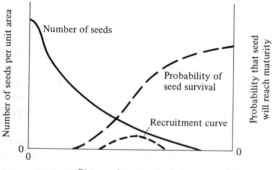

Figure 6.28   Hypothetical model of seed recruitment versus distance from the parent tree. Near the tree all seeds are eaten by seed predators; at increasing distances from the tree the probability of seed survival increases as the density of seeds and their predators decreases. Although the number of seeds drops with distance from the parental tree, recruitment is highest at some distance from the tree. [After Janzen (1970).]

found evolutionary influences upon various reproductive characteristics of conifers, including details of cone anatomy and location, the number of seeds per cone (and the variability in the number per cone), the time at which the cones shed their seeds, the thickness of seed coats, and annual fluctuations in the size of cone crops. Evolution of these defense mechanisms by the conifers has in turn forced squirrels to adapt in various ways, such as choosing cones carefully and stockpiling them.

Fluctuations in cone crops from year to year are pronounced and are best interpreted as an antisquirrel strategy, since they occur even when climatic conditions are apparently favorable for trees. Apparently the conifers withhold their products of primary production and store them for later use. Cone crops and failures are often synchronized among different tree species in a given area—a further indication that the phenomenon is directed at the squirrels. Individual trees out of synchrony presumably set fewer seeds and are thus selected against by natural selection. Smith points out that different conifer species have *diverged* from one another in evolution of cone anatomy, size, location, and time of shedding, but that these same conifer species have *converged* in their fluctuations in cone crops. Both reduce the squirrel's efficiency.

Symbiotic Relationships

Symbiosis means living together. Usually the term is used only to describe pairs of organisms that live together without harming one another, thereby excluding parasitism (+, −) and amensalism (−, 0), in which one party is affected adversely (see introduction to this chapter and Table 6.1 for explanation of symbols). Hence symbiotic relationships include neutralism (0, 0), protocooperation (+, +, nonobligatory), mutualism (+, +, obligatory), and commensalism (+, 0). As pointed out earlier, true neutralism is probably rare and therefore need not be considered. However, protocooperation, mutualism, and commensalism are fairly widespread, particularly in diverse communities. Plant–pollinator and plant–seed dispersers, as well as the ant–acacia interaction referred to above, are examples of protocooperation. Numerous other cases of protocooperation are known: some small birds ride on the backs of water buffalo (the bird obtains food while the mammal is freed of many insect pests) and other birds pick between the teeth of crocodiles (the bird obtains food while the reptile gets its teeth cleaned). A bird known as the honey guide has formed a unique alliance with the honey badger or ratel (a mammal); the honey guide locates a beehive and leads the honey badger to it, whereupon the mammal tears open the bee's nest and eats its fill of honey and bee larvae. Later the bird has its meal. The honey guide can find bee hives with relative ease but cannot open them, while the ratel is in just the opposite situation; cooperation clearly increases the efficiency of both species.

Mutualism is less common than protocooperation, probably because both populations are completely dependent upon the relationship and neither can survive without the other. Thus termites cannot themselves produce enzymes to digest the cellulose in wood, but, by harboring in their intestines a population of protozoans that can make such enzymes, the insects are able to exploit successfully wood as a food source. Neither termite nor protozoan could survive without the other. These intestinal endosymbionts are passed on from one generation of termites to the next through exchange of intestinal contents (see pp. 168–169). Another putative example of mutualism is lichens, which are composed of a fungus and an alga; the fungus provides the supportive tissue while the alga performs photosynthesis. (However, algae of some lichens can be grown without the fungi, so some lichens may represent not true mutualism but protocooperation.)

Commensalism occurs when one population is benefited but the other is unaffected (+, 0); it is probably uncommon. Small epiphytes such as bromeliads and orchids, which grow on the surfaces of large trees without obvious detriment to the tree, might be an example. A well-documented case of commensalism is the association between cattle egrets and cattle (Heatwole, 1965). These egrets follow cattle that are grazing in the sun and capture prey (crickets, grasshoppers, flies, beetles, lizards, frogs) that move as cattle approach. Heatwole found that the number of cattle egrets associated with cattle was strongly dependent on the activities of the cattle; thus he observed fewer egrets than expected on a random basis near resting cattle, but nearly twice as many egrets as expected (if the association were entirely random) accompanied cattle that were actively grazing in the sun. Since the birds seldom take prey (such as ticks and other ectoparasites) directly from the bodies of the cattle, the mammals probably do not benefit directly from their relationship with egrets. Moreover, Heatwole demonstrated that egret feeding rates and feeding efficiency were markedly higher when these birds were associated with cattle (Table 6.7).

Symbiotic relationships are easily modeled with equations similar to the Lotka-Volterra competition equations (1) and (2) simply by changing the signs of the alphas (Ks have also been changed to Xs since they no longer represent maximal densities)

$$\frac{dN_1}{dt} = r_1 N_1 \left(\frac{X_1 - N_1 + \alpha_{12} N_2}{X_1} \right) \tag{15}$$

$$\frac{dN_2}{dt} = r_2 N_2 \left(\frac{X_2 - N_2 + \alpha_{21} N_1}{X_2} \right) \tag{16}$$

Equilibrium conditions are described by a pair of linear equations and are shown graphically in Figure 6.29. Populations reach equilibrium at

Table 6.7   Various Aspects of the Association of Cattle Egrets with Cattle

Category	No. of Cattle	Percent Cattle	Number of Associated Egrets expected	observed
Grazing in sun	735	39.1	239	439
Grazing in shade	55	2.9	18	21
Standing in sun	146	7.8	48	46
Standing in shade	257	13.7	84	17
Lying in sun	503	26.8	164	69
Lying in shade	143	7.6	47	17
Walking	39	2.1	13	3
Total	1878	100.0	612	

	Mean Number per Minute	No. of Times Count Was Higher Than for Opposite Egret	Percent of Times Count Was Higher Than for Opposite Egret
FEEDINGS, $N = 84$			
Associated	2.34	58	69
Nonassociated	1.71	26	31
STEPS, $N = 62$			
Associated	20.1	7	11
Nonassociated	32.1	55	89
FEEDINGS/STEP, $N = 59$			
Associated	0.129	52	88
Nonassociated	0.051	7	12

Source: From Heatwole (1965).

Note: Upper box shows the numbers of egrets associated with cattle engaged in different activities. Lower box shows feeding rates, steps taken per prey item (energy expended in foraging), and feeding efficiencies of egrets associated with and not associated with cattle.

density X_1 or X_2 in the absence of the other species, and each population's equilibrium density is increased by increasing the density of the other species. If both X_1 and X_2 are positive and if α_{12} and α_{21} are chosen so that isoclines cross (Figure 6.29), the joint equilibrium is stable.

Complex Examples of Population Interactions

Interactions among populations often become quite intricate, particularly in diverse communities. Colwell (1973) studied interactions among four species of nectar-feeding birds, four species of flowering plants, and two species of mites (Figure 6.30) in Costa Rica. Three hummingbirds, *Colibri thalassinus*, *Eugenes fulgens*, and *Panterpe insignis*, compete for nectar and are pollinating vectors for the plants. Flowers of various species differ in corolla lengths and are visited differentially by hummingbirds that differ in beak lengths. Only *Eugenes*,

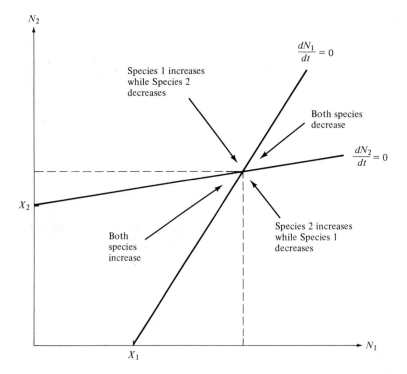

Figure 6.29   Isoclines for two mutually beneficial species with a stable joint equilibrium (see text).

which has the longest beak, can reach the nectar of *Centropogon talamancensis* through its very long corolla (Figure 6.30). However, *Panterpe* hummingbirds steal nectar from this *Centropogon* species by piercing the base of flower corollas. The fourth species of bird, *Diglossa plumbea,* is a nectar thief that obtains nectar from all four plant species by breaking their corolla bases. Two species of mites live within flowers of different plant species, moving among flowers by climbing up beaks and riding in hummingbird nostrils. Both mite species are found on *Panterpe* and *Eugenes* hummingbirds, which make legitimate pollinating visits through the corollas of flowers of three and four species of plants, respectively. One mite, *Rhinoseius colwelli,* is restricted to flowers of the two species of *Centropogon* and never occurs in either *Macleania* or *Cavendishia* flowers. The second species of mite, *R. richardsoni,* is found only in *Macleania* and *Cavendishia* flowers and never occurs in *Centropogon* flowers. Significantly, avian pollinator visits are structured so that transfers between *Macleania* and *Cavendishia* flowers are frequent while transfers between flowers of these two species and those of *Centropo-*

gon are much more infrequent. Experimental introductions of mites into flowers without mites showed that both species can live and reproduce successfully in the flowers normally occupied only by the other species. Adult male mites are extremely aggressive, particularly in interspecific encounters, and Colwell observed male *R. colwelli* killing *R. richardsoni*. Over evolutionary time, adult male mites may have reinforced the observed species-specific separation on flowers of different species by killing mites of the other species when these made the mistake of leaving their hummingbird carriers to invade a flower of the wrong species containing adult males of the other species of mite. Among these ten species, then, interactions include intense interference competition (between the two species of mites), exploitation

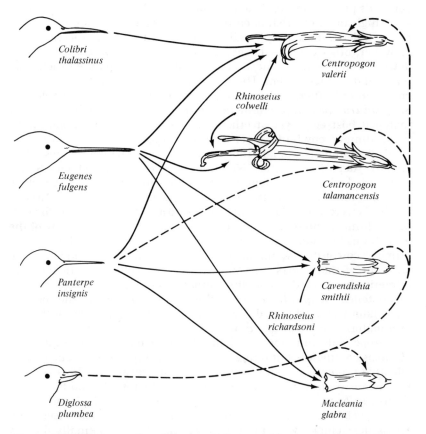

Figure 6.30 Patterns of exploitation of flowers of four species by birds and mites. Broken lines indicate illegitimate visits by nectar thieves, which pierce the base of the corolla. [From Colwell (1973). Copyright © 1973 by The University of Chicago Press.]

competition (among nectarivorous birds), protocooperation (between hummingbird pollinators and plants), parasitism (between plants and nectar thieves), and commensalism (between mites and their hummingbird carriers).

A second example of the complexity that can develop in interactions among populations concerns a Panamanian cowbird and its hosts, studied by N. Smith (1968). The giant cowbird, *Scaphidura oryzivora*, is a brood parasite, or an animal that puts its eggs into the nest of another species and lets the latter raise its young. It has four hosts in this area, three species of oropendulas and one cacique (oriole-like birds). These oropendulas and caciques build long, dangling oriole-like nests in mixed-species colonies, often with more than a hundred pairs of birds nesting together. Some colonies have been in existence for over 20 years. Two colonies, averaging 173 nests, fledged an average of 111 cowbirds per year or nearly one per nest. Local Panamanians said that such high rates of production of *Scaphidura* were of regular occurrence (indeed, these people called the cowbirds "black oropendulas"). However, other colonies fledged many fewer *Scaphidura*. Smith cut down several thousand nests, examined and manipulated their contents, and replaced them. He found striking variation from colony to colony in coloration and marking of cowbird eggs; at some colonies, *Scaphidura* produced mimetic (mimicking) eggs which closely matched host eggs, yet at other colonies cowbirds laid nonmimetic eggs. Mimetic cowbirds never deposited more than a single egg in a host's nest and always laid their eggs in nests in which the hosts had already deposited eggs. Nonmimetic cowbirds, which Smith nicknamed "dumpers," deposited their eggs in empty nests as well as full ones and often left two or three eggs in a given nest. Mimetic cowbirds were very cryptic and cautious to avoid being seen by the host birds; dumpers often laid their eggs conspicuously, in full view of the host. By using model eggs, Smith showed that hosts in colonies with mimetic cowbirds discriminated between eggs and threw mismatched ones out of their nests; in contrast, hosts in colonies with the nonmimetic dumpers did not discriminate, but readily accepted even mismatched *Scaphidura* eggs. While 73 percent of the nondiscriminator nests contained cowbird chicks, only 28 percent of the discriminators had *Scaphidura* in their nests.

Oropendula and cacique colonies are often clustered near the nests of wasps or stingless but biting bees (presumably these insects provide effective protection against vertebrate nest predators). However, the major cause of nestling mortality was not vertebrates but botflies of the genus *Philornis*, which lay their eggs on the baby birds. Upon hatching, botfly larvae (maggots) burrow into the chick's body and feed upon its tissues; chicks with more than seven maggots normally die. By examining chicks and old nests for botfly larvae and pupae, respectively, Smith found that colonies near large nests of bees or wasps had a substantially lower incidence of botfly parasitism than did colonies without bees or wasps. Even within wasp-protected colonies, nests farther than 7 meters from the wasp nest were more likely to have botflies than those closer to the wasp nests. The precise mechanism of botfly protection provided by proximity to bees or wasps is unknown, but it seems likely that the latter insects actively capture the former. Colonies protected against *Philornis* by bees or wasps were invariably composed of discriminator hosts and mimetic cowbirds, whereas colonies without bees or wasps were always composed of nondiscriminator hosts and nonmimetic cowbirds of the "dumper" type. However, even in colonies unprotected by bees or wasps, host chicks in nests containing at least one cowbird chick were seldom infested with botfly maggots (8.5 percent), whereas those in nests without *Scaphidura* chicks were heavily parasitized (90 percent of all such chicks had botflies). *Scaphidura* chicks, which are precocial, actively preen their nest mates and remove botfly eggs and maggots. (Occasionally cowbird chicks even capture adult botflies!) Thus the cowbird chicks protect the host chicks from botfly infestation.

Smith kept careful records of the composition of 4807 nests and the fledging success of hosts and cowbirds (Figure 6.31), which clearly show that nondiscriminating hosts can gain an advantage from association with cowbirds, whereas reproductive success of discriminator hosts is reduced by *Scaphidura* parasitism. The average number of

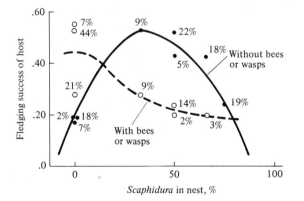

Figure 6.31 Fledging success of host oropendulas and caciques plotted against percentage of cowbird chicks in the nest for broods with different compositions of host chicks and cowbird chicks (total brood size also varies). Colonies without bees or wasps are depicted by closed circles and a solid line; those with bees or wasps are shown by open circles and a dashed line.
Percentages represent the fraction of hosts' nests observed in each of the various situations. [Data from N. Smith (1968).]

host chicks fledged per nest in colonies with and without bees or wasps was 0.39 and 0.43, respectively. Similar average numbers of young fledged per nest for cowbirds in the two types of colonies were 0.76 and 0.73. Hence reproductive success of both host and brood parasite is nearly the same under either situation, suggesting that a delicate balance has been reached (if one strategy were markedly better than the other, one would not expect to observe birds doing both). This example thus illustrates an intricate interplay between parasitism (mimetic cowbird and discriminator host, botfly and oropendula or cacique chicks), protocooperation (nonmimetic cowbird and nondiscriminator hosts), and commensalism (oropendulas or caciques and bees or wasps).

Selected References

Introduction

Haskell (1947, 1949); Krebs (1972); MacArthur (1972); MacArthur and Connell (1966); MacArthur and Wilson (1967); Odum (1959, 1971).

Competition

Birch (1957); Brian (1956); Crombie (1947); Elton (1949); Hazen (1964, 1970); Miller (1967); Milne (1961); Milthorpe (1961); Pianka (1976a).

LOTKA-VOLTERRA EQUATIONS AND COMPETITION THEORY

Andrewartha and Birch (1953); Bartlett (1960); Haigh and Maynard Smith (1972); Levins (1966, 1968); Lotka (1925); MacArthur (1968, 1972); May (1976a); Neill (1974); Pielou (1969); Schoener (1973, 1976b); Slobodkin (1962); Strobeck (1973); Vandermeer (1970, 1972, 1973, 1975); Volterra (1926a, 1926b, 1931); Wangersky and Cunningham (1956); Wilson and Bossert (1971).

COMPETITIVE EXCLUSION

Bovbjerg (1970); Cole (1960); DeBach (1966); Gause (1934); Hardin (1960); Jaeger (1971); MacArthur and Connell (1966); Miller (1964); Patten (1961); Pianka (1972).

THE BALANCE BETWEEN INTRASPECIFIC AND INTERSPECIFIC COMPETITION

Connell (1961a, 1961b); Fretwell (1972); Fetwell and Lucas (1969); MacArthur (1972); MacArthur, Diamond, and Karr (1972).

EVOLUTIONARY CONSEQUENCES OF COMPETITION

Collier *et al.* (1973); Connell (1961a, 1961b); Grant (1972); MacArthur (1972); MacArthur and Wilson (1967); Orians (1962); Pianka (1973); Ricklefs and Cox (1972).

LABORATORY EXPERIMENTS

Gause (1934, 1935); Gill (1972); Harper (1961a, 1961b); Lerner and Ho (1961); Neill (1972, 1974, 1975); Neyman, Park, and Scott (1956); Park (1948, 1954, 1962); Park, Leslie, and Mertz (1964); Vandermeer (1969); Wilbur (1972).

EVIDENCE FROM NATURE

Beauchamp and Ullyott (1932); Bovbjerg (1970); Brown and Wilson (1956); Cody (1968, 1974); Colwell and Fuentes (1975); Connell (1961a, 1961b); Crowell (1962); Dayton (1971); Elton (1946, 1949, 1958); Gadgil and Solbrig (1972); Grant (1972); Hairston (1951); Huey *et al.* (1974); Hutchinson (1959); Kohn (1959, 1968); MacArthur (1958); Menge (1972a, 1972b); Menge and Menge (1974); Orians and Willson (1964); Pianka (1969, 1973, 1974, 1975); Pittendrigh (1961); Schoener (1965, 1968, 1974a, 1974c, 1975); Terborgh and Weske (1969); Vaurie (1951); Williams (1964).

FUTURE PROSPECTS

MacArthur (1968, 1972); May (1973, 1976); Pianka (1976a); Roughgarden (1974, 1976); Schoener (1973, 1976b).

Predation

Errington (1946); Janzen (1971b); MacArthur (1972); MacArthur and Connell (1966); Wilson and Bossert (1971).

THEORIES: PREDATOR–PREY OSCILLATIONS

Elton (1942); Errington (1946); Gause (1934, 1935); Haigh and Maynard Smith (1972); Holling (1959a, 1959b, 1966); Keith (1963); Levins (1966); Lotka (1925); Maynard Smith (1974); Pielou (1969); Rosenzweig (1971, 1973a, 1973b); Rosenzweig and MacArthur (1963); Solomon (1949); Volterra (1926a, 1926b, 1931); Wangersky and Cunningham (1956); Wilson and Bossert (1971).

"PRUDENT" PREDATION AND OPTIMAL YIELD

Beverton and Holt (1957); Gilpin (1975); MacArthur (1960b, 1961); Slobodkin (1968).

SELECTED EXPERIMENTS AND OBSERVATIONS

Connell (1970); Errington (1946, 1956, 1963); Force (1972); Gause (1934, 1935); Holling (1959a, 1965); Huffaker (1958); Luckinbill (1973, 1974); Maly (1969); Menge (1972); Murdoch (1969); Neill (1972); Paine (1966); Salt (1967); Utida (1957); Wilbur (1972).

EVOLUTIONARY CONSEQUENCES OF PREDATION: PREDATOR ESCAPE TACTICS

Benson (1933); Benson (1972); Benson, Brown, and Gilbert (1975); Cott (1940); Ehrlich and Raven (1964); Feeny (1968, 1970); Fisher (1958b); Gordon (1961); Janzen (1966, 1967, 1970, 1971b); Jones (1962, 1966); Kettlewell (1956, 1958); McKey (1974); Otte and Williams (1972); Sheppard (1959); Stahl (1888); Whittaker and Feeny (1971).

Coevolution

Brower (1969); Brower and Brower (1964); Caswell *et al.* (1973); Cates (1975); Cates and Orians (1975); Chambers (1970); Dressler (1968); Ehrlich and Raven (1964); Faegri and van der Pijl (1971); Feeny (1975); Fraenkel (1959); Freeland and Janzen (1974); Futuyma (1976); Gilbert (1971, 1972, 1977); Gilbert and Raven (1975); Gordon (1961); Heinrich (1975); Heinrich and Raven (1972); Janzen (1966, 1967, 1971a, 1971b); Kircher and Heed (1970); Kircher *et al.* (1967); Lawlor and Maynard Smith (1976); Orians (1974); Rhoades and Cates (1976); C. Smith (1970); Willson (1973c).

Symbiotic Relationships

Allee (1951); Allee *et al.* (1949); Colwell (1973); Heatwole (1965); Seifert and Seifert (1976); N. Smith (1968).

The Ecological Niche

7

The concept of the niche pervades all of ecology; were it not for the fact that the ecological niche has been used in so many different ways, ecology might almost be defined as the study of niches. Many aspects of the niche have already been considered, and others are examined in Chapters 8 and 9; this chapter is closely related to both Chapters 5 and 6.

History and Definitions

Among the first to use the term niche was Grinnell (1917, 1924, 1928). He viewed the niche as the functional role and position of an organism in its community. Grinnell considered the niche essentially a behavioral unit, although he also emphasized it as the ultimate distributional unit (thereby including spatial features of the physical environment). Later Elton (1927) defined an animal's niche as "its place in the biotic environment, *its relations to food and enemies*" (his italics) and as "the status of an organism in its community." Further, he said that "the niche of an animal can be defined to a large extent by its size and food habits." Others, such as Dice (1952), use the term to refer to a subdivision of habitat; thus Dice states that "the term (niche) does not include, except indirectly, any consideration of the function the species serves in the community." Clarke (1954) distinguished two separate meanings for the term niche, the "functional niche" and the "place niche." Clarke noted that different species of animals and plants fulfill different functions in the ecological complex, and that the same functional niche may be filled by quite different species in different geographical regions. The idea of "ecological equivalents" was first stressed by Grinnell in 1924 (see also pp. 300–302).

The most influential modern treatment of niche is that of Hutchinson (1957a). Using set theory, he treats the niche somewhat more formally and defines it as the total range of conditions under which the individual (or population) lives and replaces itself. Hutchinson's examples for niche coordinates are nonbehavioral and have thus emphasized the niche as a place in space rather like a microhabitat or the "habitat niche" of Allee *et al.* (1949). This emphasis is unfortunate to the extent that it tends to exclude the "behavioral niche" from consideration. Hutchinson's distinction between the fundamental and the realized niche (p. 241) is one of the most explicit statements that an animal's potential niche is seldom fully utilized at a given moment in time or a particular place in space. This distinction has proven useful in clarifying the roles of other species, both competitors and predators, in determining the niche of an organism.

More recently Odum (1959) defined the ecological niche as "the position or status of an organism within its community and ecosystem resulting from the organism's structural adaptations, physiological responses, and specific behavior (inherited and/or learned)." He emphasized that "the ecological niche of an organism depends not only on where it lives but also on what it does." The place an organism lives, or where one would go to find it, is its habitat. For Odum the habitat is the organism's "address" while the niche is its "profession." Weatherley (1963) suggested that the definition of niche be restricted to "the nutritional role of the animal in its ecosystem, that is, its relations to all the foods available to it." Some ecologists prefer, however, to define the term niche more broadly and to subdivide it into components such as the "food niche" or the "place niche."

Because concepts of the ecological niche have taken on so many different forms, it is often difficult to be sure exactly what a particular ecologist means when this entity is invoked. Some therefore avoid using the word altogether and insist that we can get along perfectly well without it. However, no one denies that there is a broad zone of interaction between the traditional entities of "environment" and "organismic unit"; the major problem is to specify precisely in any given case just what subset of this enormous subject matter should be considered the "ecological niche."

Following the terminology of Chapter 4 (pp. 83–86), I define the ecological niche as *the sum total of the adaptations of an organismic unit*, or as all of the various ways in which a given organismic unit conforms to its particular environment. As with environment, we can speak of the niche of an individual, a population, or a species. The difference between an organism's environment and its niche is that the latter concept includes the organism's abilities at exploiting its environment and involves the ways in which an organism actually interfaces with and *uses* its environment.

The niche concept has gradually become inextricably linked to the phenomenon of interspecific competition, and it is increasingly becom-

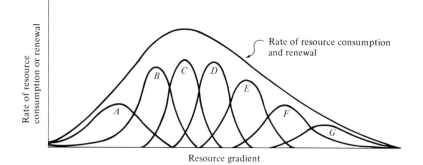

Figure 7.1   Niche relationships among members of competitive communities are usually modeled with bell-shaped utilization curves along a resource spectrum such as height above ground or prey size. Among the seven hypothetical species shown, those toward the tails have broader utilization curves because their resources renew more slowly. In such a community, consumers are at equilibrium with their resources and the rate of resource consumption is equal to the rate of renewal along the entire resource gradient.

ing identified with patterns of resource utilization. Niche relationships among competing species are frequently visualized and modeled with bell-shaped utilization curves along a continuous resource gradient, such as prey size or height above ground (Figure 7.1). Emphasis on resource use is operationally tractable and has generated a rich theoretical literature on niche relationships in competitive communities. In this chapter we consider in detail various aspects of this theory; including niche breadth, niche overlap, and niche dimensionality.

The Hypervolume Model

Building on the law of tolerance (Chapter 4), Hutchinson (1957a) and his students constructed an elegant formal definition of niche. When the tolerance or fitness of an organismic unit is plotted along a single environmental gradient, bell-shaped curves usually result (Figure 7.1). Tolerances for two different environmental variables can be plotted simultaneously (Figure 7.2). In Figure 7.3 hypothetical tolerances for three different variables are plotted in a three-dimensional space. Adding each new environmental variable simply adds one more axis and increases the number of dimensions of the plot by one. Actually Figure 7.3 represents a four-dimensional space, with the three axes shown representing the three environmental variables while the fourth axis (this one cannot be shown directly but is implicit in the

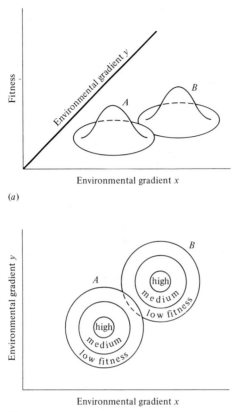

Environmental gradient x

(a)

Environmental gradient x

(b)

Figure 7.2 Two plots of the fitnesses of two organismic units, A and B, versus their position along two environmental gradients, x and y. (a) A three-dimensional plot with a fitness axis. (b) A two-dimensional plot with the fitness axis omitted; low, medium, and high fitness represented by contour lines.

figure in the same way that Figure 7.2a is implicit in Figure 7.2b) represents reproductive success or some other convenient measure of performance which we will call *fitness density*. Parts of this space with high fitness density are relatively optimal for the organism concerned; those with low fitness density are suboptimal. Conceptually, this process can be extended to any number of axes, using n-dimensional geometry. Thus Hutchinson defines an organism's niche as an n-dimensional hypervolume enclosing the complete range of conditions under which that organism can successfully replace itself

(Hutchinson's "niche" may be closer to my definition of environment). All variables relevant to the life of the organism must be included, and all must be independent of each other. An immediate difficulty with this model of the niche is that not all environmental variables can be nicely ordered linearly. To avoid this problem and to make the entire model more workable, Hutchinson translated his n-dimensional hypervolume formulation into a set theory mode of representation. Unfortunately, fitness density attributes of the n-dimensional model are lost in the conversion to a set theory model.

Hutchinson designates the entire set of optimal conditions under which a given organismic unit can live and replace itself as its *fundamental niche*, which can then be represented as a set of points in environmental space. The fundamental niche is thus a hypothetical, idealized niche in which the organism encounters no "enemies" such as competitors or predators and in which its physical environment is optimal. In contrast, the actual set of conditions under which an organism exists, which is always less than or equal to the fundamental niche, is termed its *realized niche*. The realized niche takes into account various forces that restrict an organismic unit, such as competition and perhaps predation. The fundamental niche is sometimes referred to as the *precompetitive* or *virtual* niche, whereas the realized niche is the *postcompetitive* or *actual* niche (however, this terminology neglects factors other than competition—such as predation—that might restrict the occupied region of the fundamental niche). These two concepts are thus somewhat analogous to the notions of r_{max} and r_a, discussed in Chapter 5.

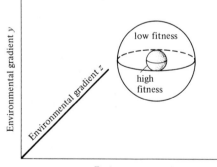

Figure 7.3   A plot (like that of Figure 7.2*b*) of fitness along three different environmental gradients, x, y, and z, showing zones of low and high fitness. A four-dimensional plot with a fitness axis analogous to Figure 7.2*a* is implicit in this graph.

Niche Overlap and Competition

Niche overlap occurs when two organismic units use the same resources or other environmental variables. In Hutchinson's terminology, each n-dimensional hypervolume includes part of the other, or some points in the two sets that constitute their realized niches are identical. Overlap is complete when two organismic units have identical niches; there is no overlap if two niches are completely disparate. Usually niches overlap only partially, with some resources being shared and others being used exclusively by each organismic unit.

Hutchinson (1957a) treats niche overlap in a simplistic way, assuming that the environment is fully saturated and that niche overlap cannot be tolerated for any period of time; hence competitive exclusion must occur in the overlapping parts of any two niches. Competition is assumed to be intense and to result in survival of only a single species in contested niche space. While this simplified approach has its shortcomings, it is useful to examine each of the logically possible cases (Figure 7.4) before considering niche overlap and competition in a more realistic way. First, two fundamental niches could be identical, corresponding exactly to one another, although such ecological identity is infinitely unlikely. In this most improbable event, the competitively superior organismic unit excludes the other. Second, one fundamental niche might be completely included within another (Figure 7.4a); given this situation, the outcome of competition depends on the relative competitive abilities of the two organismic units. If the one with the included niche is competitively inferior, it is exterminated and the other occupies the entire niche space; if the former organismic unit is competitively superior, it eliminates the latter from the contested niche space. The two organismic units then coexist with the competitively superior one occupying a niche included within the niche of the other. Third, two fundamental niches may overlap only partially, with some niche space being shared and some used exclusively by each organismic unit (Figure 7.4b and c). In this case each organismic unit has a "refuge" of uncontested niche space and coexistence is inevitable, with the superior competitor occupying the contested (overlapping) niche space. Fourth, fundamental niches might abut against one another (Figure 7.4d); although no direct competition can occur, such a niche relationship may reflect the avoidance of competition. Finally (Figure 7.4e), if two fundamental niches are entirely disjunct (no overlap), there can be no competition and both organismic units occupy their entire fundamental niche. Figure 7.5 illustrates the distinction between the fundamental and the realized niche for an organismic unit with six competitors.

A major shortcoming of the foregoing discussion is that, in nature, niches often do overlap yet competitive exclusion does not take place. Niche overlap in itself obviously need not necessitate competition. Overlap in habitats used may simply indicate that competitors have

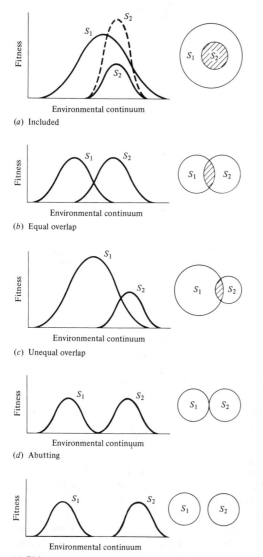

(a) Included

(b) Equal overlap

(c) Unequal overlap

(d) Abutting

(e) Disjunct

Figure 7.4 Various possible niche relationships, with fitness density models on the left and set theory models on the right. (*a*) An included niche. The niche of species 2 is entirely contained within the niche of species 1. Two possible outcomes of competition are possible: (1) if species 2 is superior (dashed line), it persists and species 1 reduces its utilization of the shared resources; (2) if species 1 is superior (solid lines), species 2 is excluded and species 1 uses the entire resource gradient. (*b*) Overlapping niches of equal breadth. Competition is equal and opposite. (*c*) Overlapping niches of unequal breadth. Competition is not equal and opposite because species 2 shares more of its niche space than does species 1. (*d*) Abutting niches. No direct competition is possible, but such a niche relationship may result from competition in the past and be indicative of the avoidance of competition. (*e*) Disjunct niches. Competition cannot occur and is not even implicit in this case.

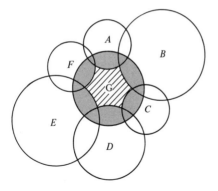

Figure 7.5 Set theory model of the fundamental niche (stippled and crosshatched) of species *G* and its realized niche (crosshatched), which is a subset of the fundamental niche, after competition and complete competitive displacement due to six superior competitors, species *A*, *B*, *C*, *D*, *E*, and *F*.

diversified in other ways. Should resources not be in short supply, two organismic units can share them without detriment to one another. In fact extensive niche overlap may often be correlated with *reduced* competition, just as disjunct niches may frequently indicate avoidance of competition in situations where it could potentially be severe (such as in cases of interspecific territoriality). For these reasons the ratio of demand to supply, or the degree of saturation, is of vital concern in the relationship between ecological overlap and competition. Indeed, much current research is designed to clarify, both theoretically and empirically, the relationship between competition and niche overlap; as pointed out in Chapter 6, modern ecologists are asking questions such as, "How much niche overlap can coexisting species tolerate?" and "How does this maximal tolerable niche overlap vary with the degree of saturation?" (Figure 7.6).

As was indicated in Chapter 6, competition is the conceptual backbone of much current ecological thought. Nonetheless competition remains surprisingly elusive to study in the field and hence is still poorly understood (probably because avoidance of competition is always advantageous when possible). Precise mechanisms by which available resources are divided among members of a community must be known before determinants of species diversity and community structure can be understood fully. Resource partitioning among coexisting species, or niche segregation, has therefore attracted considerable recent interest (for reviews, see Lack, 1971; MacArthur, 1972; Schoener, 1974; Pianka, 1976b).

The basic raw data for analysis of niche overlap is the resource matrix, which is simply an *m* by *n* matrix indicating the amount (or rate of consumption) of each of m resource states utilized by each of n different species. From this matrix one can generate an n by n matrix of overlap between all pairs of species with ones on the diagonal and values less than unity as off-diagonal elements. Overlap is sometimes equated with competition coefficients (alphas) because overlap is much easier to measure. Again, the caveat: overlap need not result in competition unless resources are in short supply. Extensive overlap may be possible when there is a surplus of resources (low demand/ supply), whereas maximal tolerable overlap may be much less in more saturated environments (see Figure 7.6). Because the principle of equal opportunity dictates that the ratio of demand over supply be constant along any *particular* resource gradient, intensity of competition should be directly proportional to the actual overlap observed along any given resource spectrum (see Figures 6.7 and 7.1). Patterns of niche overlap along different resource axes or between different communities must be compared with caution.

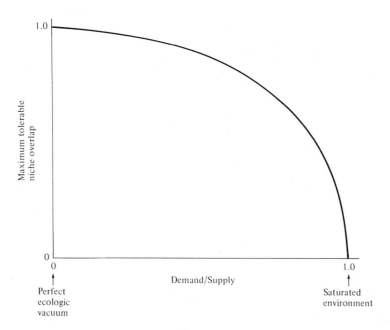

Figure 7.6   The niche overlap hypothesis predicts that maximum tolerable niche overlap must decrease as the intensity of competition increases. Maximal niche overlap in a saturated environment may not be zero, as figured, but some positive quantity related to the character displacement ratio of 1.3 (see p. 195). [From Pianka (1972). Copyright © 1972 by The University of Chicago Press.]

Niche Dynamics

Realized niches of most organisms change both in time and from place to place as physical and biotic environments vary. Temporal niche changes can be considered at two levels: (1) on a short-term basis (that is, in an ecological time scale), usually the life of a single individual or at most a few generations, and (2) on a long-term basis, over evolutionary time and many generations (see pp. 266–267). Thus the realized niche can be thought of as an ever-changing subset of the fundamental niche or, in the n-dimensional hypervolume model, as a pulsing hypervolume bounded by the hypervolume corresponding to the fundamental niche.

Some organisms, particularly insects, have entirely disjunct, nonoverlapping niches at different times in their life histories: caterpillars and butterflies, maggots and flies, tadpoles and toads, planktonic larval but sessile adult barnacles, and aquatic larval versus terrestrial adult insects (mosquitos, stoneflies, dragonflies, etc.). In all these cases, a drastic and major modification of an animal's body plan at metamorphosis allows a pronounced niche shift. Niches of other organisms change more gradually and continuously during their lifetimes. Thus juvenile lizards eat smaller prey than do adults and are often active earlier in the day at lower environmental temperatures (their smaller size and greater surface-to-volume ratio facilitates faster warming). Incomplete metamorphosis, in which an insect changes rather gradually with each molt (as in grasshoppers), is similar.

An organism's immediate neighbors in niche space, or its potential competitors, can (but need not) exert strong influences upon its ecological niche. While realized niches of relatively r-selected organisms are determined primarily by their physical environments, realized niches of more K-selected organisms are perhaps more strongly influenced by their biotic environments. Selective pressures and niches may vary during an individual's life. Thus, in temperate zones, early spring is a time when annual plants are relatively r-selected; later in the season they become progressively more K-selected (Gadgil and Solbrig, 1972). Even within a given species some individual organisms may be more r-selected than others, such as populations in different microhabitats or parts of a species' geographic range, or individuals at different positions in the rocky intertidal.

Theoretically, reduced interspecific competition should often allow niche expansion. In an attempt to observe this, Crowell (1962) examined and compared ecologies of three species of birds on Bermuda with those of mainland populations. Many fewer species of land birds occur on the island than on the mainland; the three most abundant are the cardinal, catbird, and white-eyed vireo. On Bermuda these three species have very dense populations; however, in mainland habitats where there are many more species of birds—and thus a greater variety of interspecific competitors—populations of these same three species are usually considerably smaller. Although there are, of course, inevitable differences between the habitats of Bermuda and mainland North America, the striking difference between avifaunas should nevertheless have a major effect on the ecologies of the birds. The catbird and cardinal have generally more restricted place and foraging niches on the island (perhaps available niches are more restricted), but the white-eyed vireo has expanded both its place niche and its foraging niche. All three species nest at a wider variety of heights on Bermuda (see also p. 197).

Niche Dimensionality

Although the n-dimensional hypervolume model of the niche is extremely powerful conceptually, it is too abstract to be of much practical value and is usually rather difficult to apply to the real world. Indeed, to construct such a hypervolume we would have to know essentially everything about the organism concerned. Because we can never know *all* factors impinging upon any organismic unit, the fundamental niche must remain an abstraction. Even realized niches of the majority of organisms have so many dimensions that they defy quantification. When considering relatively K-selected organisms, the number of niche dimensions can be limited to those on which competition is effectively reduced. Competition is often avoided by differences in microhabitats exploited, foods eaten, and/or times of activity (Chapter 6), and so the effective number of niche dimensions can be reduced to three: place, food, and time. We can think of a saturated community as occupying some volume in a space with these three dimensions; thus a community is something like a three-dimensional jigsaw puzzle, each piece being one species occupying only a part of the overall volume.

Most niche theory is framed in terms of a single niche dimension. Each species thus has only two neighbors in niche space, and overlap matrices contain many zeros and only two positive entries on the off-diagonal per row. Real species of plants and animals, however, seldom sort themselves out on a single resource axis; rather, pairs of species usually show moderate niche overlap along two or more niche dimensions. Complementarity of niche dimensions often occurs, with pairs that have high overlap along one niche dimension overlapping little on another dimension—and vice versa (Figure 7.7). Multidimensional niche relationships can be complex. One feature of increased niche dimensionality is that niches may be overlapping or identical along one axis and yet be separated or even disjunct along another (Figure 7.8). Thus an observer oblivious to the first niche dimension in Figure 7.8 would consider species pair A and F, pair E and B, and pair D and C as

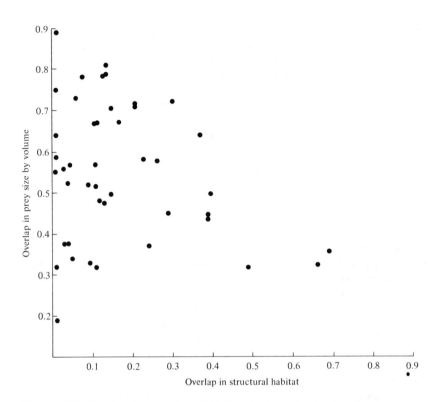

Figure 7.7   Overlap in prey size plotted against overlap in structural microhabitat among various species of *Anolis* lizards on the island of Bimini. Pairs with high dietary overlap tend to exploit different structural microhabitats; conversely, those with high spatial overlap have relatively little overlap in prey sizes eaten. [From Schoener (1968). Copyright by the Ecological Society of America.]

completely overlapping, when in fact they are partially or entirely separated along the unknown dimension!

In a one-dimensional niche space any given niche can be bounded only on two sides, whereas there can be many more neighbors in a two-dimensional niche space, and still more in three or more dimensions. As the effective number of niche dimensions rises, the potential number of neighbors in niche space increases more or less geometrically. As dimensionality increases, overlap matrices contain fewer off-diagonal elements of zero and the variance in observed overlap usually falls, both within rows and over the entire matrix. Hence niche dimensionality strongly affects the *potential* for "diffuse" competition arising from the total competitive effect of all interspecific competitors (MacArthur, 1972). The overall effect of relatively weak competitive inhibition per species summed over many other species could well be as strong or even stronger than much more intense competitive inhibition (per species) by fewer competing species. Thus an increased number of niche dimensions, by generating a greater potential for immediate neighbors in niche space, can intensify diffuse competition.

Imagine that height above ground and prey size are two such critical niche dimensions which species use differentially and thereby avoid or reduce interspecific competition. Analysis of resource utilization and niche separation along more than a single niche dimension should ideally proceed through estimation of proportional *simultaneous* utilization of all resources along each separate niche dimension. These define a three-dimensional resource matrix, with each entry representing the probability of capture of a prey item of a given size category at a particular height interval by each species present. Obtaining such multidimensional utilization data is extremely difficult, however, because most animals move and integrate over both space and time. Accurate estimates of an animal's true use of a multidimensional niche space could only be obtained by continually monitoring an individual's use of all resources. (Even then the degree to which prey individuals move between microhabitats will affect competition in obscure but vitally important ways!) As such continual observation is often ex-

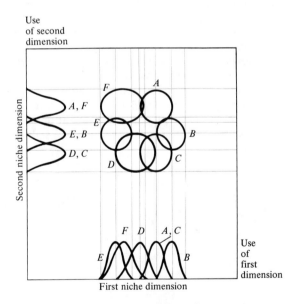

Figure 7.8   Hypothetical niches of six species differing along two niche dimensions. Although overlap is broad or complete along either single dimension alone (bell-shaped curves), niches overlap minimally or not at all when both niche dimensions are considered (circles and ellipses).

tremely tedious or even impossible, one usually approximates from separate unidimensional utilization distributions (Figures 7.8 and 7.9). Just as the three-dimensional shape of a mountain cannot be accurately determined from two of its silhouettes viewed at right angles, these "shadows" do not allow inference of the true multidimensional utilization.

The question of the degree of dependence or independence of dimensions becomes critical. Provided that niche dimensions are truly independent, with prey of any size being equally likely to be captured at any height, overall multidimensional utilization is simply the product of the separate unidimensional utilization functions (May, 1975). Under perfect independence, the probability of capture of prey item i in microhabitat j is then equal to the probability of capture of item i times the probability of being in microhabitat j. Unidimensional estimates of various niche parameters (including overlap) along component niche dimensions may then simply be multiplied to obtain multidimensional estimates. However, should niche dimensions be partially interdependent (Figure 7.9), there is no substitute for knowledge of true multidimensional utilization. True multidimensional overlap can vary greatly depending on the exact form of this dependence (see Figure 7.9). In the extreme case of complete dependence (if, for example, prey of each size are found only at one height) there is actually only a single niche dimension, and a simple average provides the best estimate of true utilization. Moreover, the arithmetic average of estimates of unidimensional niche overlap obtained from two or more separate unidimensional patterns of resource use actually constitutes an *upper bound* on the true multidimensional overlap (May, 1975).

It is difficult (or virtually impossible) to evaluate the degree of interdependence of niche dimensions for many species. However, in relatively sedentary species, the degree to which foods eaten are influenced by microhabitat can sometimes be assessed. One such study of a seden-

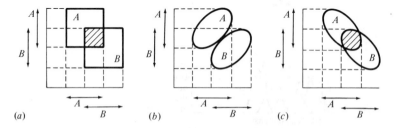

Figure 7.9   Three different cases of niche overlap on two resource dimensions with identical unidimensional projections (double-headed arrows). In (a) niche dimensions do not interact and unidimensional projections indicate multidimensional conditions accurately (dimensions are truly independent and niche axes are orthogonal). In (b) and (c) niche dimensions are partially interdependent and their unidimensional shadows are misleading.

tary legless lizard showed that most species and castes of termites are eaten in fairly similar proportions by lizards taken from different microhabitats, suggesting that these two niche dimensions are largely independent (Huey *et al.*, 1974).

In reviewing major factors leading to ecological isolation among birds, Lack (1971) concluded that the most important were differences in geographic range, habitat, and foods eaten. Schoener (1974) recently reviewed patterns of resource partitioning in over 80 natural communities ranging from simple organisms such as slime molds through various mollusks, crustaceans, insects, and other arthropods to various members of the five classes of vertebrates, including lizards. He identified and ranked five resource dimensions by degree of importance in niche segregation: macrohabitat, microhabitat, food type, time of day, and seasonality of activity. Schoener concludes that habitat dimensions are generally more important in separating niches than food type dimensions, which in turn tend to be important more often than temporal dimensions. Terrestrial poikilotherms partition food by being active at different times of day relatively often compared with other animals.* Predators partition resources by diurnal differences in time of activity more than do other groups, and vertebrates segregate less by seasonal activity differences than do lower animals. Schoener also found that segregation by food type is more important for animals feeding on large foods relative to their own size than it is among animals that feed on relatively small items.

Guild Structure

To what extent are species overdispersed in niche space? Do clusters of functionally similar species exist? Members of such a group of similar species, known as a *guild* (Root, 1967), interact strongly with one another but weakly with the remainder of their community. Although techniques of objectively defining a guild are still in their infancy, the concept is clearly of some interest because guilds presumably represent the arenas of most intense interspecific competition. An opera-

---

* Use of time of activity as a niche dimension can be justified in several ways. If resources are rapidly renewed, exploitative competition cannot occur unless individuals are active within a fairly short time of each other (otherwise resources may be replenished during the interval separating the species). When microhabitat and food categories are crude and changing in time, temporal differences in activity may in fact be associated with differential use of space and food resources that would not be reflected without much greater precision in recognition of appropriate nontemporal resource states. Thus a microhabitat such as "open sun" clearly changes with the daily march of temperature; similarly a given crude prey category such as "ants" may usually lump a series of prey species with temporal segregation. Hence it is often both useful and appropriate to treat time as a niche dimension.

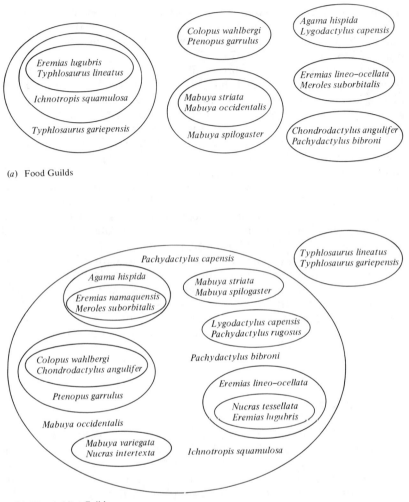

(a) Food Guilds

(b) Microhabitat Guilds

Figure 7.10   The guild structure of the lizard fauna of the Kalahari Desert, based on stomach contents (food guilds) and microhabitat usage (microhabitat guilds). Of the 21 species, 12 and 14 are members of a two-species food guild or microhabitat guild, respectively (the opposite member of each pair is the other's nearest neighbor in niche space). All but one (*Pachydactylus capensis*, a food and microhabitat generalist) of the 21 species is a member of at least one guild. Many two-species guilds are nested inside larger guilds. The microhabitat "superguild" consists of all nonsubterranean species, none of which overlap at all with either of the two subterrannean species *Typhlosaurus lineatus* and *T. gariepensis*. (Note that some species in the same food guild, such as *Eremias lineo-ocellata* and *Meroles suborbitalis*, are in different microhabitat guilds.)

tional means of delimiting the members of a guild has been developed by L. R. Lawlor (personal communication), who defines a guild as a cluster of species separated from all other such clusters by a distance greater than the largest distance between the two most disparate members of the guild concerned. Under this rather conservative definition, a community of n species whose component species are evenly spread out in niche space contains n one-species guilds (each species is its own guild) and one large n-species guild (the entire community) but no guilds of intermediate size. Real communities may usually contain many guilds of intermediate size (Figure 7.10).

Specialization versus Generalization

Some organisms have smaller niches than others. Niche breadth, also called "niche width" and "niche size," can be thought of as the extent of the hypervolume representing the realized niche of an organismic unit. Thus a koala bear, *Phascolarctos cinerus*, which eats only leaves of certain species of *Eucalyptus*, has a more specialized food niche than the Virginia opossum, *Didelphis virginianus*, which is a true omnivore that eats nearly anything. Statements about niche breadths must invariably be comparative; we can only say that a given organismic unit has a niche that is narrower or broader than that of some other organismic unit. Highly specialized organisms like the koala usually, though not always, have narrow tolerance limits along one or more of their niche dimensions. Often such specialists have very specific habitat requirements, and as a result they may not be very abundant. In contrast, organisms with broad tolerances are typically more generalized, with more flexible habitat requirements, and are usually much more common. Thus specialists are often relatively rare while generalists are more abundant. Rare organisms may, however, frequently occur in clumps so that their local density need not necessarily be low.

The only currency of natural selection is differential reproductive success. This fact raises a question: If specialization involves becoming less abundant, why have organisms become specialized at all? Since generalized organisms can usually exploit more food types, occupy more habitats, and build up larger populations, they might be expected by their very numbers to outreproduce slightly more specialized competing members of their own population and thereby swamp the populational gene pool. The answer to this apparent dilemma lies in the old adage that a jack-of-all-trades is a master of none. More specialized individuals are more efficient on their own ground than are generalists.

Under what conditions will a jack-of-all-trades win in competition with more specialized species? MacArthur and Levins (1964, 1967) considered this question and developed the following model. First imagine an ant-eating lizard in an environment that contains only a single food resource type, colonies of ants 3 mm in length, and a variable population of ant-eating lizards that exploit the ant food resource. Assume that ants are eaten whole and that lizards differ only in the size of their jaws, forming a fairly continuous phenotypic spectrum. Some phenotypes will be well adapted to use 3 mm ants and very effective at harvesting them; others will be less efficient, either because their jaws are too large or too small. Next consider the same phenotypic spectrum of lizards in another "pure" environment, this one composed solely of colonies of 5 mm ants. Almost certainly the best-adapted phenotype will differ from that in the 3 mm ant environment (Figure 7.11*a*), and the phenotype most efficient at using 5 mm ants will be one with a larger mouth. Now consider a *mixed* ant colony with equal numbers of 3 mm and 5 mm ants (say, two castes) in a *homogeneous* mixture. Which phenotype will be optimal in this new mixed environment? Assuming that the lizards encounter and use the two ant sizes in exactly equal proportions, the relationship between phenotype and harvesting effectiveness must be exactly intermediate between the two similar relationships in pure environments (Figure 7.11*a*). The dashed line can thus be drawn in Figure 7.11*a* midway between the first two (if the two resources were not in exactly equal proportions, this new line would simply be closer to one or the other of the original lines); depending on the shapes of the curves and the distance between them, this new line can take either unimodal (Figure 7.11*a*) or bimodal (Figure 7.11*b*) shape. In the former case, the phenotype of highest harvesting efficiency is intermediate between the best "pure 3 mm ant eater" and the best "pure 5 mm ant eater," and this "jack-of-both-trades" (probably the phenotype that could best exploit 4 mm ants) is competitively superior. In the latter case, because the two types of ants are very different in size, the two phenotypes with high harvesting effectiveness are separated from one another by intermediate phenotypes with *lower* efficiencies at exploiting a mixture of 2 mm and 8 mm ants, and the two specialists will eliminate the jack-of-both-trades.

Effects of interspecific competition on niche breadth are complex and under different conditions may actually favor either niche contraction or niche expansion. Thus a competitor may reduce food availability in some microhabitats but leave prey densities in other microhabitats unaltered, effectively reducing expectation of yield in some patches but not others. A competitor that is an optimal forager should restrict its patch utilization to those with higher expectation of yield, thereby decreasing the breadth of its place niche (see also pp.148–150 and pp. 260–266). Conversely, a more generalized competitor that reduces food availability more or less equally in all microhabitats by reducing the overall level of prey availability can force its competitor to expand the range of resources it uses, thereby increasing the breadth of its food niche. In a food-sparse environment, an optimal forager simply cannot afford to bypass as many potential prey items as it can in a food-dense

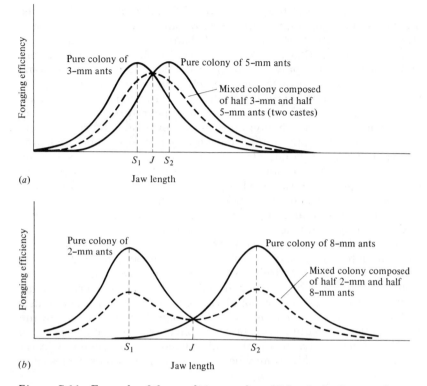

Figure 7.11   Example of the conditions under which a jack-of-two-trades
outperforms two specialists. Foraging efficiency, measured in useful calories
gathered per unit of foraging time, is plotted against a phenotypic spectrum of
jaw lengths for a hypothetical population of ant-eating lizards. (*a*) Solid curves:
performance of various lizard phenotypes in "pure" environments containing
only 3 mm and 5 mm ants, respectively. Dashed line: performance of various
phenotypes in an environment containing a homogeneous mixture of equal
numbers of both 3 mm and 5 mm ants (say, two castes). The jack-of-both-
trades, *J*, has a higher foraging efficiency than either specialist, S_1 or S_2,
which are the most efficient foragers in the respective "pure" environments.
(*b*) Foraging efficiencies of the various lizard phenotypes when the *difference*
between the two ant types is greater, say, 2 mm versus 8 mm ants. In this
case performance in the mixed environment is bimodal, and the two specialists
outperform the jack-of-both-trades. [After MacArthur and Connell (1966).]

environment; therefore more suboptimal prey must be eaten in the
former type of habitat. Reduced interspecific competition is often accompanied by an increase in the range of habitats a species uses, but
marked changes in the variety of foods eaten with changes in interspecific competition seem to be much less common (MacArthur,
1972).

A growing body of theory on optimal foraging predicts that niche breadth should generally increase as resource availability decreases (Emlen, 1966, 1968; MacArthur and Pianka, 1966; Schoener, 1971; MacArthur, 1972; Charnov, 1973, 1976a, 1976b). In an environment with a scant food supply, a consumer cannot afford to bypass many inferior prey items because mean search time per item encountered is long and expectation of prey encounter is low. In such an environment, a broad niche maximizes returns per unit expenditure, promoting generalization. In a food-rich environment, however, search time per item is low since a foraging animal encounters numerous potential prey items; under such circumstances, substandard prey items can be bypassed because expectation of finding a superior item in the near future is high. Hence rich food supplies are expected to lead to selective foraging and narrow food niche breadths.

Two fundamental components of niche breadth have only relatively recently begun to be distinguished: the "between-phenotype" versus "within-phenotype" components (Van Valen, 1965; Orians, 1971; Roughgarden, 1972, 1974b, 1974c). A population with a niche breadth determined entirely by the between-phenotype component would be composed of specialized individuals with no overlap among them in resources used; a population composed of pure generalists with each member exploiting the entire range of resources used by the total population would have a between-phenotype component of niche breadth of zero and a maximal within-phenotype component. Clearly real populations will lie somewhere between these two extremes (Figure 7.12).

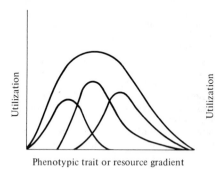

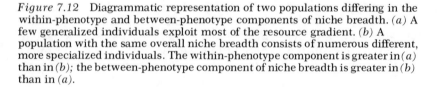

Figure 7.12   Diagrammatic representation of two populations differing in the within-phenotype and between-phenotype components of niche breadth. (*a*) A few generalized individuals exploit most of the resource gradient. (*b*) A population with the same overall niche breadth consists of numerous different, more specialized individuals. The within-phenotype component is greater in (*a*) than in (*b*); the between-phenotype component of niche breadth is greater in (*b*) than in (*a*).

Time, Matter, and Energy Budgets

As we saw in Chapter 4, any organism has a limited amount of time, matter, and energy available to devote to foraging, growth, maintenance, and reproduction. The way in which an organism allocates these resources among various conflicting demands is of fundamental interest. This apportionment determines the ways in which the organism can conform to many aspects of its environment and thus indicates a great deal about its ecological niche. Time and energy budgets vary widely among organisms; for example, relatively r-selected organisms allot more time and energy to reproduction at any instant than do more K-selected ones (Chapter 5). Varying time and energy budgeting is a potent means of coping with a changing environment while retaining some degree of adaptation to it. Thus a male marsh wren expends a great deal of energy on territorial defense during the breeding season but little at other times of the year. Similarly, in animals with parental care, an increasing amount of energy is spent on growing offspring until some point when progeny begin to become independent of their parents, whereupon the amount of time and energy devoted to them decreases. Indeed, adult female red squirrels, *Tamiasciurus*, at the height of lactation consume an average of 323 kcal of food per day, whereas the average daily energy consumption of an adult male is only about 117 kcal (C. Smith, 1968); time budgets of these squirrels also vary markedly with the seasons. In a bad dry year, many annual plants "go to seed" while still very small, whereas in a good wet year, these plants grow to a much larger size before becoming reproductive; presumably more seeds are produced in good years, but perhaps none (or very few) would be produced in a bad year if individuals grew to the sizes they reach in good years.

Organisms can be viewed as simple input-output systems (Figure 7.13), with foraging or photosynthesis providing an input of materials and energy which are in turn "mapped" into an output consisting of progeny. Fairly extensive bodies of theory now exist both on reproductive tactics and on optimal foraging. In optimal foraging theory, the "goal" usually assumed to be maximized is energy uptake per unit time (successful offspring produced during an organism's lifetime would be a more realistic measure of its foraging ability, but fitness is exceedingly difficult to measure). Similarly, among organisms without parental care, reproductive effort has sometimes been estimated by the ratio of calories devoted to eggs or offspring over total female calories at any instant (rates of uptake versus expenditure of calories have unfortunately not yet infiltrated empirical studies of reproductive tactics). To date, empirical studies of resource partitioning and niche structure have been concerned largely with "input" phenomena such as overlap in and efficiency of resource utilization and have neglected to relate these to "output" aspects. In contrast, empirical studies of reproductive tactics have done the reverse and almost entirely omitted any con-

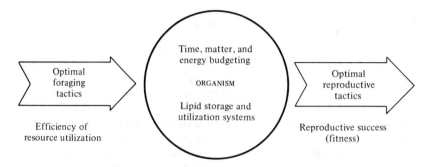

Figure 7.13 Diagrammatic view of an organism as a simple input-output system. Optimal foraging theory is concerned with efficiency of resource utilization and input phenomena; theory on optimal reproductive tactics deals largely with output aspects (reproductive success). Constraints and interactions between input and output (and vice versa) arise both from lipid storage and utilization systems and through finite time and energy budgets. [From Pianka (1976b).]

sideration of foraging. Interactions and constraints between foraging and reproduction have barely begun to be considered. A promising area for future work will be to merge aspects of optimal foraging with optimal reproductive tactics to specify rules by which input is translated into output; optimal reproductive tactics ("output" phenomena) surely must often impose substantial constraints upon "input" possibilities.

An animal's time and energy budget provides a convenient starting point for clarifying some ways in which foraging influences reproduction and vice versa. Any animal has only a certain finite period of time available in which to perform all its activities, including foraging and reproduction. This total time budget, which can be considered either on a daily basis or over the animal's lifetime, will be determined both by the diurnal rhythm of activity and by the animal's ability to "make time" by performing more than one activity at the same time (such as a male lizard sitting on a perch, simultaneously watching for potential prey and predators while monitoring mates and competing males). Provided that a time period is profitable for foraging (expected gains in matter and energy exceed inevitable losses from energetic costs of foraging), any increase in time devoted to foraging clearly will increase an animal's supply of matter and energy. Necessarily accompanying this increase in matter and energy, however, is a concomitant *decrease* in time available for nonforaging activities such as mating and reproduction. Thus profits of time spent foraging are measured in matter and energy while costs take on units of time lost. Conversely, increased time spent on nonforaging activities confers profits in time while costs take the form of decreased energy availability. Hence gains in energy correspond to losses in time, while dividends in time require reductions in energy availability. (Of course, risks of foraging and reproduction also need to be considered.)

The arguments above suggest that optimal allocation of time and energy ultimately depends on how costs in each currency vary with profits in the opposite. However, because units of costs and profits in time and energy differ, one would like to be able to convert them into a common currency. Costs and profits in time might be measured empirically in energetic units by estimation of the net gain in energy per unit of foraging time. If all potential foraging time is equivalent, profits would vary linearly with costs; under such circumstances, the loss in energy associated with nonforaging activities would be directly proportional to the amount of time devoted to such activity. Optimal budgeting of time and energy into foraging versus nonforaging activities is usually profoundly influenced by various circadian and seasonal rhythms of physical conditions, as well as those of predators and potential prey. Clearly certain time periods favorable for foraging return greater gains in energy gathered per unit time than other periods. Risks of exposure to both harsh physical conditions and predators must often figure into the optimal amount of time to devote to various activities. Ideally one would ultimately like to measure both an animal's foraging efficiency and its success in budgeting time and energy by its lifetime reproductive success, which would reflect all such environmental "risks."

Foraging and reproductive activities interact in another important way: many organisms gather and store materials and energy during time periods that are unfavorable for successful reproduction but then expend these same resources on reproduction at a later, more suitable, time. Lipid storage and utilization systems obviously facilitate such temporal integration of uptake and expenditure of matter and energy. This temporal component greatly complicates the empirical measurement of reproductive effort.

Prey density can strongly affect an animal's time and energy budget. Gibb (1956) watched rock pipits, *Anthus spinoletta*, feeding in the intertidal along the English seacoast during two consecutive winters. The first winter was relatively mild; the birds spent $6\frac{1}{2}$ hours feeding, $1\frac{3}{4}$ hours resting, and 45 minutes fighting in defense of their territories (total daylight slightly exceeded 9 hours). The next winter was much harsher and food was considerably scarcer; the birds spent $8\frac{1}{4}$ hours feeding, 39 minutes resting, and only 7 minutes on territorial defense! Apparently the combination of low food density and extreme cold (endotherms require more energy in colder weather) demanded that over 90 percent of the bird's waking hours be spent feeding and no time remained for frivolities. This example also illustrates that food is less defendable at lower densities as indicated by reduced time spent on territorial defense. Obviously food density in the second year was near the lower limit that would allow survival of rock pipits. When prey items are too sparse, encounters may be so infrequent that an individual cannot survive. Gibb (1960) calculated that, to balance their energy budget during the winter in some places, English tits must find an insect on the average once every $2\frac{1}{2}$ seconds during daylight hours.

Time and energy budgets are influenced by a multitude of other ecological factors, including body size, mode of foraging, vagility, trophic level, prey size, resource density, environmental heterogeneity, rarefaction, competition, predation, and reproductive tactics.

Foraging Tactics and Feeding Efficiency

Foraging, as we have seen, has both profits and costs. Its profits are the matter and energy gathered, which can be used in growth, maintenance, and reproduction. Costs of foraging are perhaps more elusive, but a foraging animal must expend energy in foraging and must expose itself to potential predators. Further, much of the time it spends foraging is unavailable for other activities, including reproduction. Natural selection should favor foraging behaviors that maximize the difference between foraging profits and their costs (see also pp. 148–150).

Carnivorous animals forage in two extreme ways. In the "sit-and-wait tactic," a predator waits in one place until a moving prey item comes by and then "ambushes" the prey; in the "widely-foraging tactic," the predator actively searches out its prey (Pianka, 1966b; Schoener, 1969a, 1969b). The second strategy normally requires a greater energy expenditure than the first. The success of the sit-and-wait tactic usually depends on one or more of three conditions: a fairly high prey density, high prey mobility, and low predator energy requirements. The widely-foraging tactic also depends on prey density and mobility and on the predator's energy needs, but here the distribution of prey in space and the predator's searching abilities assume paramount importance. Although these two tactics are endpoints of a continuum of possible foraging strategies (and hence somewhat artificial) foraging techniques actually employed by many organisms are rather strongly polarized. The dichotomy of sit-and-wait versus widely foraging therefore has substantial practical value. Among snakes, for example, racers and cobras forage widely when compared with boas, pythons, and vipers, which are relatively sit-and-wait foragers. Among hawks, accipiters such as Cooper's Hawks and Goshawks often hunt by ambush using a sit-and-wait strategy, whereas most buteos and many falcons are relatively more widely foraging. Web-building spiders and sessile filter feeders such as barnacles typically forage by sitting and waiting. Many spiders expend considerable amounts of energy and time building their webs rather than in moving about in search of prey; those that do not build webs forage much more widely.

Similar considerations can be applied in comparing herbivores with carnivores. Because the density of plant food almost always greatly exceeds the density of animal food, herbivores often expend little energy, relative to carnivores, in finding their prey (to the extent that secondary chemical compounds of plants, such as tannins, and other antiherbivore defenses—see pages 221–224—reduce palatability of plants or parts of plants, effective supply of plant foods may be greatly reduced). Because cellulose in plants is difficult to digest, however, herbivores must expend considerable energy in extracting nutrient from their plant food. (Most herbivores have a large ratio of gut volume to body volume, harbor intestinal organisms that digest cellulose, and spend much of their time eating or ruminating—envision a cow chewing its cud.) Animal food, composed of readily available proteins, lipids, and carbohydrates, is more readily digested; carnivores can afford to expend considerable effort in searching for their prey because of the large dividends obtained once they find it. As would be expected, efficiency of conversion of food into an animal's own tissues (*assimilation*) is considerably lower in herbivores than it is in carnivores.

Many carnivores have extremely efficient prey-capturing devices (Chapter 6); often the size of a prey object markedly influences this efficiency. Using simple geometry (Figure 7.14), Holling (1964) esti-

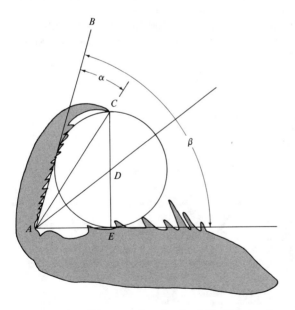

Figure 7.14   Diagram showing how Holling used geometry to calculate an estimated optimal size of a prey item from the anatomy of a mantid's foreleg. Optimal prey diameter, D, is simply $T \sin (\beta - \alpha)$, where T is the distance A to C. [From Holling (1964).]

mated the diameter of a prey item that should be optimal for a praying mantid of a particular size. He then offered five hungry mantids prey objects of various sizes and recorded percentages attacked (Figure 7.15). Mantids were noticeably reluctant to attack prey that were either much larger or much smaller than the estimated optimum! Thus natural selection has resulted in efficient predators both by producing efficient prey-capturing devices and by programming animals so that they are unlikely to attempt to capture decidedly suboptimal items. Larger predators tend to take larger prey than smaller ones (see Figure 6.13). It may in fact be better strategy for a large predator to overlook prey below some size and to spend the time that would have been spent in capturing and eating small items in searching out larger prey (see also pp. 148–150). Similarly, the effort a predator will expend on any given prey item is proportional the expected return from that item (which often increases with prey size). Thus a lizard waiting on a perch will not usually go far for a very small prey item but will often move much greater distances in attempts to obtain larger prey.

Because small prey are generally much more abundant than large prey, most animals encounter and eat many more small prey items than large ones. Small animals that eat small prey items encounter prey of suitable size much more frequently than do larger animals that rely on larger prey items; as a result, larger animals tend to eat a wider range of prey sizes. Because of such increased food niche breadths of larger animals, size *differences* between predators increase markedly with increasing predator size (MacArthur, 1972).

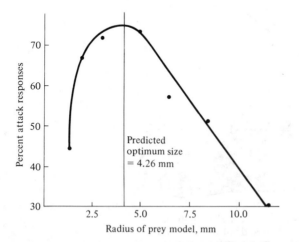

Figure 7.15   The percentage of prey items actually attacked by hungry mantids versus prey size. [From Holling (1964).]

Optimal Use of Patchy Environments

Environments that are decidedly discontinuous—that is, heterogeneous environments consisting of a patchwork of rather different resources—are termed "patchy" environments; those with similar or well-mixed resources are called homogeneous or uniform environments. Frequently a particular organism in a heterogeneous environment exploits only a part of the environmental mosaic, and different organisms are specialized to use each of the patch types. However, as the degree of difference between patches decreases, the advantage of being generalized increases. Thus we distinguish again two different extremes in the way in which organisms can utilize their environments. When an organism in the course of its daily activities encounters and uses resources in the same proportions in which they actually occur (i.e., does not select resources or patches in the environment), the organism is said to utilize its environment in a *fine-grained* manner. Organisms that spend disproportionate amounts of time in different patches are said to use their environment in a *coarse-grained* manner. An animal that encounters its prey in the actual proportions in which they occur (i.e., in a fine-grained manner), through selection of particular prey types, can exploit this fine-grained environment in a coarse-grained way. Fine-grained utilization of the environment may be forced on an organism in situations where the size of the patches in the environmental mosaic (the *grain size* of the environment) is small compared to the organism's own size. Thus larger animals, all else being equal, tend to encounter the world in a more fine-grained way than smaller ones. As an example, a meadow and an adjacent forest each has its own herbivorous mouse (often a *Microtus* and a *Peromyscus*, respectively), whereas a deer exploits both the meadow and the forest, using the former at night for feeding and the latter by day for sleeping and retreat. Small animals such as insects generally exploit their environments in a more coarse-grained way than do larger animals such as vertebrates. Sessile animals and plants, as individuals, must tend toward coarse-grained utilization of space simply by virtue of their immobility. Because individual sessile organisms may be distributed widely in an environmental mosaic, however, populations of them may approximate fine-grained utilization of the spatial environment. Statements about environmental grain size or the degree of coarse-grained versus fine-grained utilization of resources are usually comparative; when one says that a given animal "lives in a coarse-grained environment," this is relative to some other animal.

Consider now the ways in which insectivorous birds could forage in a mixed forest containing several different species of trees and a wide range of different types of insects (MacArthur and MacArthur, 1961). Assume that no single bird has the capacity to exploit everything in the forest but that some degree of coarse-grained utilization is necessary. A bird could forage only in one species of tree and fly from one such tree to another; much time and energy would then be expended in flying between trees and over unsuitable ones. Alternatively, birds could eat only one category of insect food (perhaps a particular spectrum of insect sizes) capturing these wherever they occur in any of the various tree species. In the course of searching out its particular insect prey, the bird would doubtless encounter a variety of other prey types (or sizes). The first strategy would presumably lead to different species of birds being specialized to each of the various tree species; the latter would result in each bird species having its own particular range of prey types. A third way in which birds could exploit such a patchy environment is to compromise in both of the above ways but to specialize as to exactly *where* in the trees they forage and *how* they feed. Thus a bird might select a layer in the forest and forage through this layer, taking whatever prey are available within broad limits, feeding in different trees and on many different prey types. In this case, different species of birds would differ as to where and how they forage, with each species exploiting a "natural feeding route" in the forest. A compromise strategy such as this is usually a more efficient way for birds to exploit a patchy environment than is specializing as to places or prey exploited; most birds have indeed developed their own unique patterns as to where and how they forage. Kinglets and titmice tend to forage in the crowns of trees; other species such as many warblers (see also Figure 6.9) forage in other parts of trees. Many woodpeckers exploit crevices in tree trunks in an ascending spiral to some height and then fly or glide down to a low point on the trunk of a nearby tree and repeat the process. Only in cases of extreme *concentration* of food does it pay to specialize on a particular food type. Many parrots are quite specialized as to fruits and nectars they eat; when these foods are encountered, they are usually very dense and in superabundance. Such exceptionally rich energy sources are worth searching out; once one is located dividends are relatively great.

Consider now a model for utilization of patchy environments based on optimization of an animal's time budget, although the model applies equally well to an energy budget (MacArthur and Pianka, 1966). Assume that environmental resources, say prey species, are encountered by a foraging animal in the same proportions in which they actually occur, or that the environment is fine-grained (this assumption is relaxed below). The animal is able to select from the available array of prey types and can *use* this fine-grained environment in a coarse-grained way. For convenience we also assume that the animal either does or does not eat any given kind of prey; that is, no one type is eaten only part of the time it is encountered. What number of different kinds of prey will provide the animal with maximal return per unit time (or expenditure)? Total foraging time, per item eaten, can be broken down into two components: time spent on search (search time) versus that spent on pursuit, capture, and eating (pursuit time). A fine-grained environment is searched for all types of food simultaneously, whereas prey are pursued, captured, and eaten singly. Prey types are ranked from those providing the highest harvest per unit time (and/or energy) to those of lowest yield, that is, from the prey species whose capture requires the least expenditure per calorie assimilated to that requiring the most. The diet of course includes the most rewarding item; as an animal expands its diet to include progressively less rewarding kinds of prey, more and more acceptable items are encountered and search time (and/or energy) per prey item *decreases*. However, as new and varied prey, often hard to catch or to swallow, are added to the diet, pursuit time (and/or energy) per item eaten will usually *increase*. So long as the time or energy saved in reduced search is greater than the increase in time or energy expended in increased pursuit, the diet should be enlarged to include the next, less rewarding, kind of prey. At some point losses accompanying further enlargement will balance or exceed gains, thereby marking the optimal diet.

Though no general statements about an animal's diet are possible from this model, some testable *comparative* predictions can be made. For instance, search time per item eaten should be less in a food-dense environment than in a food-sparse one. Pursuit time, a function of the relative abilities of predator and prey and of the variety of prey types and predator escape mechanisms, should however be little altered by changes in food density per se. As a result, productive environments should be used in a more specialized way than less productive ones. Similarly, animals that spend little effort searching for their prey should be more specialized than those with higher ratios of time spent in search to time spent on pursuit. Should food be scarce, foraging animals are unlikely to bypass potential prey, whereas during times or on areas with abundant food, individuals may be more selective and restrict their diets to better food types. Thus a low expectation of finding prey, or a high mean search time per item, demands generalization; a higher expectation of locating prey items (short mean search time per item) allows some degree of specialization.

So far the environment in this model has been homogeneous and fine-grained. Let us now include patchy environments in which various patches contain different arrays of prey. Types of patches are now ranked in order of decreasing *expectation of yield*, or from the patch in which the most calories are likely to be obtained from prey per unit expenditure to that with the least return. The two components of the animal's time (and/or energy) budget are now "hunting time" per item captured—time spent *within* suitable patches (hunting time is equivalent to "foraging time" above, or to the sum of search time plus pursuit time)—versus the "traveling time" per item caught—time spent traveling *between* suitable patches. Time spent traveling between patches decreases as an animal expands the number of different kinds of patches on its itinerary; time spent hunting within patches must increase as the itinerary is enlarged to include an increased variety of patches. Optimal use of a patchy environment therefore depends on the rate of decrease in traveling time (per prey item) relative to the rate of increase in hunting time (per prey item) associated with expanding the number of different patch types exploited. Suppose that food density is suddenly increased in all patches; *both* traveling time and hunting time, per item, will then be reduced. Since only search time decreases with increased food density, animals that expend greater amounts of energy on search (searchers) will have their hunting time reduced more than those which spend relatively more energy on pursuit (pursuers). Hence, under high food densities, pursuers should restrict the variety of patches they use more than searchers.

Consider now the effects of patch size on the optimal number of patch types exploited. Envision two environments differing only in the sizes of their patches and not at all in the proportions or qualities of the various patch types. Hunting time per item in identical but different sized patches is the same in both environments because patch quality is unaltered. Traveling time per item, however, *decreases* as patch size increases because distance between patches varies linearly with the linear dimension of a patch, while hunting area or volume within a patch varies as its square or cube, depending on whether the animal exploits space in two or three dimensions, respectively. Since larger patches offer smaller traveling time per unit of hunting time, they can be used in a more specialized way than can smaller patches. In the extreme, as patches become vanishingly small compared to the size of the organism, patch selection is impossible and completely fine-grained utilization must take place. Similarly, patches that are very large compared to an organism approximate a one-patch environment and an animal can (or must) spend all of its time in that one patch. Thus smaller and/or less mobile animals should use fewer different patches than larger or more mobile ones.

Competitors should normally reduce the density of some types of prey in some patches. Any prey item worth exploiting in the absence of competition is still worth eating if there is competition for it, but this is not true of patterns of patch exploitation. The decision as to whether or not to forage within a given patch type depends on an organism's expectation of yield in that patch. Should food within a given patch type become scarce (due to competitors or some other factor), inclusion of that patch type in the itinerary increases mean hunting time per item sharply and reduces efficiency. Thus the presence of competitors in some patches should cause an optimal predator to restrict the variety of patch types it exploits (see also pp. 317–318).

Evolution of Niches

Niche changes over evolutionary time are rather difficult to document, although their occurrence cannot be disputed. As new species arise from the fission of existing ones through the process of speciation, new niches come into existence. Life on Earth almost certainly arose in aquatic environments, and early organisms were doubtless very small and simple. During the evolutionary history of life over geological time, organisms have become more and more complex and diversified and Earth has been filled with an overwhelming variety of plants and animals. Some taxonomic groups of organisms, such as dinosaurs, have gone extinct and been replaced by others. Major breakthroughs in the body plans of organisms periodically open up new adaptive zones and allow bursts of evolution of new and diverse species, termed *adaptive radiations*. A major force that has lead to niche separation and diversification is interspecific competition. Thus the first terrestrial organisms found themselves in a wide open ecological and competitive vacuum, freed from competition with aquatic organisms, and they rapidly radiated into the many available new terrestrial niches. Similarly, evolution of endothermy and aerial exploitation patterns have allowed major adaptive radiations; flight has evolved independently at least four times, in insects, reptiles, birds, and mammals. Often evolutionary interactions between two or more taxa have had reciprocal effects upon one another; thus the origin and radiation of flowering plants (angiosperms) in the Mesozoic presumably allowed insects to diversify widely, while species specificity of pollinating insects in turn may well have allowed considerable diversification of plants. Indeed, Whittaker (1969) has suggested that organic diversity is self-augmenting.

Periodic Tables of Niches

In chemistry the urge of scientists to order and classify natural phenomena resulted in the well-known periodic table of the elements, which allowed chemists to predict new elements and their chemical properties and led to our understanding of electron shells. Some ecologists wonder whether something like a "periodic table of niches" might be possible. Of course nothing about niches is quite so simple or discrete as the number of electrons in the outer shell of a chemical element, but most aspects of niches have many more dimensions and are more continuous. Some patterns described earlier may be used to construct a very primitive periodic table of niches (Figure 7.16). Thus trophic niches repeat themselves in organisms of different sizes that are relatively more or less r- and K-selected. An aphid is more like a lemming and a mantid more like a weasel, in their food niche, whereas in terms of body size and position on the $r \rightarrow K$ selection continuum (Chapter 5), the aphid and mantid are relatively alike, as are the lemming and the weasel. Niche dimensions other than those used in Figure 7.16, such as diurnal and nocturnal time of activity, could also be used to construct similar but different periodic tables. A periodic table of niches for aquatic organisms would differ somewhat from that for

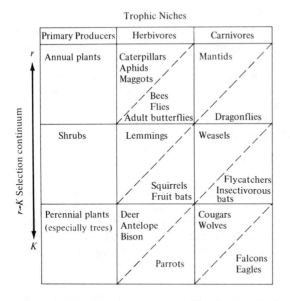

Figure 7.16 A crude "periodic table of niches" of terrestrial organisms, with examples. Dashed diagonal lines separate herbivores and carnivores that exploit space in two and three dimensions. Many other niche dimensions, such as time of activity, might profitably be used in such classifications.

terrestrial ones, especially in that there are very few relatively *K*-selected aquatic primary producers. Perhaps one day, as the young science of ecology matures, we will be able to construct something analogous to the periodic table of the elements (but much more complex) that will order niches, allow predictions, and improve our understanding of the elusive ecological niche.

Selected References

Colwell and Fuentes (1975); Hutchinson (1957a, 1965); Levins (1968, Chapter 3); MacArthur (1968, 1972); McNaughton and Wolf (1970); Odum (1959, 1971); Pianka (1976a); Van Valen (1965); Whittaker, Levin, and Root (1973); Whittaker and Levin (1975).

History and Definitions

Allee, Emerson, Park, Park, and Schmidt (1949); Clarke (1954); Dice (1952); Elton (1927); Gaffney (1975); Grinnell (1917, 1924, 1928); Hutchinson (1957a);

Odum (1959); Parker and Turner (1961); Ross (1957, 1958); Savage (1958); Udvardy (1959); Vandermeer (1972b); Weatherley (1963).

The Hypervolume Model

Green (1971); Hutchinson (1957a, 1965); Maguire (1967, 1973); Miller (1967); Shugart and Patten (1972); Vandermeer (1972b); Warburg (1965).

Niche Overlap and Competition

Abrams (1975, 1976); Case and Gilpin (1974); Colwell and Futuyma (1971); Hespenhide (1971); Horn (1966); Huey *et al.* (1974); Hutchinson (1957a); Inger and Colwell (1977); Klopfer and MacArthur (1960, 1961); MacArthur (1957, 1960a); MacArthur and Levins (1964, 1967); May (1974, 1975); May and MacArthur (1972); Miller (1964); Orians and Horn (1969); Pianka (1969, 1972, 1973, 1974, 1976a); Pielou (1972); Roughgarden (1972, 1973, 1974); Sale (1974); Schoener (1968a, 1970, 1974c); Schoener and Gorman (1968); Selander (1966); Smouse (1971); Terborgh and Diamond (1970); Vandermeer (1972b); Willson (1973b).

Niche Dynamics and Niche Dimensionality

Cody (1968); Colwell and Fuentes (1975); Colwell and Futuyma (1971); Crowell (1962); Gadgil and Solbrig (1972); Inger and Colwell (1977); Levins (1968); MacArthur (1964, 1972); May (1975b); MacArthur, Diamond, and Karr (1972); MacArthur and Pianka (1966); MacArthur and Wilson (1967); McMahon (1976); Pianka (1973, 1974, 1975, 1976a); Pianka *et al.* (1977).

Guild Structure

Cody (1974); MacMahon (1976); Pianka, Huey, and Lawlor (1977); Root (1967); Ulfstrand (1977).

Specialization versus Generalization

King (1971); MacArthur and Connell (1966); MacArthur and Levins (1964, 1967); Roughgarden (1972, 1974a, 1974b, 1974c); Van Valen (1965); Willson (1969).

Time, Matter, and Energy Budgets

Emlen (1966); Gadgil and Bossert (1970); Gibb (1960); Grodzinski and Gorecki (1967); Hickman (1975); Pianka (1976b); C. Smith (1968); Willson (1972a); Zeuthen (1953).

Foraging Tactics and Feeding Efficiency

Charnov (1976a, 1976b); Cody (1968); Emlen (1966, 1968a); Holling (1964); MacArthur (1972); MacArthur and Pianka (1966); Morse (1971); Orians and

Pearson (1977); Pianka (1966b); Pulliam (1974); Rapport (1971); Royama (1970); Schoener (1969a, 1969b, 1971); Tullock (1970); Werner and Hall (1974).

Optimal Use of Patchy Environments

Emlen (1966, 1968a); Hutchinson and MacArthur (1959); King (1971); Levins (1968); MacArthur (1972); MacArthur and Levins (1964); MacArthur and MacArthur (1961); MacArthur and Pianka (1966); Schoener (1969a, 1969b, 1971).

Evolution of Niches

Hutchinson (1965); Lawlor and Maynard Smith (1976); MacArthur (1968, 1972); MacArthur and Levins (1964, 1967); Roughgarden (1975, 1976); Whittaker (1969, 1972).

Community Structure   8

Except for the brief treatment of biomes in Chapter 3, we have until now considered the ecology of individuals and populations; here we examine community ecology. Just as populations have properties that transcend those of the individuals comprising them, communities have both structure and properties not possessed by their component populations. Thus communities have trophic structures, rates of energy fixation and flow, efficiencies, stabilities, diversities, distributions of relative importance among species, guild structures, successional stages, and so on. Moreover, these attributes of communities have profound effects on the organisms living in a given community. Community structure concerns all the various ways in which members of communities relate to and interact with one another, as well as community-level properties that emerge from these interactions. The extreme complexity of most ecosystems makes their study difficult but at the same time quite challenging. The concept of a community is itself an abstraction; communities are seldom clear-cut and distinct, but almost always grade into one another. By considering ecological systems as "open" rather than "closed," and by allowing for continual inflow and outflow of materials, energy, and organisms, this difficulty can be partially overcome and the community concept can be useful. Thus communities change both in space and in time, and the picture developed in this chapter is essentially an instantaneous view of a fairly localized portion of a larger community.

Food Webs and Trophic Levels

Any community can be represented by a *food web,* which is simply a diagram of all the trophic relationships among and between its component species. A food web is generally composed of many *food chains*, each of which represents a single pathway up to the food web. The direction of flow of matter and energy between species can be shown with arrows, as in Figure 8.1. A complete food web would include the rates of flow of energy between the various populations comprising a community.

Primary producers, or *autotrophs*, represent the first trophic level: they are the green plants that use solar energy to produce energy-rich chemicals. Primary producers are an essential part of a community in that practically all other organisms in the community are directly or indirectly dependent on them for energy. Organisms other than the primary producers, or *heterotrophs*, include *consumers* and *decomposers*. Herbivores are primary consumers and represent the second trophic level. Carnivores that eat herbivores are *secondary consumers* or *primary carnivores* and are on the third trophic level. Carnivores that eat primary carnivores in turn constitute the fourth trophic level and are termed *tertiary consumers* or *secondary carnivores*. Similarly those that prey on secondary carnivores are *quaternary consumers* or *tertiary carnivores*, and so on. Because many animals, such as omnivores that eat both plant and animal matter, prey on several different trophic levels simultaneously, it is often impossible to assign

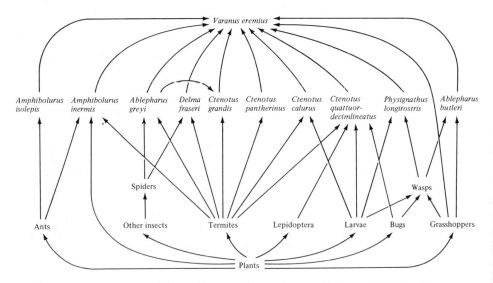

Figure 8.1 Part of the food web in an Australian sandy desert. The "top" predator, a monitor lizard, *Varanus eremius*, eats grasshoppers and ten other species of lizards, which in turn have diets dominated by various sorts of arthropods or plants. A more detailed food web would separate all food types into species and would indicate the actual rate of flow of energy up each link in the web.

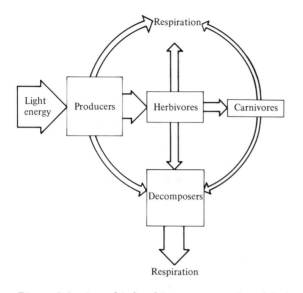

Figure 8.2   A trophic level "compartment" model of a hypothetical community, with arrows indicating the flow of energy through the system. The width of each arrow reflects the rate of flow of energy between particular parts of the system.

them to a given trophic level. Such organisms can usually be assigned partial representation in different trophic levels in proportion to the composition of their diet. The trophic level concept has proven to be an extremely useful abstraction in the study of community structure; it facilitates examination of the flow of matter and energy through communities and underscores the differences between interactions that take place within trophic levels as opposed to those which operate between trophic levels.

Another way in which the major components of an ecosystem can be diagrammed conveniently is shown in Figure 8.2, where each trophic level is treated as a "compartment" and arrows again designate the direction of flow of matter and energy. Many materials, including calcium, carbon, nitrogen, and phosphorus, move from compartment to compartment as individual organisms are consumed by others at higher trophic levels, and eventually return to the abiotic "nutrient pool," where they may be reused by primary producers; such movement of matter through ecosystems is termed *cycling* (Figure 8.3). An essential component of any ecosystem is its decomposers, or *reducers*, which function in returning materials to the nutrient pool. Whereas matter continually circulates through the various compartments of an ecosystem and is always being reused, energy can be used only once.

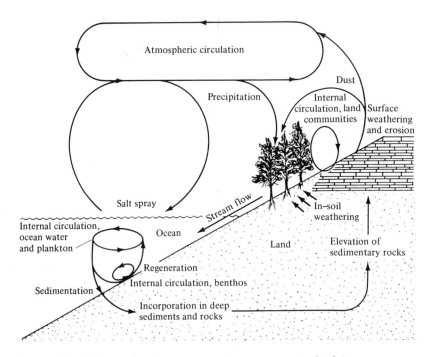

Figure 8.3   A biogeochemical cycle: major movements of calcium are represented diagrammatically. Many other materials circulate within and between ecosystems in a somewhat similar fashion. [From Whittaker (1970). *Communities and Ecosystems*. Reprinted with permission of Macmillan Publishing Co., Inc. © Copyright Robert H. Whittaker, 1970.]

All ecosystems thus depend on a continual inflow of energy (see also pp. 276–278).

The Community Matrix

A table of numbers with rows and columns is known as a matrix. Building on the Lotka-Volterra competition equations, Levins (1968) formulated the concept of an alpha matrix, or (more generally) a community matrix. For a community composed of n species, the community matrix is an $n \times n$ matrix that gives the sign and degree of interaction between each pair of species. Figure 8.4 illustrates the community matrix of trophic relationships among ten hypothetical species with a food web as shown. In pairs of competing species, α_{ij} and α_{ji} values both have positive signs, indicating that each inhibits the other's population. The magnitude of the alpha values indicates the intensity of competitive inhibition, which need not be equal and opposite. A prey–predator relationship is indicated by α_{ij} and α_{ji} values with opposite signs, with the predator benefiting from the relationship and the prey suffering. Again the degree of benefit or detriment is represented by the magnitude of the alpha values. Thus, if α_{ij} is negative and α_{ji} is positive,

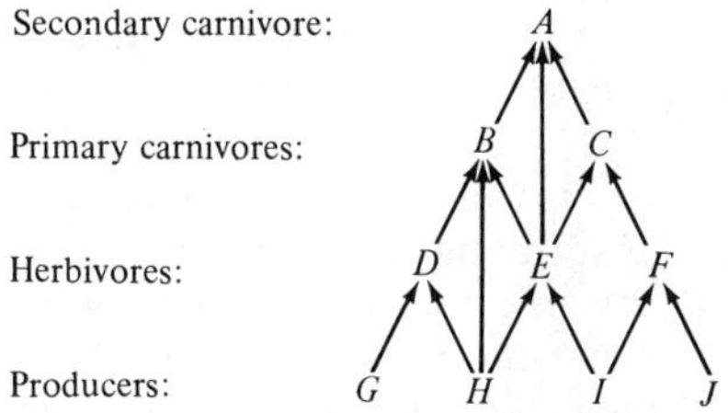

Species having the effect

		A	B	C	D	E	F	G	H	I	J
	A	1	−	−	0	−	0	0	0	0	0
	B	+	1	+	−	−	0	0	−	0	0
	C	+	+	1	0	−	−	0	0	0	0
Species affected	D	0	+	0	1	+	+	α_{DG}	−	0	0
	E	+	+	+	+	1	+	−	−	−	0
	F	0	0	+	+	+	1	0	0	−	−
	G	0	0	0	α_{GD}	+	0	1	+	+	+
	H	0	+	0	+	+	0	+	1	+	+
	I	0	0	0	0	+	+	+	+	1	+
	J	0	0	0	0	0	+	+	+	+	1

Figure 8.4   A community matrix for a hypothetical ten-species community. Only the signs of the alphas are given and only *direct* interactions between and within two trophic levels are considered. All members of a given trophic level are assumed to be in competition. Higher order interactions, such as the effect of a plant's antiherbivore defenses on a carnivore that depends on the plant's herbivores for food, or the indirect effects of the predator on the plant through reducing herbivore densities, are ignored for simplicity. (A community matrix showing the *net* effect of each species on every other could also be constructed.) One pair of alphas is identified within the matrix, that between species D (predator) and species G (prey). Note that α_{DG} is negative whereas α_{GD} is positive. The community matrix can also be formulated in terms of the sensitivity of each species' population density to changes in the density of the others using partial derivatives, in which case the sign structure is reversed.

species i eats species j. Mutualistic relationships are represented by pairs of negative alphas. Thus the sign structure of such a matrix specifies the trophic structure of a community.

In competitive communities, a greater number of effective niche dimensions will result in fewer off-diagonal elements of zero and more intense diffuse competition (community "connectedness" increases). Moreover, the guild structure of a community is implicit in its community matrix, since clusters of species that interact strongly will have high alphas while those that interact weakly will have lower coefficients (see also pp. 251–253).

A virtue of the community matrix idea is that it facilitates abstraction and quantification of the interactions among members of a community. The concept of the community matrix is actually independent of the Lotka-Volterra competition equations since it can be formulated equally well in terms of the sensitivities of each species' population density to changes in the densities of other species (using this convention reverses the sign structure of the matrix—see also p. 199). Although one species can affect another in both positive and negative ways simultaneously, the appropriate coefficient in the matrix represents an overall effect. Alpha values in community matrices traditionally represent only the *direct* pairwise effects of each species on every other; however, an exactly analogous matrix of the *net* effects between all pairs could just as easily be constructed. Such net community matrices might well provide useful insights into community structure. For example, a pair of competing species can actually have a mutualistic relationship provided that each reduces sufficiently the densities of the other species' other competitors. In fact alphas are presumably neither constant nor independent but change both in time and in space and with community composition. Nevertheless the concept of an ever-changing community matrix is an extremely useful abstraction that helps us to visualize what presumably is actually happening in a complex real community.

Principles of Thermodynamics

An important facet of community ecology is the energy relationships between and among members of a community. Before studying community energetics, however, we must review some fundamentals of thermodynamics.

All organisms require energy to persist and to replace themselves, and the ultimate source of practically all Earth's energy is the sun. One can think of the sun as "feeding" Earth via its radiant energy. But 99 percent or more of this incident solar radiation goes unused by organisms and is lost as heat and heat of evaporation; only about 1 percent is actually captured by plants in photosynthesis and stored as chemical energy. Moreover, energy available from sunlight varies widely over Earth's surface both in space and in time (see Chapters 2 and 3).

Physics and chemistry have given us two basic laws of thermodynamics that are obeyed by *all* forms of matter and energy, including living organisms. The first law is that of "conservation of matter and energy," which states that matter and energy cannot be created or destroyed. Matter and energy can be transformed, and energy can be converted from one form into another, but the total of the equivalent amounts of both must always remain constant. Light can be changed into heat, kinetic energy, and/or potential energy. Each time energy is converted from one form into another, some of it is given off as heat, which is the most random form of energy. Indeed, the only energy conversion that is 100 percent efficient is conversion to heat, or burning. Burning aliquots of dried organisms in "bomb calorimeters" is a common method of determining how much energy is stored in their tissues (Paine, 1971). Energy can be measured in a variety of different units such as ergs and joules, but the common denominator used in ecology is heat energy or calories.

The second law of thermodynamics states that energy of all sorts, whether it be light, potential, chemical, kinetic, or whatever, tends to change itself spontaneously into a more random, or less organized, form. This law is sometimes stated as "entropy increases"—entropy being random, unavailable energy. Suppose I heat a skillet to cook an egg, and after finishing I leave it on the stove. At first, heat energy is concentrated near the skillet, which is, relative to the rest of the room, quite nonrandom. But by the next morning the skillet has cooled to air temperature, and the heat energy has radiated throughout the room. That heat energy is now dispersed and unavailable for cooking; the system of the skillet, the room, and the heat has gone toward equilibrium, become more random, and entropy has increased. Unless an outside source of energy such as a stove, with fuel or electricity, is continually at work to maintain a nonequilibrium state, dispersion of heat results in a random equilibrium state. The same is true for all kinds of energy. According to this law, our solar system and presumably the entire universe should theoretically become a completely random array of molecules and heat in the far distant future.

Life has sometimes been called "reverse entropy" because organisms maintain complex organized states compared to their surroundings. But they obey the second law just as any other system of matter and energy; all organisms must work continually to build and maintain nonrandom assemblages of matter and energy. This process requires energy, and organisms use the energy of the decaying sun (which of course also obeys the second law of thermodynamics and tends toward increasing disorder) to "oppose" the second law within their own tissues by producing order out of increasing disorderliness. Wherever there is a live plant or animal, there must be an energy source. Without a continued influx of energy, no organism can survive for very long. Again, this "reverse entropy" occurs only *within* each organism, and the overall energy relations of the entire solar system are in accord with the second law of thermodynamics, with the overall system continually becoming more and more random.

Pyramids of Energy, Numbers, and Biomass

The rate of flow of energy through a given trophic level decreases with increasing trophic level for several reasons. Because energy transfers are never 100 percent efficient, not all the energy contained in any given prey item is actually available to a predator; some is lost in converting prey tissue into predator tissue, and some is not even assimilated but passes through the predator's intestinal tract unchanged and is then decomposed by reducers. The efficiency of transfer of matter and energy from prey to predator is often greatly reduced by predator-avoidance tactics of prey, such as the chemical defenses of plants. In addition, each organismic unit (and each trophic level) expends some of its available energy on its own activities, further reducing the amount of energy available at higher trophic levels. Ultimately, of course, at equilibrium all the energy captured by primary producers must be expended and dissipated back into space as heat; that is, the amount of energy entering the system must exactly balance that leaving it.

The reduction in the rate of flow of energy from each trophic level to the next higher one determines many of a community's properties, including the total number of trophic levels as well as the proportion of predators to prey. Ecologists estimate that, after standardization per unit area and unit time, approximately 10 to 20 percent of the energy at any given trophic level is available to the next higher trophic level. Hence, if a thousand calories are available to primary producers, usually only a few of the thousand are actually available to a secondary carnivore three trophic levels away. A result of this rapid reduction in

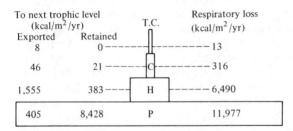

Figure 8.5 Pyramid of energy for Silver Springs, Florida. [From Phillipson (1966) after Odum.]

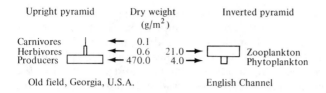

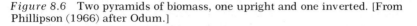

Figure 8.6   Two pyramids of biomass, one upright and one inverted. [From Phillipson (1966) after Odum.]

the availability of energy is that animals at higher trophic levels are generally much rarer than those at lower ones. Moreover, decreasing availability of energy places a distinct upper limit on the number of trophic levels possible, about five or six being the normal maximum.

A convenient means of expressing the energetic structure of a community is the *pyramid of energy*, which consists simply of the rates of energy flow between various trophic levels (Figure 8.5). The laws of thermodynamics and above considerations dictate that the pyramid of energy can never be inverted; that is, the flow of energy through each trophic level must always decrease with increasing trophic level.

Two other types of ecological "pyramids" are the *pyramid of numbers* and the *pyramid of biomass*. These are instantaneous measures rather than rates; they have no time dimension (units of the pyramid of energy are calories/m^2/year, but the units of the pyramid of numbers are number/m^2, and the units of the pyramid of biomass are grams/m^2). The pyramid of numbers consists of a set of the densities of individuals in each trophic level; the pyramid of biomass is the biomass (usually measured in grams dry weight) per square or cubic meter in each trophic level. Pyramids of numbers and biomass measure only the *standing crop* (the amount present at an instant) of each trophic level, not their turnover rate. Because they lack a time dimension, these two pyramids may be inverted, with, for example, lower densities and/or smaller biomasses at lower trophic levels. Thus one tree may support many insects (an inverted pyramid of numbers); likewise, a rapid rate of turnover allows a small biomass of prey to support a larger biomass of predators with a slower turnover rate. Such an inverted pyramid of biomass often characterizes aquatic ecosystems, where primary producers (phytoplanktonic algae) are small and rapidly dividing while their zooplanktonic predators are larger and longer lived (Figure 8.6).

Energy Flow and Ecological Energetics

The energy content of a trophic level at any instant (that is, its standing crop in energy) is usually represented by capital lambda, Λ, subscripted to indicate the appropriate trophic level: Λ_1 = primary producers, Λ_2 = herbivores, Λ_3 = primary carnivores, and so on. Similarly the rate of flow of energy between trophic levels is designated by lower case lambdas, λ_{ij}, where the i and j subscripts indicate the two trophic levels involved, with i representing the level receiving and j the level losing energy. Subscripts of zero denote the world external to the system; subscripts of 1, 2, 3, etc., indicate trophic level as above.

Using these conventions, an ecosystem can be represented by a compartment model, as in Figure 8.7. At equilibrium the amount of energy contained in every compartment (trophic level) must be constant, which in turn requires that the rate of flow of energy into every compartment be exactly balanced by the flow of energy out of the compartment concerned. At equilibrium $(d\Lambda_i/dt = 0$ for all $i)$ energy flow in the system portrayed in the figure may thus be represented by a set of simple equations (with inputs on the left and rate of outflow to the right of the equals signs):

$$\lambda_{10} = \lambda_{01} + \lambda_{02} + \lambda_{03} + \lambda_{04}$$
$$\lambda_{10} = \lambda_{21} + \lambda_{01} + \lambda_{41}$$
$$\lambda_{21} = \lambda_{32} + \lambda_{02} + \lambda_{42}$$
$$\lambda_{32} = \lambda_{03} + \lambda_{43}$$
$$\lambda_{41} + \lambda_{42} + \lambda_{43} = \lambda_{04}$$

The rate at which energy is actually captured by plants (λ_{10}) is estimated at only about 1 percent of the total solar energy hitting Earth's surface. This rate of uptake of solar energy by primary producers, λ_{10}, is termed the gross productivity. It is usually given in

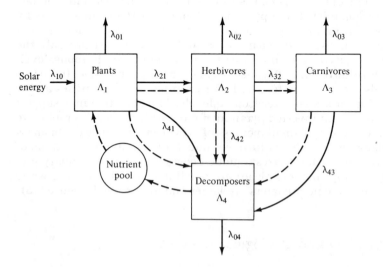

Figure 8.7 Another compartment model of an ecosystem. Energy flow is shown with solid arrows and the flow of matter with dashed arrows (see text).

calories/m²/year, which represents the *gross annual production* (GAP). Since plants use some of this energy in their own respiration (i.e., λ_{01}), only a part of the gross annual production is actually available to animals and decomposers; this fraction, λ_{21}, plus the energy used by decomposers, λ_{41}, is termed the net productivity or (on an annual areal basis) *net annual production* (NAP). Net production may be considerably less than gross production; in some tropical rain forests plants use as much as 75 to 80 percent of their gross production in respiration.

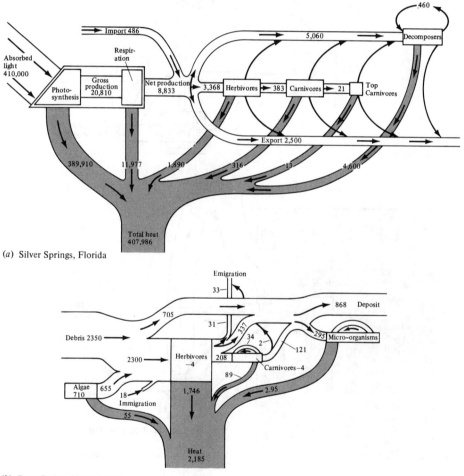

(a) Silver Springs, Florida

(b) Root Spring, Massachusetts

Figure 8.8   Two energy flow diagrams from actual communities. Figures are in kcal/m²/yr; numbers in boxes in (b) represent changes in standing crops. [From Phillipson (1966) after Teal and Odum.]

In temperate deciduous forests respiration is usually 50 to 75 percent of gross primary production, while in most other communities it is about 25 to 50 percent of gross production. Whittaker (personal communication) estimates that perhaps 7 percent of the plant food on land is actually harvested by animal consumers: the remainder of the net

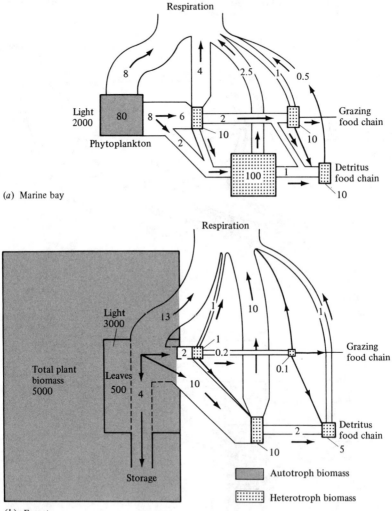

Figure 8.9   Energy flow diagrams for two ecosystems with very different standing crops. (*a*) A marine bay. (*b*) A forest. Standing crop biomass measured in kcal/m²; energy flow in kcal/m²/day. [From Phillipson (1966) after Odum, Connell, and Davenport.]

primary production is consumed by decomposers. Efficiency of transfer of energy from one trophic level to the next higher trophic level, say level i to level j, may be estimated as $\lambda_{ji}/\lambda_{ih}$, where $j = i + 1$ and $h = i - 1$. Thus the ratio $\lambda_{21}/\lambda_{10}$ is a measure of the efficiency with which primary producers pass the solar energy they capture on to herbivores and, indirectly, to consumers at higher trophic levels. Such efficiencies of transfer of energy from one trophic level to the next are generally estimated to be between about 5 and 30 percent, and a reasonable average figure is about 10 to 15 percent (Slobodkin, 1960, 1962).

Energy flow diagrams have been constructed for some natural communities (Figures 8.8 and 8.9); these diagrams underscore the relatively minor energetic importance of carnivores in ecological systems. Decomposers often play a major energetic role, especially in terrestrial ecosystems where much of the primary production is not consumed by herbivores but instead falls to the ground as dead leaves and other plant material. Indeed, as much as 90 percent of the net annual production in some communities may be consumed by their decomposers. Community ecologists are currently investigating energy flow and efficiency of transfer of energy in natural ecosystems, and much remains to be learned about ecological energetics; such studies have obvious practical significance to human exploitation of ecological systems.

Systems Ecology

Ecological communities are exceedingly complex, with a myriad of intricate and often quite subtle interactions between and among their component resources, individuals, and populations. Even a process as fundamental as the interaction between a predator and its prey may require elaborate analysis in terms of numerous subcomponents: these include hunger, search, pursuit and capture, functional and numerical response, predator escape tactics, spatial and temporal distribution of prey, predator learning and interference between predators, etc. (Holling, 1965, 1966). A young and vigorous branch of ecology, termed "systems ecology," has been developed to deal with such complexity. Systems ecologists rely heavily on computers to build models of complex ecological systems that allow for various sorts of interactions between and among components, which components may themselves have many interacting subcomponents. Using actual data on how each component, at each level, affects others (in practice such data are *very* difficult to obtain), systems ecologists hope to model ecological systems accurately enough to predict their responses to particular perturbations. Because the systems approach is basically descriptive and deductive, it is limited in that the behavior of a system usually cannot be predicted with accuracy at states outside those used in the original data describing interactions between compartments.

Saturation with Individuals and with Species

In any closed ecosystem at equilibrium, all the energy of net production must be used up by consumers and decomposers in order for the system to have a balanced energy budget (Figure 8.7 and equations on p. 280). Such an idealized system can be thought of as being saturated with individual organisms because all available energy is used and no more organisms could be supported. However, predators, by reducing densities of organisms at lower trophic levels, can prevent their prey populations from reaching *otherwise maximal sustainable densities* and thereby effectively preclude true saturation within that lower trophic level. If this is the case, only top predator populations reach a sort of "complete" saturation. Furthermore, antiherbivore defenses of plants render much of the net primary productivity unusable by animal consumers and thus require that many plant tissues be routed directly through a community's decomposers.

Communities, or portions thereof, can be kept from reaching saturation with individuals in other ways. Real ecological systems are almost never truly closed; rather they usually both receive materials and energy from other systems and lose them to others. A community or a portion of a community that is not closed may be rarefied by continual or sporadic removal of organisms. As a hypothetical example, consider a lake along a river: both lake and river contain communities of phytoplankton and zooplankton, but the lake receives an inflow of river water containing no members of the lake community while losing water containing members of its community. Such a system can never become truly saturated because of the continual removal of organisms from it. Moreover, since the physical environment is changing continually (Chapters 2 and 3), and because it takes time to respond to these changes, populations and communities probably seldom reach equilibrium, although some K-selected organisms may occasionally approach it.

How much does the degree of saturation with individuals vary within and between communities? And how does efficiency of transfer of energy change with degree of saturation with individuals? It is known (Chapter 6) that the turnover rate of prey is usually highest at intermediate prey densities; moreover, such prey populations often support higher predator densities than larger prey populations. The immense practical value of such knowledge is readily apparent.

Can communities be saturated with *species*? That is, is there a maximum number of *different* species that can exist within an ecological system? If so, a new species introduced into such a community must either go extinct or cause the extinction of another species (which it then replaces). Conversely, the successful invasion of a new species into a community without the extermination of an existing species would imply that the original community was not saturated with species.

Limited evidence suggests that portions of some communities may indeed be saturated with species, at least within habitats. R. H. MacArthur and his colleagues demonstrated that bird species diversity is strongly correlated with foliage height diversity (Figure 8.10) in a remarkably similar way on three continents: North America, South America, and Australia. Habitats with equal amounts of foliage (measured by leaf surface) in three layers (0–2, 2–25, and over 25 feet above ground) are richer in bird species than are habitats with unequal pro-

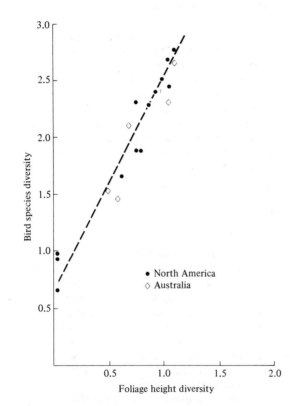

Figure 8.10 Correlation between foliage height diversity and bird species diversity. North American habitats are shown as solid circles; Australian habitats are represented by diamonds. This is perhaps the best evidence that communities may sometimes be saturated with species. [After Recher (1969).]

portions of foliage in the three layers. The diversity of bird species is lowest in habitats with only one of these layers of vegetation, such as a grassland. Interestingly enough, knowledge of plant species diversity does not allow an improvement in the prediction of bird species diversity (MacArthur and MacArthur, 1961), which suggests that birds recognize the structure rather than type of the vegetation. Despite the fact that avian niche space is partitioned in a fundamentally different way in Australia, bird species diversity within a given habitat on that continent is very close to what it is in a habitat of similar structure in North America (Recher, 1969). In addition to illustrating that spatial heterogeneity regulates bird species diversity, the convergence of these data suggests that these avifaunas are saturated with species. However, such neat convergences in species densities of plants, insects, and desert lizards do not occur, which suggest that these groups may not always be saturated with species (Whittaker, 1969, 1970, 1972; Pianka, 1973).

The number of species that can coexist at any point in space may have a distinct upper limit, as suggested above; however, there is no obvious limitation on the number of species that can occur in a given *area* because horizontal replacement of species can allow coexistence of many more species than actually share the use of a common point in space within that area. Indeed, MacArthur (1965) has suggested that the horizontal component of diversity ("between habitat" diversity) may be increasing continually during evolutionary time, whereas point diversities remain nearly constant. Some upper limit on horizontal turnover of species also seems likely.

Species Diversity

Why does one community contain more species than another? Some complex communities, such as tropical rain forests, consist of many thousands of different plant and animal species while other communities, such as tundra communities, support perhaps only a few hundred species. Number of species often varies greatly even at a local level; thus a grassland habitat typically contains many fewer species of birds than does an adjacent forest. Indeed, different forest communities in the same general region usually vary in numbers as well as types of plant and animal species. The number of species is referred to as "species richness" or, more frequently, as "species density."

Communities with similar species densities often differ in yet another way: some contain a few very common species and many rare ones, whereas others support no very common species but many of intermediate abundance. Abundance is only one way of estimating the *relative importance* of various component species within a community; other measures frequently employed include both the biomass of, and the energy flow through, various species' populations. The relative importance of species varies within and between communities, and considerable effort has been expended in attempts to document such differences and to understand why they occur. Importances of different species within a community (or a portion of one) can be depicted conveniently by "species importance curves" (Figure 8.11). Several hypothetical distributions of the relative importance of species within communities have been suggested that generate different shaped species importance curves (see Figure 8.11 and Whittaker, 1970).

Species density and relative importance have been combined in the concept of *species diversity*, which increases both with increasing species density and with increasing equality of importance among the members of a community. Species diversity is high when it is difficult to predict the species of a randomly chosen individual organism and low when an accurate prediction can be made. For example, an or-

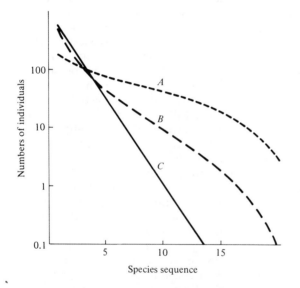

Figure 8.11 Species importance curves, with species ranked from most important to least important. Lines illustrate three different hypothetical curves based on (A) random niche boundaries (MacArthur, 1957, 1960a), (B) multiple niche dimensions, which generate a lognormal distribution of species importances (Preston, 1948, 1960, 1962a, 1962b; Whittaker, 1970, 1972), and (C) niche preemption, which leads to a geometric series (Motomura, 1932; Whittaker, 1970, 1972). Data from various real communities fit each hypothetical distribution reasonably well. For a given number of species, diversity would be highest in A and lowest in C. [After Whittaker (1972).]

ganism chosen at random from a cornfield would probably be a stalk of corn, whereas one would not venture to guess the probable identity of a randomly chosen organism from a tropical rain forest. We are currently trying to determine not only why different communities contain different numbers of species with differing relative importances, but how such differences in species richness and importance affect other community properties such as trophic structure and community stability.

Communities can differ in species diversity in several ways; more diverse communities may (1) contain a greater range of available resources (that is, a larger total niche hypervolume space) and/or (2) their component species may, on the average, have smaller niche breadths (that is, each species might exploit a smaller fraction of the total niche hypervolume). The former corresponds roughly to "more niches" and the latter to "smaller niches." (3) Two communities with identical niche space and mean niche breadth can still differ in species diversity if they differ in the average degree of *niche overlap* because greater niche overlap means that more species can exploit any particular resource (this situation is described as "smaller exclusive niches"). (4) Alternatively, in communities that do not contain all the species they could conceivably support (i.e., those that are "unsaturated" with species), species diversity can vary with the extent to which all available resources are exploited by as many *different* species as possible (that is, with the degree of saturation with species, or with the number of "empty niches"). Resources are seldom if ever wasted, even in communities that do not contain their full quota of species, because those species that do occur in such communities generally expand their activities and exploit nearly all the available resources, although their efficiency of exploitation may be less than that of some better adapted species. (Thus most communities are probably effectively saturated with individuals even if they are not saturated with species.)

Because thorough understanding of community species diversity inevitably requires analysis of niche structure and diversification in component populations, investigations of species diversity usually go hand in hand with the study of niches. In practice, one is seldom able to study the species diversity of an entire community, and usually attention is focused on a portion of a community (an "assemblage") such as trees, ants, lizards, or birds. Using the three major niche dimensions (Chapter 7), total species diversity of an area can be partitioned into its spatial, temporal, and trophic components. Species replace one another along each of these three niche dimensions, and diversity is generated by separation along each.

I have studied species diversity and niche relations of desert lizards in certain deserts of North America (the Sonoran and Mojave deserts), southern Africa (the Kalahari Desert), and Western Australia (the Great Victoria Desert) (Pianka, 1973, 1975). Lizard niches in these deserts differ in the three fundamental niche dimensions (place, time, and food). Moreover, the variety of resources actually used by lizards along each niche dimension, as well as the amount of niche overlap along them, differs markedly among desert systems. Food is a major dimension separating niches of North American lizards; in the Kalahari food niche separation is slight and differences in the place and time niches are considerable. In the most diverse Australian lizard "communities," all three niche dimensions are important in separating niches and niche overlap is distinctly reduced. Differences between deserts in lizard species diversity are *not* accompanied by conspicuous differences in niche breadths; rather they stem primarily from differences in the variety of resources used by lizards. Moreover, niche overlap does not increase with diversity but varies inversely and is lowest in the most diverse lizard communities of Australia (Pianka, 1973, 1974, 1975).

The spatial component of diversity is due to differential use of space by different populations; for convenience it can be broken down into horizontal and vertical components. At a gross geographic level, species replace one another horizontally as one moves from one habitat to another (this is the between-habitat component of overall diversity). Similar replacement of species occurs both horizontally and vertically *within* habitats. Birds, for instance, often tend to partition a given habitat by occupying different vertical strata, such as low bushes, tree trunks, lower foliage, and high canopy. Ground-dwelling mammals and lizards generally partition microhabitats horizontally, with some using open spaces between shrubs and others exploiting the ground beneath or near specific types of vegetation such as grasses, shrubs, and trees. Different populations, by occupying different microhabitats, are thus able to coexist within a given habitat and contribute to within-habitat diversity. The within-habitat component of diversity is most easily distinguished from the between-habitat component in relatively homogeneous communities (heterogeneous communities, such as edge communities and ecoclines—see Figure 3.17—include both components). However, even a homogeneous community has an internal structure in that it consists of a mosaic of repeatable horizontal and vertical patches. Because communities and habitats frequently blend into one another, it is often difficult to distinguish between-habitat from within-habitat diversity. Where does one habitat "stop" and another "begin"? A sandridge gradually gives way to a sand plain and the intertidal grades into the deeper benthic zone. The problem of defining a habitat can be overcome by the use of "point diversities," which consist of the species diversity occurring at a point in space. Point diversities are difficult to estimate (one might have to wait a very long time to see *all* the species that use a particular point!). However, they should invariably be lower than any areal estimate of diversity because the different species in a community have each specialized somewhat as to the microhabitats they use.

Provided that different resources are utilized, temporal separation, both daily and seasonally, of species' populations can allow coexistence of more species and hence may add to community diversity. Many instances of subtle differences in times of activity between populations are known, in addition to such conspicuous distinctions as that between nocturnal and diurnal animals.

Yet another means by which community diversity may be enhanced is by trophic differences. Here again, in addition to the conspicuous differences between trophic levels (such as herbivores, omnivores, and carnivores), there are more subtle but nevertheless important differences between species even within a given trophic level in the prey they eat. Thus different species of predators living in the same area tend to eat prey of different sizes and types, with the larger species taking larger prey items (this generalization applies to most fish, lizards, carnivorous mammals, and hawks). Moreover, the composition of the diet often varies markedly among potential competitors (see Table 6.4). Finally, diversity of plant defensive chemicals doubtless creates numerous different potential food niches for herbivores, especially insects (Whittaker, 1969; Whittaker and Feeny, 1971), thereby greatly facilitating trophic diversity at higher trophic levels.

A prevalent global pattern of species diversity is of some interest. The diversity of living organisms is usually high near the equator and decreases rather gradually with increasing latitude, both to the north and to the south (Figures 8.12 and 8.13). Such "latitudinal gradients"

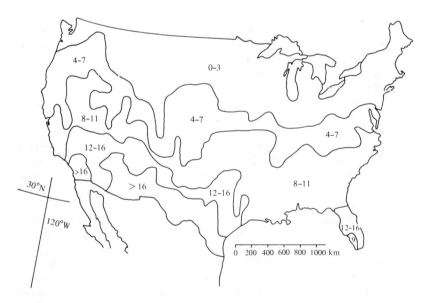

Figure 8.12   Numbers of species of lizards known to occur per square degree of latitude and longitude within the continental United States. [Courtesy of J. J. Schall.]

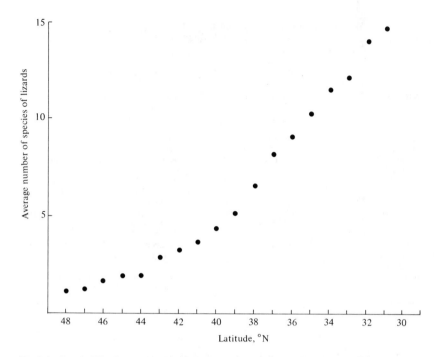

Figure 8.13   The latitudinal gradient in the average number of species of lizards per degree square in the continental United States. [Courtesy of J. J. Schall.]

in diversity are widespread among different plant and animal groups, and it is likely that a general explanation underlies these ubiquitous patterns. One reason species diversity is higher at lower latitudes than it is in the temperate zones is that there are often more habitats in the tropics. At high altitudes in the tropics, habitats similar to, but richer in species than, those in temperate zones often occur, whereas true tropical habitats are seldom found in temperate areas. However, the fact that there is a greater variety of species where there are more habitats is neither surprising nor theoretically very interesting. Diversity differences *within* a given habitat type are of much greater interest, for they reflect the partitioning of available niche space within habitats.

Considerable speculation about the causes of both local and latitudinal patterns in species diversity has generated many theories and hypotheses for their explanation (Table 8.1), all of which probably operate in some situations (Pianka, 1966a). Various mechanisms for determination of diversity are clearly not independent, and several may often act in concert or in series in any given case. Each hypothesis or theory is briefly outlined below, and some ways in which they could interact are considered.

Table 8.1   Various Hypothetical Mechanisms for the Determination of Species
Diversity and Their Proposed Modes of Action upon Ecological
Niches

Level	Hypothesis or Theory	Mode of Action
Primary	1. Evolutionary time	Degree of unsaturation with species
Primary	2. Ecological time	Degree of unsaturation with species
Primary	3. Climatic stability	Mean niche breadth
Primary	4. Climatic predictability	Mean niche breadth
Primary or secondary	5. Spatial heterogeneity	Range of available resources
Secondary	6. Productivity	Especially mean niche breadth, but also range of available resources
Secondary	7. Stability of primary production	Mean niche breadth and range of available resources
Tertiary	8. Competition	Mean niche breadth
Primary, secondary, or tertiary	9. Rarefaction	Degree of allowable niche overlap and level of competition
Tertiary	10. Predation	Degree of allowable niche overlap and level of competition

These mechanisms can be classified as primary, secondary, or tertiary, depending on whether they act respectively, mainly through the physical environment alone, through a mixture of both the physical and the biotic environments, or through the biological environment alone (Poulson and Culver, 1969). Ultimately a thorough understanding of patterns in diversity requires knowledge of primary-level mechanisms.

1. EVOLUTIONARY TIME. This theory assumes that diversity increases with the age of a community, although the validity of this assumption is still open to question. Temperate habitats are thus considered to be impoverished with species because their component species have not had time enough to adapt to, or to occupy completely, their environment since the recent glaciations and other geological disturbances. More "mature" tropical communities, however, are more diverse because there has been a longer period without major disturbances for organisms to speciate and diversify within them. The evolutionary time theory does not necessarily imply that temperate communities are unsaturated with individuals; niche expansion may often allow nearly full utilization of available resources even in a habitat that is impoverished with species.

2. ECOLOGICAL TIME. This theory is similar to the evolutionary time theory but deals with a shorter, more recent, time span. Here we are concerned primarily with time available for dispersal, rather than

with time for speciation and evolutionary adaptation. Newly opened or remote areas of suitable habitat, such as a patch of forest burned by lightning, an isolated lake, or a patch of sand dunes, may not have their full complement of species because there has been inadequate time for dispersal into these areas. Dispersal powers of most organisms are good enough that this mechanism may be of relatively minor importance in most communities (see also Chapter 9).

3. CLIMATIC STABILITY. A stable climate is one that does not change much with the seasons. Successful exploitation of environments with unstable climates often requires that organisms have broad tolerance limits to cope with the wide range of environmental conditions they encounter. Thus, by demanding generalization, such variable environments favor organisms with broad niches. Conversely, environments with more constant climates allow finer specialization and narrower niches. Plants and animals in the relatively constant tropics, for example, are often highly specialized in both the places they forage and the foods they eat. Obviously, in two habitats with the same range of available resources, the one whose component species each use a smaller fraction of these resources will support more species. By these means the number of species should increase with climatic stability.

4. CLIMATIC PREDICTABILITY. Many aspects of climate, although temporally variable, are nevertheless highly predictable in that they repeat themselves fairly exactly from day to day and year after year. Such cyclical predictability can allow organisms to evolve some degree of dependence on, as well as to specialize on, particular environmental conditions and temporal patterns of resource availability, thereby enhancing daily and/or seasonal replacement of species and the temporal component of total diversity. Deep freshwater lakes in the temperate zones, for example, typically have a consistent annual succession of primary producers; a major causal factor is nutrient availability, which changes markedly during the year, being greatest during the spring and fall turnovers (pp. 68–71). Thus different species of phytoplankton have adapted to exploit lakes under particular environmental conditions that recur regularly every year. Annual plants in Arizona's Sonoran Desert have adapted to the marked bimodal annual precipitation pattern (Figure 2.11*b*) with distinct rainfall peaks in winter and summer: there are two distinct sets of species, one whose seeds germinate under wet and cooler conditions (winter annuals) and another set that germinates when conditions are wet but warmer (summer annuals that bloom in late summer after the flash floods).

5. SPATIAL HETEROGENEITY. A forest contains more different species of birds than a grassland, an arboreal desert generally supports more species of lizards than one without trees, and a tidal flat with a great variety of particle sizes and substrate types has more species of

mud-dwelling invertebrates than a homogeneous mudflat. Structurally complex habitats obviously offer a greater variety of different microhabitats than do simple ones. Because there are more different ways of exploiting them, such spatially heterogeneous habitats usually support more species than homogeneous ones; thus species replace one another in space with greater frequency and the spatial component of diversity is higher. Correlations between the structural complexity of a habitat and the species diversity of its biota are widespread (one discussed earlier in this chapter is illustrated in Figure 8.10).

6. PRODUCTIVITY. In habitats with little food, foraging animals cannot afford to bypass many potential prey items; where food is abundant, however, individuals can be more selective and confine their diets to better prey items (see also pp. 263–266). Thus more productive habitats, or those where food is dense, offer more prey choice and hence allow greater dietary specialization than do less productive habitats. Because each species uses less of the total range of available foods, the same spectrum of food types will support more species in a more productive environment (Figure 8.14). In addition, productive habitats may support more species than similar, less productive ones, by virtue of the fact that certain resources, too sparse to support a species in unproductive habitats, are dense enough to be successfully exploited in productive habitats (MacArthur, 1965). An open desert with only one ant nest per hectare might not support a population of specialized ant-eating lizards, whereas another, richer, area with several ant nests per hectare would.

7. STABILITY OF PRIMARY PRODUCTION. Just as more stable and more predictable climates support more species, areas with temporally stable and/or predictable patterns of production should often allow coexistence of more species than would be possible in areas with more variable and/or erratic productivity. This secondary- or tertiary-level mechanism differs from those of climatic stability and climatic predictability in that plants themselves react to climatic conditions and hence can alter temporal variability with their own homeostatic adaptations and storage capacities. Plants both buffer and enhance physical fluctuations either by releasing the products of primary production (flowers or seeds) gradually and continuously or by expending them in temporally erratic blooms (see also p. 227).

Mechanisms 3, 4, and 7 (climatic stability, climatic predictability, and stability of primary production) could be combined under a more inclusive heading of "temporal heterogeneity" to parallel "spatial heterogeneity" (see also Menge and Sutherland, 1975).

8. COMPETITION. In many diverse communities, such as tropical rain forests, populations are thought to be often near their maximal sizes (equilibrium populations of Chapter 5), with the result that intraspecific and interspecific competition are frequently keen. Selection for

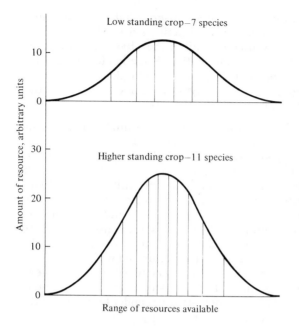

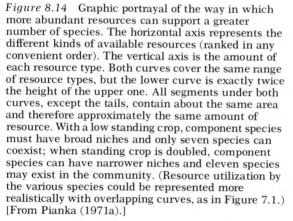

Figure 8.14 Graphic portrayal of the way in which more abundant resources can support a greater number of species. The horizontal axis represents the different kinds of available resources (ranked in any convenient order). The vertical axis is the amount of each resource type. Both curves cover the same range of resource types, but the lower curve is exactly twice the height of the upper one. All segments under both curves, except the tails, contain about the same area and therefore approximately the same amount of resource. With a low standing crop, component species must have broad niches and only seven species can coexist; when standing crop is doubled, component species can have narrower niches and eleven species may exist in the community. (Resource utilization by the various species could be represented more realistically with overlapping curves, as in Figure 7.1.) [From Pianka (1971a).]

competitive ability (K selection) is therefore strong, and most success-ful organisms in these communities have their own zone of competi-tive superiority. Thus organisms that have specialized as to foods and/or habitats are at a competitive advantage (see pp. 253–255), and the resulting small niches make high diversity possible. By way of contrast, populations in many less diverse communities, such as tem-perate and polar ones, are thought to be less stable and often well below their maximal sizes (opportunistic populations of Chapter 5). As a result, portions of the community are often unsaturated with individuals and intraspecific and interspecific competition are frequently relatively lax. In such communities it is often the physical world rather than the biotic one that demands adaptation. Selection for competitive ability is weak, whereas selection for rapid reproduction (r selection) is strong. Such r-selected populations typically have broad tolerance limits and relatively large niches.

9. RAREFACTION. Rarefaction is the continual density-independent removal of organisms from a community (see pp. 284–285). The rarefaction hypothesis is essentially an alternative to the competition hypothesis; these two mechanisms would seem to be mutually exclusive. In communities that are not fully saturated with individuals, competition is reduced and coexistence is possible *without competitive exclusion*. Thus this hypothesis suggests that communities (or portions of communities) can in some sense be oversaturated with species in that more species coexist than would be possible if the system were allowed to become truly saturated with individuals. Rarefaction may operate through primary-, secondary-, or tertiary-level mechanisms (see also Predation). Disturbances such as catastrophic winter cold snaps and subsequent density-independent kills (see Table 5.3) illustrate primary-level rarefaction.

10. PREDATION. By either selective or random removal of individual prey organisms, predators can act as rarefying agents and effectively reduce the level of competition among their prey. Indeed, as described in Chapter 6 (pp. 213–214), predators can allow the local coexistence of species that are eliminated by competitive exclusion in the absence of the predator. Because many predators prey preferentially upon more abundant prey types, predation is often frequency dependent, which promotes prey diversity.

Clearly several of these mechanisms may often act together to determine the diversity of a given community, and the relative importance of each mechanism doubtless varies widely from community to community. A multitude of various possible ways in which these mechanisms could interact have been suggested; two are shown in Figure 8.15.

Tree Species Diversity in Tropical Rain Forests

In the lowland tropics, between 50 and 100 different species of trees usually occur together on a single hectare. While tree species diversity is thus exceedingly high,* many of these trees are phenotypically almost identical with broad evergreen leaves and smooth bark (indeed, only an expert can distinguish among the majority of species).

---

* Such high plant species diversity, of course, fosters diversification of animal species, especially insects and birds.

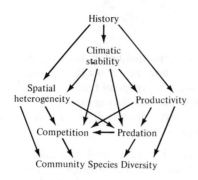

(a)

Rate of Production of New Species

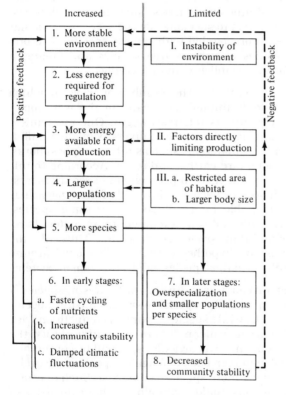

(b)

Figure 8.15 Two ways in which various mechanisms might interact to determine community diversity. (a) Simple schematic diagram showing possible interactions of hypothetical processes. [From Pianka (1971a).] (b) A cybernetic model from Connell and Orias (1964) proposed for the production and regulation of diversity, which includes several of the ten proposed mechanisms in Table 8.1.

Many species are quite rare with densities below one tree per hectare. How can so many similar species, apparently all light limited, coexist at such low densities? Explaining why tropical tree diversity is so high is among the most challenging questions facing ecologists. Numerous hypotheses have been proposed but relevant data on this fascinating phenomenon remain distressingly scant.

SEED PREDATION HYPOTHESIS. Because seed predation is intense in the tropics, Janzen (1970) argued that seedlings cannot establish themselves in the vicinity of parental trees since the high densities of seeds attract many seed predators (see Figure 6.28, p. 226). This argument predicts that successful recruitment of seeds to seedlings will occur in a ring around (but at some distance from) the parental tree. Inside and outside this ring, other tree species can establish themselves. The variety of seed protection tactics (such as toxic matrices) has forced many seed predators to specialize on the seeds of particular species. Heavy seed predation coupled with species-specific seed predators holds down densities of various tree species and creates a mosaic of conditions for the establishment of seedlings.

NUTRIENT MOSAIC HYPOTHESIS. The number of ways in which plants can differ is decidedly limited, especially in the wet tropics where variation in soil moisture is relatively slight. One mechanism that could help to maintain high plant species diversity involves differentiation in the use of various materials, such as nitrogen, phosphorus, potassium, calcium, various rare earths, and so on. According to this argument, each tree species has its own peculiar set of requirements; the soil underneath each species becomes depleted of those particular resources making it unsuitable for seedlings of the same species. (Eventually after the tree falls and is decomposed, these materials re-enter the nutrient pool and that species grows again.) Thus, like the seed predation hypothesis, this hypothesis predicts a "shadow" around a parent tree where seedlings of that species will be rare or nonexistent.

CIRCULAR NETWORKS HYPOTHESIS. In this mechanism, species A is envisioned as being competitively superior to species B while species B in turn excludes species C, while species C wins in competition with species A. Under such a circular hierarchy of competitive ability, the identity of the species occurring at a particular spot will be continually changing from A to B to C and then back to A again repeating the cycle. Circular networks with many more species could exist. Such nontransitive competitive interactions could maintain the high diversity of tropical trees.

DISTURBANCE HYPOTHESES. Frequent disturbances by fires, floods, and storms might interrupt the process of competitive exclusion locally and allow maintenance of high diversity (Connell, 1977). D. Strong (pers. comm.) has developed a variant on this hypothesis involving epiphytes as agents of disturbance (many more epiphytes occur in the tropics than in the temperate zones); Strong suggests that tree falls (due to epiphyte loads) are frequent in the tropics, continually opening up patches in the forest and fostering local secondary succession.

The preceding hypotheses barely begin to address the question of why hardly any temperate forests support more than a dozen species of trees. If there are more species-specific seed predators in the tropics, why? Why should nutrient differentiation be more pronounced in the tropics? Why doesn't the circular network mechanism foster higher diversity in temperate forests? Are disturbances more frequent in the tropics and if so, why? Ultimately latitudinal variation in any of the previously proposed mechanisms will have to be related to underlying variation in physical variables such as climate.

Community Stability

Numerous concepts of stability can be and have been applied to communities (for reviews, see Lewontin, 1969; Holling, 1973; and/or Orians, 1975). All too often the term "stability" is left vague and undefined. Without going into great detail, theorists claim various relationships between community diversity and stability. Traditional ecological "wisdom" holds that more diverse communities are in some sense more stable than simpler ones.* MacArthur (1955) suggested that the stability of populations in a community should increase both with the number of different trophic links between species and with the equitability of energy flow up the various food chains. He reasoned that a community with many trophic links provides greater possibilities for checks and balances to operate between and among various species' populations: Should any one population begin to increase markedly, its predators, by changing their diets and feeding selectively upon this abundant prey type, would exert disproportionate effects on its population growth.

May (1973) challenged the "commonsense" generalization that diversity begets stability, noting that mathematical stability decreases with complexity in randomly constructed model communities. However, Lawlor (1977) points out that real communities are not constructed randomly but must obey various constraints: there can be no more than five to seven trophic levels, no three-species food chain "loops" can occur, there must be at least one producer in the system, and so on. Lawlor argues that it is extremely unlikely that *any* of May's

---

* This argument is often used to "explain" why our traditional method of agriculture, the monoculture of stands of a single species of plant, tends to lead to unstable ecological systems. Pest populations easily grow at an exponential rate and spread rapidly through the fields (little wonder we depend upon massive applications of pesticides to control such pests!).

random systems even remotely resemble real ecological systems, and he asserts that the relationship between complexity and stability in large-scale model ecosystems remains unresolved.

Empirical studies have also produced conflicting results (for a review, see Goodman, 1975). Watt (1968) found that herbivorous insect species that feed on a wide variety of tree species in Canada actually have *less* stable populations than do similar insects with more restricted diets. Watt does not indicate whether insects with stable populations have a greater variety of potential predators, but population stability did increase with increasing numbers of competing species as would be expected. Clearly the relationship between diversity and stability remains an important but unresolved problem in community ecology.

Evolutionary Convergence and Ecological Equivalence

Organisms evolving independently of one another under similar environmental conditions have sometimes responded to similar selective pressures with nearly identical adaptations. Thus flightless birds such as the emu, ostrich, and rhea fill very similar ecological niches on different continents. Arid regions of South Africa support a wide variety of euphorbeaceous plants, some of which are strikingly close to American cacti phenotypically. A bird of some African prairies and grasslands, the African yellow-throated longclaw *(Macronix croceus)*, looks and acts so much like an American meadowlark *(Sturnella magna)* that a competent bird watcher might mistake them for the same species; and yet they belong to different avian families (Figure 8.16a). Such convergent phenotypic responses by different stocks of plants or animals are known as evolutionary convergence. Products of convergent evolution, organisms that have evolved independently and yet occupy roughly similar niches in various communities in different parts of the world, are known as *ecological equivalents*. More striking examples of evolutionary convergence (Figure 8.16) usually fall into either or both of two categories: They sometimes occur in relatively simple communities in which biotic interactions are highly predictable and the resulting number of different ways of exploiting the environment are few, and/or they occur under unusual conditions where selective forces for achievement of a particular mode of existence are particularly strong. Examples of the latter include the independently evolved marsupial and placental "saber-toothed tigers" (Figure 8.16f) and the fusiform shapes of sharks, ichthyosaurs, and dolphins. Evolutionary convergence can easily be read into a situation by placing undue emphasis upon superficial similarities but failing to appreciate fully the inevitable dissimilarities between pairs of supposed ecological equivalents.

Figure 8.16 Examples of convergent evolution in animals. Pairs of independently evolved but ecologically similar species which occupy similar niches in different communities are known as "ecological equivalents." (*a*) A tropical Asian cyprinid fish, *Rasbora* sp. (above), and an African characin, *Neolobias* sp. (*b*) An Australian agamid lizard, *Amphibolurus cristatus* (left), and a North American iguanid, *Callisaurus draconoides*. (*c*) An African civet (left) and an American weasel. (*d*) Another Australian agamid lizard, *Moloch horridus* (left), and another North American iguanid, *Phrynosoma coronatum*. (*e*) An Australian marsupial, a wombat (left), with its skull and an American placental, a woodchuck, with its skull. (*f*) Skulls of two fossil (but not contemporary) saber-tooth carnivores, the South American marsupial "cat," *Thylacosmilus* (left), and the North and South American placental saber-toothed tiger, *Smilodon*. (*g*) An American icterid, the Eastern Meadowlark, *Sturnella neglecta* (above), and an African motacillid, the yellow-throated longclaw, *Macronix croceus*. (*h*) A North American little auk (above) and a Magellan diving petrel, which belong to two different avian orders. [*a*, *c*, *e*, *f* after Salthe (1972), *Evolutionary Biology*. Copyright © 1972 by Holt, Rinehart and Winston, Inc. Reprinted by permission of Holt, Rinehart and Winston, Inc. *b*, *d* after Pianka (1971a). *g*, *h* after Fisher and Peterson (1964).]

Often roughly similar ecological systems support relatively few conspicuous ecological equivalents but instead are composed largely of distinctly different plant and animal types. For instance, although the bird species diversities of temperate forests in eastern North America and eastern Australia are similar (Recher, 1969, and pp. 284–286), many avian niches appear to be fundamentally different on the two continents. Honeyeaters and parrots are conspicuous in Australia while hummingbirds and woodpeckers are entirely absent. Apparently, different combinations of the various avian ecological activities are possible; thus an Australian honeyeater might combine aspects of the food and place niches exploited in North America by both warblers and hummingbirds. An analogy can be made by comparing the "total avian niche space" to a deck of cards: There are a limited number of ways this niche space can be exploited, and each bird population or species has its own ways of doing things, or its own "hand of cards," determined in part by what other species in the community are doing.

Community Evolution

Many communities change during the lifetimes of the individuals that comprise them. In addition to relatively short-term changes during ecological time, community characteristics are affected by evolution of and coevolution among the species' populations that are available to form the community over evolutionary time. At the same time, the community itself is a major determinant of the selective milieu of its component populations, and its characteristics presumably dictate many of their adaptations. The so-called taxon cycle (p. 326) is thought to be driven by biotic responses to competition and predation, or "counter adaptations" of the other species in a community (Ricklefs and Cox, 1972). Competition within, between, and among species results in the evolution of niche differences, which in turn assures that the resources of a given community, including plants and animals, are utilized more or less in proportion to their effective supply (Chapters 6 and 7). As pointed out earlier in this chapter, evolution of the species within a community has still other effects upon community structure. Evolution of prey reduces the efficiency of transfer of energy from one trophic level to the next but increases stability, whereas evolution of predators acts to increase the efficiency of this transfer but reduces stability. The diversity of prey eaten by a predator as well as the predator's ability to alter its diet with changes in prey availability probably influences the stability of prey populations, and therefore of the community.

Can natural selection operate between entire communities? The notions of selection at the levels of communities and ecosystems (Dunbar, 1960, 1968; Lewontin, 1970) constitute apparent extremes of the idea of group selection (pp. 12–13). Selection is quite unlikely to occur at these levels, in view of both the limited number of communities and ecosystems and their low rate of turnover. Most important, selection acts only by *differential reproduction* (pp. 9–12), and it is most difficult to envision reproduction by a community or an ecosystem. Organisms comprising a community are not bound together by obligate relations; instead each evolves in a manner independent of, and often antagonistic to, other members of the community, such as its prey, competitors, and predators. Indeed, community stability may even be incompatible with efficient transfer of energy to higher trophic levels because of the antagonistic interactions between predators and their prey.

Selected References

Food Webs and Trophic Levels

Allee *et al.* (1949); Elton (1927, 1949, 1966); Gallopin (1972); Hairston, Smith, and Slobodkin (1960); Hubbell (1973a, 1973b); Kozlovsky (1968); Murdoch (1966a); Odum (1959, 1963, 1971); Paine (1966); Phillipson (1966).

The Community Matrix

Levins (1968); May (1973); Neill (1974); Parker and Turner (1961); Seifert and Seifert (1976); Vandermeer (1970, 1972a, 1972b).

Principles of Thermodynamics

Bertalanffy (1957); Brody (1945); Gates (1965); Odum (1959, 1971); Paine (1971); Phillipson (1966); Wiegert (1968).

Pyramids of Energy, Numbers, and Biomass

Elton (1927); Kormondy (1969); Leigh (1965); Odum (1959, 1963, 1971); Phillipson (1966); Slobodkin (1962).

Energy Flow and Ecological Energetics

Bertalanffy (1969); Bormann and Likens (1967); Engelmann (1966); Gates (1965); Golley (1960); Hairston and Byers (1954); Hubbell (1971); Lindemann (1942); Mann (1969); Margalef (1963, 1969); Odum (1959, 1963, 1968, 1969, 1971); Paine (1966, 1971); Patten (1959); Phillipson (1966); Reichle (1970); Schultz (1969); Slobodkin (1960, 1962); Teal (1962).

Systems Ecology

Bertalanffy (1969); Bormann and Likens (1967); Caswell *et. al.* (1972); Chorley and Kennedy (1971); Clark *et al.* (1967); Dale (1970); Foin (1972); Forrester

(1971); Holling (1959a, 1959b, 1963, 1964, 1965, 1966); Hubbell (1971, 1973a, 1973b); Huffaker (1971); H. Odum (1971); Patten (1971, 1972, 1975, 1976); Reichle (1970); Van Dyne (1966); Waterman (1968); Watt (1966, 1968, 1973).

Saturation with Individuals and with Species

Cody (1970, 1973); Levins (1968); MacArthur (1965, 1970, 1971, 1972); MacArthur and MacArthur (1961); Pianka (1966a, 1973); Recher (1969); Vandermeer (1972a); Whittaker (1969, 1972).

Species Diversity

Arnold (1972); Baker (1970); Connell and Orias (1964); Fischer (1960); Fisher, Corbet, and Williams (1943); Futuyma (1973); Gleason (1922); Harper (1969); Hutchinson (1959); Janzen (1971a); Johnson, Mason, and Raven (1968); Klopfer (1962); Klopfer and MacArthur (1960, 1961); Lack (1945); Leigh (1965); Loucks (1970); MacArthur (1960a, 1965, 1972); MacArthur and MacArthur (1961); MacArthur, MacArthur, and Preer (1962); MacArthur, Recher, and Cody (1966); Margalef (1958a, 1958b, 1963, 1968); Menge and Sutherland (1976); Murdoch *et al.* (1972); Odum (1969); Orians (1969a); Paine (1966); Patten (1962); Pianka (1966a, 1973, 1975); Pielou (1975); Poulson and Culver (1969); Preston (1948, 1960, 1962a, 1962b); Recher (1969); Ricklefs (1966); Schoener (1968a); Schoener and Janzen (1968); Shannon (1948); Simpson (1949); Simpson (1969); F. E. Smith (1970a, 1970b, 1972); Tramer (1969); Vandermeer (1970); Watt (1973); Whiteside and Hainsworth (1967); Whittaker (1965, 1969, 1970, 1972); Whittaker and Feeny (1971); Williams (1944, 1953, 1964); Woodwell and Smith (1969).

TREE SPECIES DIVERSITY IN TROPICAL RAIN FORESTS

Black *et al.* (1950); Cain (1969); Connell (1977); Dobzhansky (1950); Eggeling (1947); Gilpin (1975b); Janzen (1970); Jones (1956); Richards (1952); Ricklefs (1977).

Community Stability

Frank (1968); Futuyma (1973); Goodman (1975); Hairston *et al.* (1968); Harper (1969); Holling (1973); Hurd *et al.* (1971); Lawlor (1977); Leigh (1965); Lewontin (1969); Loucks (1970); MacArthur (1955, 1965); Margalef (1969); May (1971, 1973, 1975); Milsum (1973); Murdoch (1969); Orians (1975); Peterson (1975); F. E. Smith (1972); Sutherland (1974); Usher and Williamson (1974); Watt (1964, 1965, 1968, 1973); Whittaker (1972); Woodwell and Smith (1969).

Evolutionary Convergence and Ecological Equivalence

Grinnell (1924); MacArthur and Connell (1966); Raunkaier (1934); Recher (1969); Salthe (1972).

Community Evolution

Darlington (1971); Dunbar (1960, 1968, 1972); Futuyma (1973); Kormondy (1969); Lewontin (1970); Odum (1969); Ricklefs and Cox (1972); Whittaker (1972); Whittaker and Woodwell (1971).

Biogeography 9

A major goal of ecology is to understand various factors influencing the distribution and abundance of animals and plants (Andrewartha and Birch, 1954; Krebs, 1972; MacArthur, 1972). Factors affecting abundances and microgeographic distributions (including habitat selection) were considered in earlier chapters; here we examine more gross geographic distributions—the spatial distributions of plants and animals over large geographic areas such as major land masses (continents and islands). The study of the gross geographical distributions of plants and animals, respectively, are termed phytogeography and zoogeography. Biogeography encompasses the geography of *all* organisms and involves a search for patterns in the distributions of plants and animals and an attempt to explain how such patterns arose during the geological past. In addition to classifying present distributions, biogeographers seek to interpret and to understand past movements of organisms. Ecology and biogeography are closely related and overlapping disciplines, and they have profoundly affected one another.

Classical Biogeography

When early naturalists traveled to different parts of the world they soon discovered distinctly different assemblages of species; as data were gathered on these patterns, six major biogeographic "realms" or regions were recognized, three of which correspond roughly to the continents of Australia (Australian), North America north of the Mexican escarpment (Nearctic), and South America south of the Mexican escarpment (Neotropical). (The Neotropical region also includes the Antilles.) Africa south of the Sahara is known as the Ethiopian region. Eurasia is divided into two regions, the Palearctic north of the Himalayas (which includes Africa north of the Sahara) and the Oriental south of the Himalayas (India, southern China, Indochina, the Phillipines, and Borneo, Java, Sumatra, and other islands of Indonesia east to, and including, the Celebes). Each of the six biogeographic regions (Figure 9.1) is separated from the others by a major barrier to the dispersal of plants and animals, such as a narrow isthmus, high mountains, a desert, an ocean, or an oceanic strait. There is generally a high degree of floral and faunal consistency within regions and a marked shift in higher taxa such as genera and families in going from one region to another. Although biogeographers familiar with different plant and animal groups often disagree on the exact boundaries between regions (Figure 9.2), there is broad agreement on the usefulness of recognizing these six major regions.

High species diversities in the tropics (see pp. 290–291), among other things, have led to the notion that speciation rates in these areas must be extremely high and that such regions are often "source areas" for production of new species, many of which then migrate into less hospitable areas such as the temperate zones. Thus Darlington (1957, 1959) proposed the "area–climate hypothesis," which states that the majority of dominant animal species have arisen in geographically extensive and climatically favorable areas; he considers the Old World Tropics, which includes the tropical portions of the Ethiopian and Oriental regions, the major source area for most vertebrate groups, and he argues that such dominant forms have migrated centrifugally to other smaller and less favorable areas, including Europe, North and South America, and Australia.

Much of this classical biogeography assumed some permanence in the locations of continents; as a result interpretations of faunal similarities between them often rely on hypothetical mechanisms of

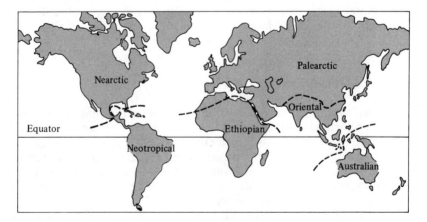

Figure 9.1 The six major biogeographic regions of the world.

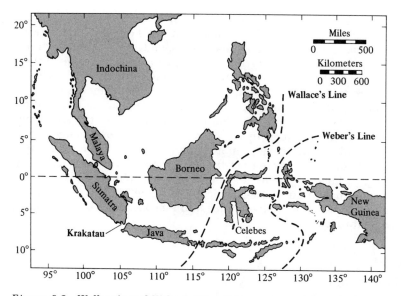

Figure 9.2   Wallace's and Weber's "lines" in southeast Asia, which separate the Oriental from the Australian regions. The position of the volcanic island of Krakatau between Sumatra and Java is indicated at lower left.

transport from one continent to another, such as "rafting" of organisms across water gaps. Recent advances in geology have forced a partial revision of this view. There is now strong evidence (Dietz and Holden, 1970) that the continents were once joined in a large southern landmass (Pangaea) and have "drifted" apart (see also pp. 37–40), with a gradual breakup that began in the early Mesozoic era (about 200 million years ago). Geological evidence that the continents have drifted (and are now drifting) is accumulating rapidly (J. T. Wilson, 1973). Much of classical biogeography will have to be reinterpreted in light of new findings; for example, certain very ancient groups of freshwater lungfishes, amphibians, and insects that had spread before the breakup of the continents now occur on several continents, whereas many other more recently evolved groups of plants and animals, such as mammals and birds, are restricted to a particular biogeographic region. These latter, more recent groups follow the regional divisions much more closely than the older groups (Kurtén, 1969).

Classical biogeography has produced several so-called biogeographic rules based on recurring patterns of adaptation of organisms. Thus homoiotherms living in cold climates are often larger than those from warmer regions; such a trend or *cline* can even be demonstrated within some wide-ranging species. This tendency, termed Bergmann's rule after its discoverer, has a probable causal basis in that large animals have less surface per unit of body volume than small ones (see p.

78), resulting in more efficient retention of body heat. Many other biogeographic rules have also been proposed, all of which are basically descriptive. Allen's rule states that the appendages and/or extremities of homoiotherms are either longer or have a larger area in warmer climates; a jackrabbit, for example, has much longer and larger ears than an arctic hare. The presumed functional significance is that large appendages, having a larger relative surface area, are better heat dissipaters than smaller ones. Another rule (Gloger's) asserts that animals from hot, dry areas tend to be paler than those from colder, wetter regions. Still another biogeographic rule is that fish from cold waters often have more vertebrae than those from warmer waters. The adaptive significance of many of these biogeographic trends remains obscure, although such geographically variable phenotypic traits are frequently developmentally flexible and respond more or less directly to temperature.

Island Biogeography

Ecosystems are usually very difficult to manipulate experimentally; hence much of modern ecology has had to rely on exploitation of "natural" experiments—situations in which one (or a few) factor(s) affecting a community differ between two (or more) ecosystems. For this reason ecologists have long been especially interested in islands, which constitute some of the finest natural ecological experiments. Different islands in an archipelago often contain different combinations of the mainland species, allowing an investigator to observe both ecological and evolutionary responses, such as niche shifts, of various component species to the presence or absence of other species (Figure 9.3). Islands can be exploited as natural ecological experiments in numerous other ways as well. Thus, because islands support fewer predatory species than comparable mainland habitats, they can be used to study the effects of predator exclusion (see also p. 138). Moreover, reduced species densities on islands, such as the land birds of Bermuda (see p. 246), allow partial analysis of the effects of interspecific competition on the ecologies of those species which have populated an island.

Islands of a sort are widespread in the terrestrial landscape as well; a patch of forest separated from a larger stand of trees can be considered a "habitat island." Similarly, isolated lakes and mountain tops (see p. 197) represent "islands." To a nonflying insect, plants in the desert or trees within an open forest may approximate islands in that they are separated from one another by relatively vast open spaces of a different and relatively inhospitable environment. Likewise, cattle droppings scattered about a field are islands to the animals that inhabit them (Mohr, 1943). Even a teaspoon of water or the body of an insect may be an island to a bacterium.

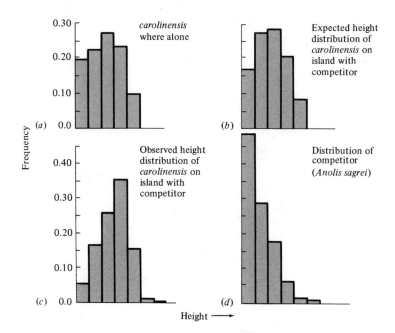

Figure 9.3 Observed and expected frequency distributions of perch heights of *Anolis* lizards. (*a*) Observed height distribution of *A. carolinensis* where it occurs on an island alone without competitors. (*b*) Expected distribution of perch heights of *A. carolinensis* on another island with different availabilities of various perch heights, assuming no niche shift. (*c*) Observed distribution of *A. carolinensis* on the second island with a competitor (compare with *b*). (*d*) Height distribution of the competing species, *A. sagrei*. [From Schoener (1975a).]

Species–Area Relationships

Larger islands generally support more species of plants and animals than smaller ones. In fact, when plotted on a double log scale, the number of species in a given taxon typically increases more or less linearly with island size (Figure 9.4). In most cases, a tenfold increase in area corresponds to an approximate doubling of the number of species. The slope of a linear regression line through such points is designated as that taxon's z-value in the particular island system. In a variety of taxa on many different island systems (Table 9.1), z-values generally range from about 0.24 to about 0.33. The z-value is the exponent in the equation

$$S = CA^z \tag{1}$$

where S is the number of species, C is a constant which varies between taxa and from place to place, and A is the area of the island(s) con-

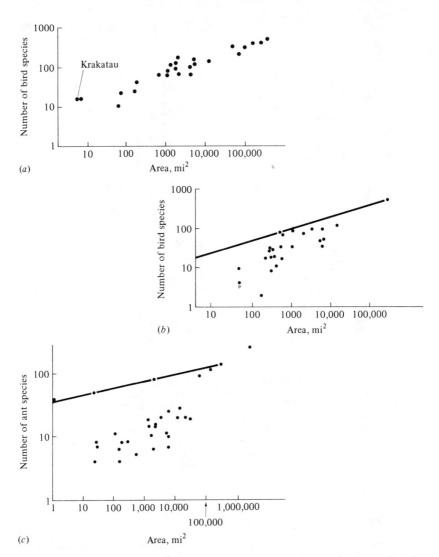

Figure 9.4 Various species–area relationships. (*a*) Numbers of species of land and freshwater birds on islands of the Sunda group in southeast Asia (Figure 9.2), with the Phillipines and New Guinea. Krakatau is plotted at the extreme left. (*b*) Numbers of species of land and freshwater birds on various often remote islands in the south Pacific, including the Moluccas, Melanesia, Micronesia, Polynesia, and Hawaii. The line is drawn through the two islands (Kei and New Guinea) nearest to source regions to demonstrate the degree of departure of species densities on the more remote islands. (*c*) Number of species of ponerine ants in the faunas of various Molluccan and Melanesian islands. The line represents the number of species with increasing areas in subsamples of New Guinea; points represent smaller islands. Note that the islands support fewer species than a comparable sized portion of New Guinea but that the *rate* of increase of species with area is greater among the islands than it is within New Guinea. [(*a, b*) from MacArthur and Wilson (1967). *The Theory of Island Biogeography.* Copyright © 1967 by Princeton University Press. Reprinted by permission of Princeton University Press. (*c*) from Wilson (1961).]

Table 9.1   Estimated z-Values for Various Terrestrial Plants and
Animals on Different Island Groups

Fauna or Flora	Island Group	z
Carabid beetles	West Indies	0.34
Ponerine ants	Melanesia	0.30
Amphibians and reptiles	West Indies	0.301
Breeding land and fresh-water birds	West Indies	0.237
Breeding land and fresh-water birds	East Indies	0.280
Breeding land and fresh-water birds	East-Central Pacific	0.303
Breeding land and fresh-water birds	Islands of Gulf of Guinea	0.489
Land vertebrates	Islands of Lake Michigan	0.239
Land plants	Galápagos Islands	0.325

Source: From MacArthur and Wilson (1967). *The Theory of Island Biogeography.* Reprinted by
permission of Princeton University Press. Copyright © 1967 by Princeton University Press.

cerned. Taking logarithms and rearranging, one obtains a linear equation in which z is the slope:

$$\log S = \log C + z \cdot \log A \tag{2}$$

Large values of z result from topographic diversity and spatial replacement of species, or "islands within islands"; lower values arise with reduced replacement of species in space, as on very homogeneous islands, continents, or subsamples of large islands (see below). Area in itself is probably not the primary factor affecting species density in most situations, but it presumably operates indirectly through increasing the variety of available habitats. Area can, however, directly affect species densities in some situations.

An area of mainland habitat comparable to and equal in size to an offshore island almost invariably supports more species, especially those at higher trophic levels, than does the island. The number of species in samples of a continental system also increases with the size (area) of the subsample, although not as rapidly as on islands (Figure 9.4). Typically, z-values in mainland situations range from about 0.12 to about 0.17. This difference arises because an island is a true "isolate" while a similar sized patch of mainland habitat is only a "sample"; rare species can occur in the mainland sample both due to migration from other areas and because areas immediately adjacent to the subsample also support other members of broad-ranging species. A mountain lion requiring a 20-square-kilometer territory would be unlikely to maintain a viable population on a small island of, say, less than 30 to 40 square kilometers, whereas these same cats are able to survive and replace themselves in a similar sized subsample of a larger landmass. This is why islands tend to support fewer species at higher trophic levels than mainland areas.

Equilibrium Theory

For many years islands were considered to be in some sense "impoverished" with species both because of the obvious problems species have in colonizing them and because islands typically support fewer species than a comparable area of mainland habitat. Recently, however, the regularity of species–area patterns led MacArthur and Wilson (1963, 1967) to examine the possibility that islands might in fact be supporting as many species as possible.

MacArthur and Wilson reasoned that the rate of immigration of *new* species to an island should decrease as the number of species on that island increases. Immigration rate must drop to zero as the species density of the island reaches the total number of species in the "species pool" available for colonization of the island at which no immigrant can be a new species. (The species pool corresponds to the total number of species in source areas surrounding the particular island system.) MacArthur and Wilson argue that the rate of extinction of species already present on an island should *increase* as the number of species on an island increases; this seems likely because, as more species invade an island, average population size must decrease and both the intensity of interspecific competition and the incidence of competitive exclusion should increase. Moreover, there are more species to go extinct.

When rate of immigration equals rate of extinction (Figure 9.5),

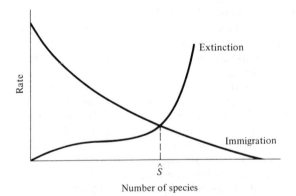

Number of species

Figure 9.5 Illustration of the equilibrium theory for island species densities, with the immigration rate of new species falling and the rate of extinction of existing species rising as the total number of species on an island increases. At equilibrium, immigration just balances extinction and $\hat{S}$ different species exist on the island. The composition of the island's biota may change as some of the existing species go extinct and are replaced by other, different, species.

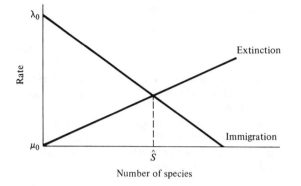

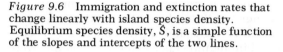

Figure 9.6   Immigration and extinction rates that
change linearly with island species density.
Equilibrium species density, $\hat{S}$, is a simple function
of the slopes and intercepts of the two lines.

existing species go extinct at the same rate that new ones invade; thus
species density reaches a dynamic equilibrium. Although species density stays constant, the continual turnover of species means that the
species composition of the island can be changing.

MacArthur and Wilson's equilibrium theory is somewhat analogous to the model of the Verhulst-Pearl logistic equation for growth
processes within a population (Chapter 5); thus number of individuals
(density), N, is replaced by the number of species (species density), S,
and density-dependent birth and death rates, b_N and d_N, are replaced by
a falling immigration rate, λ, and a rising rate of extinction, μ, as the
species density of an island increases. As a first approximation, we
might assume that rates of immigration (λ) and extinction (μ) vary
linearly with species density according to the equations

$$\lambda_s = \lambda_0 - \alpha S \tag{3}$$

$$\mu_s = \beta S \tag{4}$$

where λ_0 is the rate of immigration with no species on the island and α
and β represent rates of change in rates of immigration and extinction,
respectively, as species density increases (Figure 9.6). (MacArthur and
Wilson point out that this assumption of linearity is not as stringent as
it might at first seem, since transformations of the ordinate may allow
simultaneous straightening of immigration and extinction curves.) At
equilibrium, or $\hat{S}$, the rate of immigration must exactly equal rate of
extinction—that is, λ_s must equal μ_s. Setting equation (3) equal to (4),

$$\lambda_0 - \alpha \hat{S} = \beta \hat{S} \tag{5}$$

and rearranging, one obtains an expression for the number of species at equilibrium

$$\hat{S} = \frac{\lambda_0}{\alpha + \beta} \tag{6}$$

Equation (6) is of course identical in form to the expression for carrying capacity, K, in the logistic equation, which is $K = r/(x + y)$ [see also equation (24) in Chapter 5, p. 118].

At equilibrium, total rate of immigration of species must equal the total extinction rate. However, because species going extinct will undoubtedly often differ from those that successfully invade an island, the *composition* of an island's biota will be continually changing, even at equilibrium.

As developed above, λ_x and μ_x represent total rates of immigration and extinction and thus indicate little about the relative rates *per species* either already present on the island or available in the species pool (P). The average rate of immigration per species, $\bar{\lambda}$, and the average rate of extinction per species, $\bar{\mu}$, can be obtained by dividing by, respectively, the number of species not yet on the island $(P - S)$ and the number already present on the island (S):

$$\bar{\lambda} = \frac{\lambda_x}{P - S} \quad \text{or} \quad \lambda_x = \bar{\lambda}(P - S) \tag{7}$$

$$\bar{\mu} = \frac{\mu_x}{S} \quad \text{or} \quad \mu_x = \bar{\mu}S \tag{8}$$

Again, at equilibrium, total extinction rate (μ_x) must equal the total rate of immigration (λ_x), that is $\lambda_x = \mu_x$, or in terms of the average rates per species (which are the rates with which an ecologist will usually be working):

$$\bar{\lambda}(P - \hat{S}) = \bar{\mu}\hat{S} \tag{9}$$

Solving for the equilibrium number of species, $\hat{S}$, gives

$$\hat{S} = \frac{\bar{\lambda}P}{\bar{\mu} + \bar{\lambda}} \tag{10}$$

Equation (10) demonstrates that $\hat{S}$ increases with increasing P and $\bar{\lambda}$ and decreases with increased $\bar{\mu}$. Notice also that $\bar{\lambda}P$ is λ_0 [compare equation (10) with equation (6)] and that $\bar{\lambda}$ is identical to α in equation (3), whereas μ is β in equation (4).

Because dispersal falls off more or less exponentially with distance (Figure 9.7), MacArthur and Wilson reasoned that immigration rates should decrease with increasing distance from source areas (Figure 9.8). Further, they argued that rates of extinction should be largely unaffected by distance from source areas per se but should instead generally increase with decreasing island size because smaller islands support smaller, more tenuous, populations (Figure 9.9). [Because they present a smaller "target" for potential invaders, smaller islands might

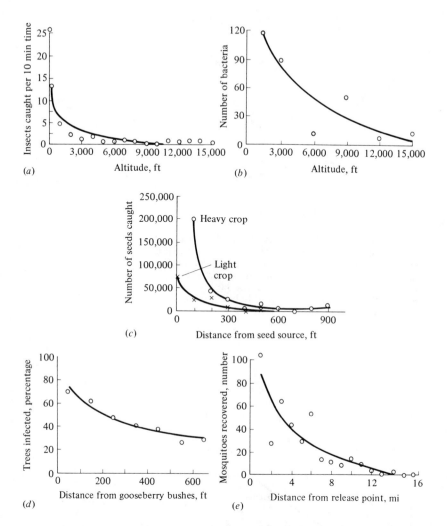

Figure 9.7 Some actual patterns of dispersal, both vertical and horizontal. The number of organisms decays rapidly at first and then more and more slowly with increasing distance. [From Odum (1959) after Wolfenbarger.]

also have slightly lower immigration rates than other equivalent but larger islands. But this change should be minor compared with the expected decline due to the exponential decay in the number of immigrants with distance—(Figure 9.7).] Note also that, in islands equidistant from source areas, rates of species turnover should be higher on small islands than on large ones (turnover rate should thus vary inversely with equilibrium species density). Simple Islands with little

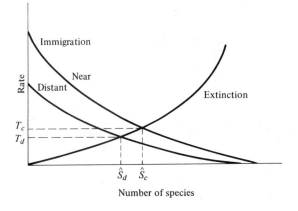

Figure 9.8 Immigration rates should decrease with increasing distance from source areas so that distant islands should reach equilibrium with fewer species, $\hat{S}_d$, than close-in islands, $\hat{S}_c$, all else being equal. Moreover, turnover rates should also be higher on nearby islands than on comparable but more distant islands ($T_c > T_d$).

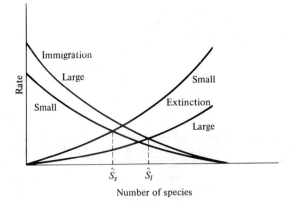

Figure 9.9 Extinction rates should be little affected by distances from source areas, but they should often vary inversely with island size and/or complexity. Immigration rates may also be slightly higher on larger islands because they present a larger "target" for potential invaders. Thus, all else being equal, a small island should equilibrate with fewer species, $\hat{S}_s$, than a larger island, $\hat{S}_l$.

topographic relief and relatively few different habitats should have generally higher extinction rates than more complex and more diverse islands with a greater variety of habitats, because the latter would provide a greater variety of immigrants with suitable opportunities for successful invasion and persistence on the island. Finally, clumped islands such as archipelagos should have higher rates of immigration than more scattered or isolated islands due to interchanges of plants and animals between islands.

Some predictions of equilibrium theory have now been supported by facts; others, especially those involving turnover rates, have proven to be very difficult to test.

The Compression Hypothesis

When faced with more intense competition from another species, many organisms restrict their utilization of shared microhabitats and/or other resources (see Chapters 6 and 7). These adjustments are those that take place in ecological time, during the lifetime of the organism concerned. The fact of such niche contractions, coupled with theoretical considerations, has led to the so-called compression hypothesis (MacArthur and Wilson, 1967), which states that, as more species invade a community, place niches are compressed while food niches either remain constant or expand (Figure 9.10). Any prey item worth eating should be acceptable no matter what the intensity of competition, but an animal must choose the places it forages on the basis of its *expectation* of yield, which will usually be markedly decreased in some patches of habitat by heightened competition (see also pp. 264–266). Thus the compression hypothesis predicts that during short-term, nonevolutionary time habitats used should shrink with increased competition while the range of foods eaten should expand or remain the same. Moreover, should competitors reduce overall levels of available foods more or less equally among all patches on a species' itinerary, food niche expansion will be favored.

The Morphological Variation–Niche Breadth Hypothesis

A species can be a generalist in two ways: (1) A population can contain a variety of different phenotypes, each using a smaller range of resources than the overall population, and/or (2) each individual in a population can itself be relatively flexible and generalized, with the resources utilized by any individual being similar to those exploited by the entire population. Roughgarden (1972) refers to these two components of niche breadth as the between-phenotype and within-phenotype components of niche breadth (see Figure 7.12). Phenotypic variability within a population, by allowing different phenotypes to exploit different resources, should thus increase the overall range of resources exploited by the population. Moreover, by reducing niche

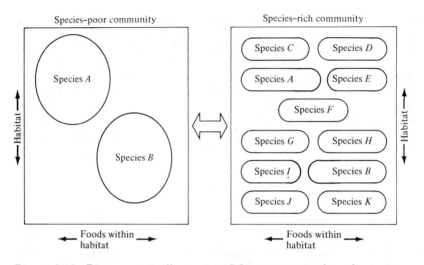

Figure 9.10 Diagrammatic illustration of the compression hypothesis. As more species invade a habitat, interspecific competition forces any single species to decrease the range of habitats it exploits; however, the range of foods eaten should either increase or remain more or less constant. Conversely, if a species invades a habitat that is impoverished with species, reduced interspecific competition should often allow an expansion in the variety of habitats (or microhabitats) exploited. The compression hypothesis applies only in ecological time and does not address evolutionary niche shifts. [After MacArthur and Wilson (1967). *The Theory of Island Biogeography.* Copyright © 1967 by Princeton University Press. Reprinted by permission of Princeton University Press.]

overlap among members of a population, between-phenotype niche expansion might be expected to reduce the average degree of inter-phenotypic competition.

Island species, freed from some interspecific competition, often tend to exploit a wider range of habitats than do mainland species—a phenomenon often referred to as "ecological release." Van Valen (1965) reasoned that reduced competition from other species should also favor an increased morphological variability because this would promote niche expansion. He postulated that island species should often be morphologically more variable than their mainland counterparts; moreover, Van Valen found evidence for just such an increased phenotypic variability in five out of six species of birds known to have broader niches on certain islands. Grant (1967), however, found less morphological variation (in length of wing, tail, tarsus, and bill) in some Mexican insular bird populations than in mainland ones. In a later study, Grant (1971) found no consistent trends in tarsal length variation between mainland and island populations of Mexican birds. He speculates that, under spatially uniform environmental conditions, selection favors little variation among individuals in feeding ecology and associated morphology by acting strongly against individuals that depart widely from the average phenotype, and that, under spatially varied (patchy) conditions, the opposite may be true.

Soulé and Stewart (1970) restate and somewhat reverse the hypothesis: generalized, broad-niched species should be phenotypically and morphologically more variable than more specialized and more narrow-niched species (they call this the niche-variation hypothesis). Soulé and Stewart, however, were unable to find any evidence that generalized African bird species such as crows are in fact morphologically more variable than more specialized species. Van Valen and Grant (1970) point out that the broad niche of these crows could well be primarily due to within-phenotype flexibility in resource utilization and that one would not necessarily expect great morphological variability in such a situation.

Also of interest here is the fact that, although tropical species are often considered to be more specialized than temperate species, bill dimensions of some tropical birds are at least as variable as those of some North Temperate species (Willson, 1969). Clearly much more information is needed before the relationships between morphological variation and niche breadth can be adequately assessed.

The Gene Flow–Variation Hypothesis

Among island populations of the lizard *Uta stansburiana* in the Gulf of California, both genetic and morphologic variability are greater on larger islands than on smaller ones (Soulé, 1971, and Figure 9.11).

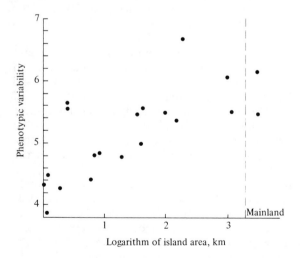

Figure 9.11   An index of overall variation in eight phenotypic characters of the lizard *Uta stansburiana* plotted against the logarithm of the area of 18 islands in the Gulf of California and 2 mainland areas. Variation increases with area. [After Soulé (1971).]

To explain these results, Soulé proposed an alternative to the niche-variation hypothesis. The gene flow–variation hypothesis states that gene flow between habitats with differing selective milieus generates genetic variability in spatially intermediate populations (see also p. 166). An important distinction between the niche-variation hypothesis and the gene flow-variation hypothesis is that the former predicts that increased genotypic and phenotypic variation is adaptive, whereas the latter suggests that it is nonadaptive and merely an unavoidable byproduct of environmental heterogeneity and movement of organisms. Not enough is known at present to assess the importance of gene flow in maintaining variability of natural populations.

Islands as Ecological Experiments: Some Examples

Darwin's Finches

The Galápagos Islands, an archipelago of relatively small and remote, deep water, volcanic islands located about 600 miles west of the Ecuadoran coast (Figure 9.12), support a remarkable group of birds that nicely illustrate a number of evolutionary and ecological principles. Named Darwin's finches after the first evolutionist to appreciate and study them, these birds dominate the avifauna of the Galápagos. Only 26 species of land birds occurred in the archipelago naturally (i.e., before human introductions), and 13 of these are finches (the islands also support 4 species of mockingbirds, 2 flycatchers, 2 owls, a hawk, a dove, a cuckoo, a warbler, and a swallow).

This archipelago (16 major islands and a sprinkling of tiny islets) was formed from volcanic eruptions of the ocean floor about a million years ago; thus, originally, there were no organisms on the islands, and their entire biota has been derived from mainland species. Because of their remoteness, relatively few different plant and animal stocks have been able to colonize the Galápagos. (The position of these islands on the equator, however, presumably makes them particularly vulnerable to invasions from rafts and floating islands carried out to sea in the equatorial current.)

The 13 species of finches are thought to have evolved from a single mainland finch ancestor that reached the islands long ago. (The birds are similar enough to each other that they are classified as a distinct subfamily of finches, endemic to the Galápagos and Cocos islands.) Archipelagos are ideal for geographic isolation and speciation, especially in land birds like finches that do not readily fly across wide stretches of water. In such effectively isolated populations, different selective pressures on different islands lead to divergent evolution and adaptations; moreover, occasional interchanges between islands result

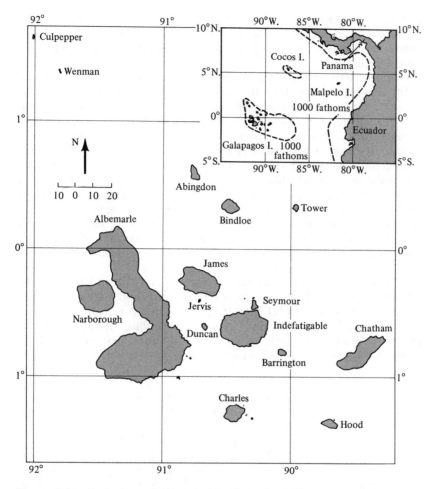

Figure 9.12 Two maps of the main islands in the Galápagos Archipelago.
[Inset from Lack (1947). Larger map from Bowman (1961), originally published
by the University of California Press. Reprinted by permission of the regents of
the University of California.]

in heightened interspecific competition, which promotes niche
diversification.

The necessity of geographic isolation and subsequent interisland
colonization for the occurrence of speciation and adaptive radiation is
nicely demonstrated by the Cocos Island finch, *Pinaroloxias inornata*.
Cocos Island is a remote but solitary island several hundred miles
north of the Galápagos and about the same distance from the mainland
(Figure 9.12, inset). Although there are a fair variety of habitats on it,
Cocos supports only one species of finch. There has been no opportu-

nity for geographic isolation or reduced gene flow, and the *Pinaroloxias* gene pool has never split. One would predict that this finch should be a generalist, probably with a high degree of phenotypic variability between individuals.

Adaptive radiation of these finches in the Galápagos has produced 3 different genera that differ in where they forage, how they forage, and what they eat. So-called ground finches *(Geospiza)* include 6 ground-foraging species with broad beaks which eat seeds of different sizes and types as well as flowers of *Opuntia* cactus. The genus *Camarhynchus*, termed tree finches because they forage in trees, also contains 6 species with generally somewhat narrower beaks; one species is a vegetarian, while the other 5 eat different sized insects in different ways (one of these, *Camarhynchus pallidus*, the "woodpecker finch," uses sticks and catus spines to probe cracks and crevices for insects, much like a woodpecker uses its long pointed tongue). One very distinctive and monotypic genus, the so-called warbler finch, *Certhidea olivacea*, occurs on almost all the islets and islands in the archipelago and breeds throughout most habitats.

From 3 to 10 species of finches occur on any given island (Table 9.2) in various combinations. Beak lengths and depths are highly variable from island to island (Figure 9.13), presumably reflecting different environmental conditions among islands, including interspecific competitive pressures. Indeed, Figure 9.13 illustrates character displacement in beak depths; the tiny islets of Crossman and Daphne support

Table 9.2 Distributions of Darwin's Finches on Various Major Islands in the Galápagos Archipelago

Species	Abingdon	Albemarle	Barrington	Bindloe	Charles	Chatham	Culpepper	Duncan	Hood	Indefatigable	James	Jervis	Narborough	Seymour	Tower	Wenman
Geospiza magnirostris	×	×	×	×	×	—	×	×	—	×	×	×	×	×	×	×
Geospiza fortis	×	×	×	×	×	×	—	×	—	×	×	×	×	×	—	—
Geospiza fuliginosa	×	×	×	×	×	×	—	×	×	×	×	×	×	×	—	×
Geospiza difficilis	×	—	—	—	—	—	—	—	—	×	×	—	×	—	×	×
Geospiza scandens	×	×	×	×	×	×	—	×	—	×	×	×	—	×	—	—
Geospiza conirostris	—	—	—	—	—	—	×	—	×	—	—	—	—	—	×	—
Camarhynchus crassirostris	×	×	—	×	×	×	—	×	—	×	×	×	×	—	—	—
Camarhynchus psittacula	×	×	×	×	×	—	—	×	—	×	×	×	×	—	—	—
Camarhynchus pauper	—	—	—	—	×	—	—	—	—	—	—	—	—	—	—	—
Camarhynchus parvulus	×	×	×	—	×	×	—	×	—	×	×	×	×	×	—	×
Camarhynchus pallidus	—	—	—	—	—	×	—	×	—	×	×	×	—	×	—	—
Camarhynchus heliobates	—	×	—	—	—	—	—	—	—	—	—	—	×	—	—	—
Certhidea olivacea	×	×	×	×	×	×	×	×	×	×	×	×	×	×	×	—
Total number of species per island	9	10	7	7	9	7	4	9	3	10	10	9	9	8	4	5

Source: After Bowman (1961). Originally published by the University of California Press. Reprinted by permission of The Regents of the University of California.

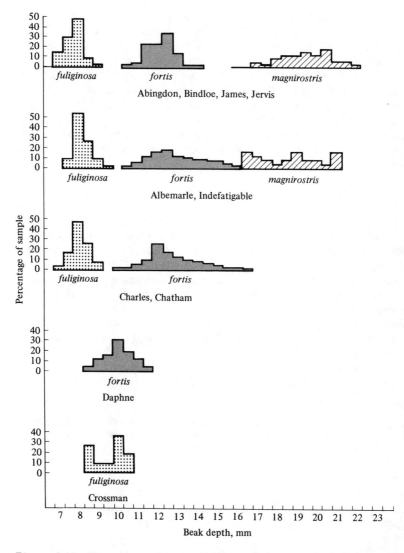

Figure 9.13 Histograms of the beak depths of several species of Darwin's finches, genus *Geospiza*, on different islands. In allopatry on the islets of Daphne and Crossman *G. fortis* and *G. fuliginosa* have beaks of very similar size, whereas in sympatry (upper three sets of histograms) their beak depths are entirely nonoverlapping. [From Lack (1947).]

only one member of a pair of very similar species, either *Geospiza fuliginosa* or *G. fortis*, respectively. On these two small islands, beaks of both species are nearly the same size (about 8.5 to 11 mm deep), whereas on larger islands where the two species occur together in sympatry (Abingdon, Bindloe, James, Jervis, Albemarle, Indefatigable,

Charles, and Chatham—upper part of Figure 9.13) beak depths are
completely nonoverlapping with *fuliginosa* having a small beak (about
7 to 9 mm deep) and *fortis* a larger beak (about 10 to 14 or 16 mm
deep). Beak dimensions, of course, determine in large part the size of
the food items the birds eat (beaks are also used in species recognition).

Larger islands in the Galápagos archipelago contain a greater variety of habitat types and as a result support more species of finches than
do smaller islands. Moreover, the total number of finch species decreases with "average isolation," or the mean distance from other islands, while the number of endemic species increases with isolation
(Hamilton and Rubinoff, 1963, 1967).

Krakatau

In 1883 the small volcanic island of Krakatau, located between
Java and Sumatra (see Figure 9.2), erupted repeatedly over a three-
month period. All of Krakatau and two adjacent islands were covered
with red-hot lava, pumice, and ash to a depth of 100 feet or more. The
islands were so hot that months afterward falling rain turned to steam
on contact. It is most unlikely that any organisms survived. Repopulation from adjacent Sumatra (about 15 miles away) and Java proceeded
rapidly and by 1921 the number of resident species of birds was comparable to that expected on a small island of eight square miles (the
size of Krakatau after the eruptions) in the general region (see Figure
9.4). Total number of bird species did not change much between 1921
and 1933, although the composition of the avifauna did (Table 9.3).

Table 9.3 Number of species of Land and Freshwater Birds on Krakatau and
Verlaten During Three Collection Periods, and the Number of
Species "Lost" Between Intervals

	1908			1919–1921		
	Non-migrant	Migrant	Total	Non-migrant	Migrant	Total
Krakatau	13	0	13	27	4	31
Verlaten	1	0	1	27	2	29

	1932–1934			Number "lost"	
	Non-migrant	Migrant	Total	1908 to 1919–1921	1919–1921 to 1932–1934
Krakatau	27	3	30	2	5
Verlaten	29	5	34	0	2

Source: From MacArthur and Wilson (1967) after Dammerman.

This example suggests that mobile organisms like birds rapidly reach an equilibrium species density. Plant species, on the other hand, were still being added rapidly at last count in 1934 (Figure 9.14). Clearly it is time for another expedition to census Krakatau's current flora and fauna.

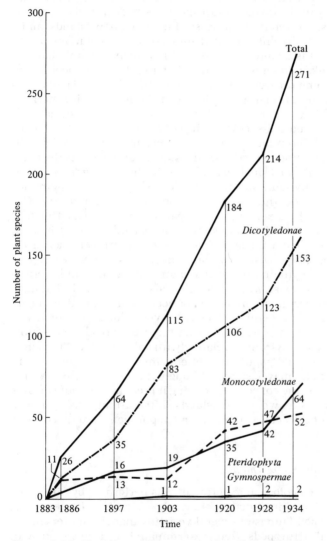

Figure 9.14   Numbers of plant species recorded on the three islands of the Krakatau group from 1883 through 1934. [From MacArthur and Wilson (1967). *The Theory of Island Biography.* Copyright © 1967 by Princeton University Press. Reprinted by permission of Princeton University Press.]

The Taxon Cycle

Many island species are thought to progress through a series of evolutionary changes, termed a taxon cycle, that may eventually greatly increase their probability of going extinct (Wilson, 1961; MacArthur and Wilson, 1967; Ricklefs and Cox, 1972). Under this hypothesis, early in the taxon cycle a species is widespread and occurs on many islands, is often in the process of invading new islands, and is only slightly (if at all) differentiated into distinct populations on the various islands. It is adapted to marginal, relatively unstable habitats such as riverbanks and forest clearings. Later in the cycle, populations of a species become progressively more and more differentiated on different islands, at first remaining widespread. At this stage it penetrates more stable habitats, such as old forests, where it must coexist with larger numbers of native species. Still later, after local extinction on some islands, a differentiated species becomes more restricted and its geographic range is fragmented. Finally, species at the end of the taxon cycle are found only on a single island (that is, they are *endemic* to one island). Occasionally some species are able to shift back into marginal, species-poor habitats, thus restarting the cycle. Ricklefs and Cox (1972) argue that the taxon cycle for a particular species is driven by counteradaptations of other members of an island biota against the species concerned. There is a shift from r selection to K selection as the taxon cycle progresses. Dispersal ability decreases as species depend less and less on the ability to colonize marginal habitats and more on their capacity to coexist with competing species in stable habitats. Such reduced dispersal and more pronounced local adaptation in turn favors speciation and endemism. Newly arrived colonists are relatively free of a counteradaptive load, allowing them to spread successfully throughout an island system in relatively unstable habitats. On small and remote islands, old populations of endemics (p. 197), such as the Cocos Island finch, may persist. Although the taxon cycle may not apply to all species, the concept underlying it presumably could operate in mainland faunas as well as on islands. Little attempt has yet been made to interpret the ecology of mainland populations in terms of such counteradaptations.

A Defaunation Experiment

An interesting ecological experiment was performed by Simberloff and Wilson (1970 and included references). After carefully censusing the entire arthropod faunas of several very tiny mangrove islets in the Florida keys, all arthropods were exterminated by fumigation with methyl bromide. The process of recolonization was then monitored over a two-year period. Arthropods rapidly recolonized; within a mere 200 days numbers of species on the islets had stabilized (Figure 9.15). Although turnover rates remained quite high throughout the exper-

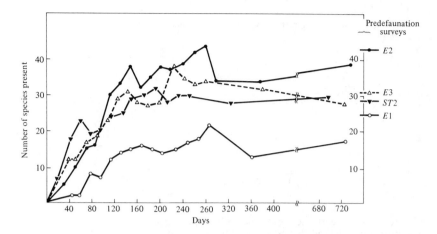

Figure 9.15   Colonization curves (number of arthropod species through time) of four tiny mangrove islets in the Florida keys following complete extermination of the arthropod fauna but relatively little damage to the vegetation. Numbers of species on each islet first rose rapidly, then soon leveled off to species densities fairly close to those before defaunation (plotted on the right vertical axis). [After Simberloff and Wilson (1969). By permission of Duke University Press.]

imental period, numbers of species on the islets remained relatively constant over a period of nearly two years, providing strong evidence that the islands have indeed reached an equilibrium. Two islands, $E1$ and $E2$, seem to have reached equilibrium at a slightly lower species density after defaunation. The apparent depression in the number of species at equilibrium may indicate that the species which reinvaded these islands interfere with one another more than the members of the original communities did; alternatively, this reduction in species density may be due to the fact that new immigrants are not as well adapted to exploit the island's resources as were the original inhabitants. In any case, these results demonstrate that the actual composition of an island's fauna may itself partially determine the equilibrium number of species an island will support.

Wilson (1969) suggests that an island community may experience a sequence of several distinct sorts of equilibria through time (Figure 9.16). First, a "non-interactive" equilibrium in the number of species present may be reached even before component populations come to equilibrium demographically and with one another. Then, because competitive and predatory interactions are intensified as populations saturate the island with individuals, a second "interactive" equilibrium is reached. Both stages should occur relatively rapidly. Wilson envisions two other types of equilibria, which would require considerably greater time spans to attain. As various species go extinct and

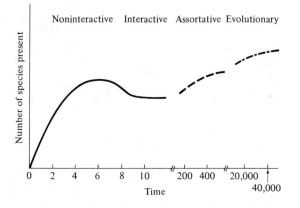

Figure 9.16 Hypothetical sequence of community equilibria on an island through time. The time scale is arbitrary and intended only to emphasize the vastly greater time spans required for the assortative and evolutionary equilibria (see text.) [From Wilson (1969).]

others invade, the composition of an island's biota may gradually change until a certain set of species from the available species pool is reached that is composed of species with particularly low extinction rates; Wilson terms this the "assortative" equilibrium. Finally, given a much longer time span, the component species could actually evolve minimal extinction rates and an "evolutionary" equilibrium might be reached (Wilson, 1969).

Applied Biogeography: Design of Nature Preserves

Tall grass prairie covered hundreds of thousands of square kilometers in the midwestern United States a few hundred years ago; today this natural community has virtually disappeared. Lowland tropical rain forest is now being destroyed at an alarming rate. Natural communities of all sorts are rapidly being replaced by overgrazed pastures, eroded fields, artificial lakes, golf courses, roads, parking lots, and housing developments. None of Earth's natural communities remain pristine; all have been disturbed either with pesticides and other pollutants or by way of introductions and extinctions of species. Even the disturbed remnants of Earth's biomes are continually being broken up into smaller and smaller isolated patches or habitat islands (Figure 9.17). As would be expected from the equilibrium theory of island biogeography, faunal and floral diversity are decreasing in such isolates as species go extinct locally (some species, such as the passenger pigeon and probably the ivory-billed woodpecker, have been entirely eradicated). Larger species of animals at higher trophic levels disappear before smaller species and those at lower trophic levels. Unfortunately, much remains to be learned about these vanishing natural communities and their inhabitants.

Biogeographic principles can be used profitably in designing natural preserves to protect endangered habitats and species. Assume that

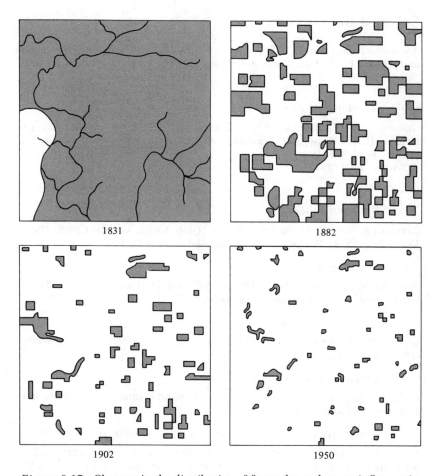

Figure 9.17 Changes in the distribution of forest due to human influence in the past century and a half. The total area depicted, in Wisconsin, is about 10 km on a side. Fragmentation of the forest has created many very small habitat islands. [From J. T. Curtis, "The Modification of Mid-Latitude Grasslands and Forests by Man," in W. L. Thomas, Jr., ed., *Man's Role in Changing the Face of the Earth*. Copyright © 1956 by The University of Chicago.]

it is desirable to maintain as great a diversity of plants and animals as possible. Clearly a single large and contiguous reserve will generally be superior to a number of smaller reserves covering an equivalent area. All else being equal, protected areas should be as diverse as possible. Further, the ratio of edge to area should be minimized. Provision of dispersal corridors or "stepping stones" of natural habitat between larger reserves enhances migration and increases diversity (a species can go extinct in one preserve but reinvade from an adjacent one).

Selected References

Andrewartha and Birch (1954); Krebs (1972); MacArthur (1972); MacArthur and Wilson (1967); Udvardy (1969); Watts (1971).

Classical Biogeography

Cracraft (1974); Dansereau (1957); Darlington (1957, 1959, 1965); Dietz and Holden (1970); Hesse, Allee, and Schmidt (1951); Kurtén (1969); MacArthur (1959); Newbigin (1936); Terborgh (1971); Udvardy (1969); Wallace (1876); J. T. Wilson (1971, 1973).

Island Biogeography

Carlquist (1965); MacArthur and Wilson (1936, 1967); Maguire (1963, 1971); Simberloff (1974); Wilson (1969); Wilson and Bossert (1971).

SPECIES–AREA RELATIONSHIPS

Gleason (1922, 1929); Krebs (1972); MacArthur and Wilson (1967); May (1975); Odum (1959, 1971); Preston (1948, 1960, 1962a, 1962b).

EQUILIBRIUM THEORY

Brown (1971); MacArthur and Wilson (1963, 1967); Simberloff (1974); Wilson (1969); Wilson and Bossert (1971).

THE COMPRESSION HYPOTHESIS

Crowell (1962); MacArthur (1972); MacArthur and Pianka (1966); MacArthur, Diamond, and Karr (1972); Schoener (1974b).

THE MORPHOLOGICAL VARIATION–NICHE BREADTH HYPOTHESIS

Grant (1967, 1971); Orians (1974); Rothstein (1973); Roughgarden (1972); Soulé and Stewart (1970); Van Valen (1965); Van Valen and Grant (1970); Willson (1969).

THE GENE FLOW–VARIATION HYPOTHESIS

Ehrlich and Raven (1969); Ford (1964); Gilbert and Singer (1973); Levins (1964); Soulé (1971); Wilson and Bossert (1971).

Islands as Ecological Experiments: Some Examples

Carlquist (1965); MacArthur (1972); MacArthur and Wilson (1967).

DARWIN'S FINCHES

Bowman (1961); Hamilton and Rubinoff (1963, 1967); Lack (1947).

KRAKATAU

Dammerman (1948); Docters van Leeuwen (1936); MacArthur and Wilson (1967).

THE TAXON CYCLE

MacArthur and Wilson (1967); Ricklefs and Cox (1972); Wilson (1961).

A DEFAUNATION EXPERIMENT

Simberloff and Wilson (1970); Wilson (1969).

Applied Biogeography: Design of Nature Preserves

Terborgh (1974a, 1974b); Wilson and Willis (1975).

Bibliography

ABRAMS, P. 1975. Limiting similarity and the form of the competition coefficient. *Theoret. Pop. Biol.* 8:356–375.

ABRAMS, P. 1976. Niche overlap and environmental variability. *Math. Biosci.* 28:357–372.

ALCOCK, J. 1975. *Animal behavior. An evolutionary approach.* Sinauer, 547 pp.

ALEXANDER, R. D. 1974. The evolution of social behavior. *Ann. Rev. Ecol. Syst.* 5:325–383.

ALLEE, W. C. 1951. *Cooperation among animals with human implications.* Schuman, New York (Revised Edition of *Social Life of Animals*, Norton, New York, 1938). 233 pp.

ALLEE, W. C., A. E. EMERSON, O. PARK, T. PARK, and K. P. SCHMIDT. 1949. *Principles of animal ecology.* Saunders, Philadelphia. 837 pp.

ANDERSON, W. W. 1971. Genetic equilibrium and population growth under density-regulated selection. *Amer. Natur.* 105:489–498.

ANDREWARTHA, H. G. 1961. *Introduction to the study of animal populations.* Methuen, London. 281 pp.

ANDREWARTHA, H. G. 1963. Density dependence in the Australian thrips. *Ecology* 44:218–220.

ANDREWARTHA, H. G., and L. C. BIRCH. 1953. The Lotka-Volterra theory of interspecific competition. *Aust. J. Zool.* 1:174–177.

ANDREWARTHA, H. G., and L. C. BIRCH. 1954. *The distribution and abundance of animals.* University of Chicago Press, Chicago. 782 pp.

ARNOLD, S. J. 1972. Species densities of predators and their prey. *Amer. Natur.* 106:220–236.

ASHMOLE, N. P. 1963. The regulation of numbers of tropical oceanic birds. *Ibis* 103b:458–473.

AYALA, F. J. 1968. Genotype, environment, and population numbers. *Science* 162:1453–1459.

AYALA, F. J., M. E. GILPIN, and J. G. EHRENFELD. 1973. Competition between species: theoretical models and experimental tests. *Theoret. Pop. Biol.* 4:331–355.

BAILEY, I. W., and E. W. SINNOTT. 1916. The climatic distribution of certain types of angiosperm leaves. *Amer. J. Bot.* 3:24–39.

BAKER, H. G. 1970. Evolution in the tropics. *Biotropica* 2:101–111.

BAKER, J. R. 1938. The evolution of breeding systems. In *Evolution, essays presented to E. S. Goodrich.* Oxford Univ. Press, London.

BARTHOLOMEW, G. A. 1972. Body temperature and energy metabolism. Chapter 8 (pp. 298–368) M. S. Gordon (Ed.), *Animal physiology: Principles and Adaptations.* Macmillan, New York.

BARTLETT, M. S. 1960. *Stochastic population models in ecology and epidemiology.* Methuen, London. 90 pp.

BARTLETT, P. N., and D. M. GATES. 1967. The energy budget of a lizard on a tree trunk. *Ecology* 48:315–322.

BEARD, J. S. 1955. The classification of tropical American vegetation types. *Ecology* 36:89–100.

BEAUCHAMP, R. S. A., and P. ULLYOTT. 1932. Competitive relationships between certain species of fresh-water triclads. *J. Ecol.* 20:200–208.

BENSON, S. B. 1933. Concealing coloration among some desert rodents of the southwestern United States. *Univ. Calif. Publ. Zool.* 40:1–70.

BENSON, W. W. 1972. Natural selection for Müllerian mimicry in *Heliconius erato* in Costa Rica. *Science* 176:936–939.

BENSON, W. W., K. S. BROWN, JR., and L. E. GILBERT. 1975. Coevolution of plants and herbivores: Passion flower butterflies. *Evolution* 29:659–680.

BERNAL, J. D. 1967. *The origin of life.* World, Cleveland. 345 pp.

BERTALANFFY, L. 1957. Quantitative laws in metabolism and growth. *Quart. Rev. Biol.* 32:217–231.

BERTALANFFY, L. (Ed.). 1969. *General systems theory: foundations, development, applications.* Braziller, New York. 290 pp.

BEVERTON, R. J. H., and S. J. HOLT. 1957. On the dynamics of exploited fish populations. *Great Brit. Min. Agr. Fish, Food, Fish. Invest. Ser. 2,* 19:1–533.

BILLINGS, W. D. 1964. *Plants and the ecosystem.* Wadsworth, Belmont, Calif.

BIRCH, L. C. 1948. The intrinsic rate of natural increase of an insect population. *J. Anim. Ecol.* 16:15–26.

BIRCH, L. C. 1953. Experimental background to the study of the distribution and abundance of insects. III. The relations between innate capacity for increase and survival of different species of beetles living together on the same food. *Evolution* 7:136–144.

BIRCH, L. C. 1957. The meanings of competition. *Amer. Natur.* 91:5–18.

BIRCH, L. C., and P. R. EHRLICH. 1967. Evolutionary history and population biology. *Nature* 214:349–352.

BLACK, C. A. 1968. *Soil-plant relationships.* (2nd ed.) Wiley, New York. 792 pp.

BLACK, G. A., TH. DOBZHANSKY, and C. PAVAN. 1950. Some attempts to estimate the species diversity and population density of trees in Amazonian forests. *Bot. Gaz.* 111:413–425.

BLAIR, T. A., and R. C. FITE. 1965. *Weather elements.* Prentice-Hall, Englewood Cliffs, N.J.

BLAIR, W. F. 1960. *The rusty lizard. A population study.* Univ. Texas Press, Austin. 185 pp.

BLIGH, J. 1973. *Temperature regulation in mammals and other vertebrates.* Elsevier, North Holland. 436 pp.

BLUM, H. F. 1968. *Time's arrow and evolution.* Princeton Univ. Press, Princeton, N.J. 232 pp.

BLUMENSTOCK, D. I., and C. W. THORNTHWAITE. 1941. Climate and the world pattern. pp. 98–127. In *Climate and man*. U.S. Department of Agriculture Yearbook. Washington, D.C. 1248 pp.

BOGUE, D. J. 1969. *Principles of demography*. Wiley, New York.

BONNER, J. T. 1965. *Sze and cycle: an essay on the structure of biology*. Princeton Univ. Press, Princeton, N.J. 219 pp.

BOORMAN, S. A., and P. R. LEVITT. 1972. Group selection on the boundary of a stable population. *Proc. Nat. Acad. Sci., USA* 69:2711–2713.

BOORMAN, S. A., and P. R. LEVITT. 1973. Group selection at the boundary of a stable population. *Theoret. Pop. Biol.* 4:85–128.

BORMANN, F. H., and G. E. LIKENS. 1967. Nutrient cycling. *Science* 155:424–429.

BOTKIN, D. B., and R. S. MILLER. 1974. Mortality rates and survival of birds. *Amer. Natur.* 108:181–192.

BOVBJERG, R. V. 1970. Ecological isolation and competitive exclusion in two crayfish *(Orconectes virilis* and *Orconectes immunis)*. *Ecology* 51:225–236.

BOWMAN, R. L. 1961. Morphological differentiation and adaptation in the Galápagos finches. *Univ. Calif. Publ. Zoology*. Vol. 58. Univ. of Calif. Press, Berkeley. 326 pp.

BRAUN-BLANQUET, J. 1932. *Plant sociology: the study of plant communities* (translated and edited by G. D. Fuller and H. C. Conard). McGraw-Hill, New York. 439 pp.

BRIAN, M. V. 1956. Exploitation and interference in interspecies competition. *J. Anim. Ecol.* 25:339–347.

BROCKELMAN, W. Y. 1975. Competition, the fitness of offspring, and optimal clutch size. *Amer. Natur.* 109:677–699.

BROCKELMAN, W. Y., and R. M. FAGEN. 1972. On modeling density-independent population change. *Ecology* 53:944–948.

BRODY, S. 1945. *Bioenergetics and growth*. Van Nostrand Reinhold, New York. 1023 pp.

BROOKS, G. R., JR. 1967. Population ecology of the ground skink, *Lygosoma laterale* (Say). *Ecol. Monogr.* 37:71–87.

BROWER, L. P. 1969. Ecological chemistry. *Sci. Amer.* 220(Mar.):22–29.

BROWER, L. P., and J. BROWER. 1964. Birds, butterflies, and plant poisons: a study in ecological chemistry. *Zoologica* 49:137–159.

BROWN, J. H. 1971. Mammals on mountaintops: nonequilibrium insular biogeography. *Amer. Natur.* 105:467–478.

BROWN, J. H., and C. R. FELDMETH. 1971. Evolution in constant and fluctuating environments: thermal tolerances of desert pupfish *(Cyrpinodon)*. *Evolution* 25:390–398.

BROWN, J. H., and R. C. LASIEWSKI. 1972. Metabolism of weasels: the cost of being long and thin. *Ecology* 53:939–943.

BROWN, J. L. 1964. The evolution of diversity in avian territorial systems. *Wilson Bull.* 76:160–169.

BROWN, J. L. 1966. Types of group selection. *Nature* 211:870.

BROWN, J. L. 1969. Territorial behavior and population regulation in birds. *Wilson Bull.* 81:293–329.

BROWN, J. L. 1975. *The evolution of behavior*. Norton, New York. 761 pp.

BROWN, J. L., and G. H. ORIANS. 1970. Spacing patterns in mobile animals. *Ann. Rev. Ecol. Syst.* 1:239–262.

BROWN, W. L., and E. O. WILSON. 1956. Character displacement. *Syst. Zool.* 5:49–64.

BURGES, A., and F. RAW (Eds.). 1967. *Soil biology.* Academic Press, New York. 532 pp.

BYERS, H. G. 1954. The atmosphere up to 30 kilometers. In G. P. Kuiper (Ed.), *The earth as a planet.* Univ. Chicago Press, Chicago.

CAIN, A. J. 1969. Speciation in tropical environments: summing up. *Biol. J. Linn. Soc.* 1:233–236.

CAIN, S. A. 1950. Life-forms and phytoclimate. *Bot. Rev.* 16:1–32.

CALDWELL, M. M., and O. A. FERNANDEZ. 1975. Dynamics of great basin shrub root systems. Pp 38–51 in N. F. Hadley (Ed.), *Environmental physiology of desert organisms.* Halstead, New York. 283 pp.

CALVIN, M. 1969. *Chemical evolution.* Oxford Univ. Press, New York. 278 pp.

CARLQUIST, S. 1965. *Island life: a natural history of the islands of the world.* Natural History Press, Garden City, N.Y.

CARPENTER, C. R. 1958. Territoriality: a review of concepts and problems. Pp. 224–250 in A. Roe and G. G. Simpson (Eds.), *Behavior and evolution.* Yale Univ. Press, New Haven. 557 pp.

CASE, T. J., and M. E. GILPIN. 1974. Interference competition and niche theory. *Proc. Nat. Acad. Sci., USA* 71:3073–3077.

CASWELL, H., H. E. KOENIG, J. A. RESH, and Q. E. ROSS. 1972. An introduction to systems science for ecologists. Pp. 4–78 in B. Patten (Ed.), *Systems analysis and simulation in ecology.* Vol. II. Academic Press, New York. 592 pp.

CASWELL, H., F. REED, S. N. STEPHENSON, and P. A. WERNER. 1973. Photosynthetic pathways and selective herbivory: a hypothesis. *Amer. Natur.* 107:465–480.

CATES, R. G. 1975. The interface between slugs and wild ginger: some evolutionary aspects. *Ecology* 56:391–400.

CATES, R. G., and G. H. ORIANS. 1975. Successional status and the palatability of plants to generalized herbivores. *Ecology* 56:410–418.

CAUGHLEY, G. 1966. Mortality patterns in mammals. *Ecology* 47:906–918.

CHAMBERS, K. L. (Ed.). 1970. *Biochemical coevolution.* 29th Biology Colloquium, Oregon State Univ. Press, Eugene. 117 pp.

CHARLESWORTH, B. 1971. Selection in density-regulated populations. *Ecology* 52:469–474.

CHARNOV, E. L. 1973. Optimal foraging: some theoretical explorations. Ph. D. dissertation. Univ. of Washington, Seattle. 95 pp.

CHARNOV, E. L. 1976a. Optimal foraging: attack strategy of a mantid. *Amer. Natur.* 110:141–151.

CHARNOV, E. L. 1976b. Optimal foraging: the marginal value theorem. *Theoret. Pop. Biol.* 9:129–136.

CHARNOV, E. L., and J. R. KREBS. 1973. On clutch size and fitness. *Ibis* 116:217–219.

CHARNOV, E. L., and J. R. KREBS. 1975. The evolution of alarm calls: altruism or manipulation? *Amer. Natur.* 109:107–112.

CHARNOV, E. L., G. H. ORIANS, and K. HYATT. 1976. Ecological implications of resource depression. *Amer. Natur.* 110:247–259.

CHITTY, D. 1960. Population processes in the vole and their relevance to general theory. *Canad. J. Zool.* 38:99–113.

CHITTY, D. 1967a. The natural selection of self-regulatory behavior in animal populations. *Proc. Ecol. Soc. Australia* 2:51–78.

CHITTY, D. 1967b. What regulates bird populations? *Ecology* 48:698–701.

CHORLEY, R. J., and B. A. KENNEDY. 1971. *Physical geography: a systems approach.* Prentice-Hall, London. 370 pp.

CHRISTIAN, J. J., and D. E. DAVIS. 1964. Endocrines, behavior, and population. *Science* 146:1550–1560.

CLAPHAM, W. B. 1973. *Natural ecosystems.* Macmillan, New York. 248 pp.

CLARK, L. R., P. W. GEIER, R. D. HUGHES, and R. F. MORRIS. 1967. *The ecology of insect populations in theory and practice.* Methuen, London. 232 pp.

CLARKE, B. C. 1972. Density-dependent selection. *Amer. Natur.* 106:1–13.

CLARKE, G. L. 1954. *Elements of ecology.* Wiley, New York. 560 pp.

CLEMENTS, F. E. 1920. *Plant succession: an analysis of the development of vegetation.* Publication No. 290. Carnegie Institute, Washington, D.C. 388 pp.

CLEMENTS, F. E. 1949. *Dynamics of vegetation.* Hafner, New York. 296 pp.

CLOUDSLEY-THOMPSON, J. L. 1971. *The temperature and water relations of reptiles.* Merrow. 159 pp.

CODY, M. L. 1966. A general theory of clutch size. *Evolution* 20:174–184.

CODY, M. L. 1968. On the methods of resource division in grassland bird communities. *Amer. Natur.* 102:107–147.

CODY, M. L. 1970. Chilean bird distribution. *Ecology* 51:455–463.

CODY, M. L. 1971. Ecological aspects of reproduction. Chapter 10 (pp. 461–512) in D. S. Farner and J. R. King (Eds.), *Avian Biology.* Vol. I. Academic Press, New York. 586 pp.

CODY, M. L. 1974. *Competition and the structure of bird communities.* Princeton Univ. Press, Princeton, N.J. 318 pp.

COLE, G. A. 1975. *Textbook of limnology.* Mosby, St. Louis. 283 pp.

COLE, L. C. 1951. Population cycles and random oscillations. *J. Wildl. Manage.* 15:233–251.

COLE, L. C. 1954a. Some features of random cycles. *J. Wildl. Manage.* 18:2–24.

COLE, L. C. 1954b. The population consequences of life history phenomena. *Quart. Rev. Biol.* 29:103–137.

COLE, L. C. 1958. Sketches of general and comparative demography. *Cold Spring Harbor Symp. Quant. Biol.* 22:1–15.

COLE, L. C. 1960. Competitive exclusion. *Science* 132:348–349.

COLE, L. C. 1965. Dynamics of animal population growth. Pp. 221–241 in M. C. Sheps and J. C. Ridley (Eds.), *Public health and population change.* Univ. Pittsburgh Press, Pittsburgh.

COLINVAUX, P. A. 1973. *Introduction to ecology.* Wiley, New York. 621 pp.

COLLIER, B., G. W. COX, A. W. JOHNSON, and P. C. MILLER. 1973. *Dynamic ecology.* Prentice-Hall, Englewood Cliffs, N.J. 563 pp.

COLWELL, R. K. 1973. Competition and coexistence in a simple tropical community. *Amer. Natur.* 107:737–760.

COLWELL, R. K., and E. R. FUENTES. 1975. Experimental studies of the niche. *Ann. Rev. Ecol. Syst.* 6:281–310.

COLWELL, R. K., and D. J. FUTUYMA. 1971. On the measurement of niche breadth and overlap. *Ecology* 52:567–576.

CONNELL, J. H. 1961a. The effects of competition, predation by *Thais lapillus,* and other factors on natural populations of the barnacle *Balanus balanoides. Ecol. Monogr.* 31:61–104.

CONNELL, J. H. 1961b. The influence of interspecific competition and other factors on the distribution of the barnacle, *Chthamalus stellatus*. *Ecology* 42:710–723.

CONNELL, J. H. 1970. A predator–prey system in the marine intertidal region. I. *Balanus glandula* and several predatory species of *Thais*. *Ecol. Monogr.* 40:49–78.

CONNELL, J. 1977. Diversity in tropical rain forests and coral reefs. *Science* 197:in press.

CONNELL, J. H., and E. ORIAS. 1964. The ecological regulation of species diversity. *Amer. Natur.* 98:399–414.

CONNELL, J. H., D. B. MERTZ, and W. W. MURDOCH. 1970. *Readings in ecology and ecological genetics*. Harper & Row, New York. 397 pp.

COTT, H. B. 1940. *Adaptive coloration in animals*. Oxford Univ. Press, London. 508 pp.

COWLES, R. B., and C. M. BOGERT. 1944. A preliminary study of the thermal requirements of desert reptiles. *Bull. Amer. Mus. Natur. Hist.* 83:261–296.

COWLES, R. B., and C. E. BRAMBEL. 1936. A study of the environmental conditions in a bog pond with special reference to the diurnal vertical distribution of *Gonyostomum semen*. *Biol. Bull. Mar. Biol. Lab. Woods Hole* 71:286–298.

CRACRAFT, J. 1974. Continental drift and vertebrate distribution. *Ann. Rev. Ecol. Syst.* 5:215–261.

CROCKER, R. L. 1952. Soil genesis and the pedogenic factors. *Quart. Rev. Biol.* 27:139–168.

CROCKER, R. L., and J. MAJOR. 1955. Soil development in relation to vegetation and surface age at Glacier Bay, Alaska. *J. Ecol.* 43:427–448.

CROMBIE, A. C. 1947. Interspecific competition. *J. Anim. Ecol.* 16:44–73.

CROOK, J. H. 1962. The adaptive significance of pair formation types in weaver birds. *Symp. Zool. Soc. London* 8:57–70.

CROOK, J. H. 1963. Monogamy, polygamy and food supply. *Discovery* (Jan.): 35–41.

CROOK, J. H. 1964. The evolution of social organization and visual communication in the weaver birds (Ploceinae). *Behaviour* 10:1–178.

CROOK, J. H. 1965. The adaptive significance of avian social organization. Pp. 181–218 in P. E. Ellis (Ed.), *Social Organization of animal communities*. Symp. Zool. Soc. London, Vol. 14. Zoological Society of London.

CROOK, J. H. 1972. Sexual selection, dimorphism, and social organization in the primates. Pp. 231–281 in B. G. Campbell (Ed.), *Sexual selection and the descent of man (1871–1971)*. Aldine-Atherton, Chicago.

CROWELL, K. L. 1962. Reduced interspecific competition among the birds of Bermuda. *Ecology* 43:75–88.

CURTIS, J. T. 1956. The modification of mid-latitude grasslands and forests by man. In W. L. Thomas, Jr. (Ed.) *Man's role in changing the face of the earth*. Univ. Chicago Press, Chicago.

DALE, M. B. 1970. Systems analysis and ecology. *Ecology* 51:2–16.

DAMMERMAN, K. W. 1948. The fauna of Krakatau 1883–1933. *Verhandel. Kon-Inkl. Ned. Akad. Wetenschap. Afdel. Natuurk.* 44:1–594.

DANSEREAU, P. 1957. *Biogeography: an ecological perspective*. Ronald, New York. 394 pp.

DARLINGTON, C. D., and K. MATHER. 1949. *The elements of genetics*. Allen and Unwin, London. 446 pp.

DARLINGTON, P. J. 1957. *Zoogeography: the geographical distribution of animals.* Wiley, New York. 675 pp.

DARLINGTON, P. J. 1959. Area, climate, and evolution. *Evolution* 13:488–510.

DARLINGTON, P. J. 1965. *Biogeography of the southern end of the world.* Harvard Univ. Press, Cambridge. 236 pp.

DARLINGTON, P. J. 1971. Nonmathematical models for evolution of altruism, and for group selection. *Proc. Nat. Acad Sci.* 69:293–297.

DARNELL, R. M. 1970. Evolution and the ecosystem. *Amer. Zool.* 10:9–15.

DARWIN, C. 1859. *The origin of species by means of natural selection* (numerous editions). Murray, London.

DARWIN, C. 1871. *The descent of man, and selection in relation to sex* (numerous editions). Murray, London.

DAUBENMIRE, R. F. 1947. *Plants and environment.* Wiley, New York. 424 pp.

DAUBENMIRE, R. F. 1956. Climate as a determinant of vegetation distribution in eastern Washington and northern Idaho. *Ecol. Monogr.* 26:131–154.

DAUBENMIRE, R. F. 1968. *Plant communities.* Harper & Row, New York. 300 pp.

DAVIDSON, J., and H. G. ANDREWARTHA. 1948. Annual trends in a natural population of *Thrips imaginis* (Thysanoptera). *J. Anim. Ecol.* 17:193–222.

DAWKINS, R. 1976. *The selfish gene.* Oxford Univ. Press. 224 pp.

DAWKINS, R., and T. R. CARLISLE. 1976. Parental investment, mate desertion and a fallacy. *Nature* 262:131–133.

DAWSON, P. S., and C. E. KING (Eds.). 1971. *Readings in population biology.* Prentice-Hall, Englewood Cliffs, N.J.

DAWSON, W. R. 1975. On the physiological significance of the preferred body temperatures of reptiles. Pp. 443–473 in D. M. Gates and R. B. Schmerl (Eds.), *Perspectives of biophysical ecology.* Ecological Studies, vol. 12, Springer-Verlag.

DAYTON, P. K. 1971. Competition, disturbance, and community organization: the provision and subsequent utilization of space in a rocky intertidal community. *Ecol. Monogr.* 41:351–389.

DEBACH, P. 1966. The competitive displacement and coexistence principles. *Ann. Rev. Entomol.* 11:183–212.

DEEVEY, E. S., JR. 1947. Life tables for natural populations of animals. *Quart. Rev. Biol.* 22:283–314.

DEEVEY, E. S. (Ed.) 1972. *Growth by intussusception. Trans. Conn. Acad. Arts. Sci.* 44:1–443.

DICE, L. R. 1952. *Natural communities.* Univ. Michigan Press, Ann Arbor. 547 pp.

DIETZ, R. S., and J. C. HOLDEN. 1970. The breakup of pangaea. *Sci. Amer.* 223(Oct.):30–41.

DOBZHANSKY, T. 1950. Evolution in the tropics. *Amer. Sci.* 38:208–221.

DOBZHANSKY, T. 1970. *Genetics of the evolutionary process.* Columbia Univ. Press, New York. 505 pp.

DOCTERS VAN LEEUWEN, W. M. 1936. Krakatau, 1833 to 1933. *Ann. Jard. Botan. Buitenzorg* 56–57:1–506.

DOEKSEN, J., and J. VAN DER DRIFT. 1963. *Soil organisms.* North-Holland, Amsterdam. 453 pp.

DOWNHOWER, J. F., and K. B. ARMITAGE. 1971. The yellow-bellied marmot and the evolution of polygamy. *Amer. Natur.* 105:355–370.

DRAKE, E. T. (Ed.). 1968. *Evolution and environment.* Yale Univ. Press, New Haven, Conn. 478 pp.

DRESSLER, R. L. 1968. Pollination by euglossine bees. *Evolution* 22:202–210.

DUNBAR, M. J. 1960. The evolution of stability in marine environments: natural selection at the level of the ecosystem. *Amer. Natur.* 94:129–136.

DUNBAR, M. J. 1968. *Ecological development in polar regions.* Prentice-Hall. Englewood Cliffs, N.J. 119 pp.

DUNBAR, M. J. 1972. The ecosystem as unit of natural selection. Pp. 113–130 in E. S. Deevey (Ed.), *Growth by intussusception. Trans. Conn. Acad. Arts Sci.* 44:1–443.

EBERHARD, M. J. W. 1975. The evolution of social behavior by kin selection. Q. *Rev. Biol.* 50:1–33.

EGGELING, W. J. 1947. Observations on the ecology of the Budongo rain forest, Uganda. *J. Ecol.* 34:20–87.

EHRLICH, P. R., D. E. BREEDLOVE, P. F. BRUSSARD, and M. A. SHARP. 1972. Weather and the "regulation" of subalpine populations. *Ecology* 53:243–247.

EHRLICH, P. R., and L. C. BIRCH. 1967. "The balance of nature" and "population control." *Amer. Natur.* 101:97–107.

EHRLICH, P. R., and R. W. HOLM. 1963. *The process of evolution.* McGraw-Hill, New York. 347 pp.

EHRLICH, P. R., and P. H. RAVEN. 1964. Butterflies and plants: a study in coevolution. *Evolution* 18:586–608.

EHRLICH, P. R., and P. H. RAVEN. 1969. Differentiation of populations. *Science* 165:1228–1231.

ELTON, C. S. 1927. *Animal ecology.* Sidgwick and Jackson, London. 209 pp.

ELTON, C. S. 1942. *Voles, mice and lemmings: problems in population dynamics.* Oxford Univ. Press, London. 496 pp.

ELTON, C. S. 1946. Competition and the structure of ecological communities. *J. Anim. Ecol.* 15:54–68.

ELTON, C. S. 1949. Population interspersion: an essay on animal community patterns. *J. Ecol.* 37:1–23.

ELTON, C. S. 1958. *The ecology of invasions by animals and plants.* Methuen, London. 181 pp.

ELTON, C. S. 1966. *The pattern of animal communities.* Methuen, London.

EMERSON, A. E. 1960. The evolution of adaptation in population systems. Pp. 307–348 in S. Tax (Ed.), *Evolution after Darwin.* Vol. I. Univ. Chicago Press, Chicago.

EMLEN, J. M. 1966. The role of time and energy in food preference. *Amer. Natur.* 100:611–617.

EMLEN, J. M. 1968a. Optimal choice in animals. *Amer. Natur.* 102:385–390.

EMLEN, J. M. 1968b. A note on natural selection and the sex ratio. *Amer. Natur.* 102:94–95.

EMLEN, J. M. 1970. Age specificity and ecological theory. *Ecology* 51:588–601.

EMLEN, J. M. 1973. *Ecology: an evolutionary approach.* Addison-Wesley, Reading, Mass. 493 pp.

ENGELMANN, M. D. 1966. Energetics, terrestrial field studies, and animal productivity. *Adv. Ecol. Res.* 3:73–115.

ERRINGTON, P. L. 1946. Predation and vertebrate populations. *Quart. Rev. Biol.* 21:144–177.

ERRINGTON, P. L. 1956. Factors limiting higher vertebrate populations. *Science* 124:304–307.

ERRINGTON, P. L. 1963. *Muskrat populations.* Iowa State Univ. Press, Ames. 665 pp.

ESHEL, I. 1972. On the neighbor effect and the evolution of altruistic traits. *Theoret. Pop. Biol.* 3:258–277.

ESSER, M. H. M. 1946a. Tree trunks and branches as optimal mechanical supports of the crown. I. The trunk. *Bull. Math. Biophys.* 8:65–74.

ESSER, M. H. M. 1946b. Tree trunks and branches as optimal mechanical supports of the crown. II. The branches. *Bull. Math. Biophys.* 8:95–100.

EVANS, F. C. and F. E. SMITH. 1952. The intrinsic rate of natural increase for the human louse, *Pediculus humanus* L. *Amer. Natur.* 86:299–310.

EVANS, H. E. 1977. Extrinsic versus intrinsic factors in the evolution of insect sociality. *BioScience* 27:613–617.

EYRE, S. R. 1963. *Vegetation and soils: a world picture.* Aldine, Chicago. 324 pp.

FAEGRI, K., and L. VAN DER PIJL. 1971. *The principles of pollination ecology.* Pergamon Press, London. 248 pp.

FALLS, J. B. 1969. Functions of territorial songs in the white-throated sparrow. Pp. 207–232 in R. A. Hinde (Ed.), *Bird vocalizations.* Cambridge Univ. Press, Cambridge, England. 394 pp.

FEENY, P. P. 1968. Effects of oak leaf tannins on larval growth of the winter moth *Operophtera brumata. J. Insect Physiol.* 14:805–817.

FEENY, P. P. 1970. Seasonal changes in oak leaf tannins and nutrients as a cause of spring feeding by winter moth caterpillars. *Ecology* 51:565–581.

FEENY, P. 1975. Biochemical coevolution between plants and their insect herbivores. Pp 3–19 in L. E. Gilbert and P. H. Raven (Eds.), *Coevolution of animals and plants.* Univ. Texas Press, Austin. 246 pp.

FEENY, P. 1976. Plant apparency and chemical defense. *Rec. Adv. Phytochemistry* 10:1–40.

FENCHEL, T. 1974. Intrinsic rate of natural increase: the relationship with body size. *Oecologia* 14:317–326.

FENCHEL, T. 1975. Character displacement and coexistence in mud snails (Hydrobiidae). *Oecologia* 20:19–32.

FINCH, V. C., and G. T. TREWARTHA. 1949. *Physical elements of geography.* McGraw-Hill, New York.

FISCHER, A. G. 1960. Latitudinal variations in organic diversity. *Evolution* 14:64–81.

FISHER, J., and R. T. PETERSON. 1964. *The world of birds.* Doubleday, New York.

FISHER, R. A. 1930. *The genetical theory of natural selection.* Clarendon Press, Oxford. 272 pp.

FISHER, R. A. 1958a. *The genetical theory of natural selection.* (2nd ed.) Dover, New York. 291 pp.

FISHER, R. A. 1958b. Polymorphism and natural selection. *J. Anim. Ecol.* 46:289–293.

FISHER, R. A., A. S. CORBET, and C. B. WILLIAMS. 1943. The relation between the number of species and the number of individuals in a random sample of an animal population. *J. Anim. Ecol.* 12:42–58.

FITZPATRICK, L. C. 1973. Energy allocation in the Allegheny Mountain salamander. *Desmognathus ochrophaeus. Ecol. Monogr.* 43:43–58.

FLOHN, H. 1969. *Climate and weather.* World Univ. Lib., McGraw-Hill, New York.

FLOREY, E. 1966. *An introduction to general and comparative physiology.* Saunders, Philadelphia. 713 pp.

FOLK, G. E., JR. 1974. *Textbook of environmental physiology*. (2nd ed.) Lea & Febiger, Philadelphia. 465 pp.

FONS, W. L. 1940. Influence of forest cover on wind velocity. *J. Forestry* 38:481–486.

FORCE, D. C. 1972. *r*- and *K*-strategists in endemic host-parasitoid communities. *Bull. Entomol. Soc. Amer.* 18:135–137.

FORD, E. B. 1931. *Mendelism and evolution*. Methuen, London. 122 pp.

FORD, E. B. 1964. *Ecological genetics*. Methuen, London. 335 pp.

FORD, R. F., and W. E. HAZEN. 1972. *Readings in aquatic ecology*. Saunders, Phildelphia. 379 pp.

FOX, S. W., and K. DOSE. 1972. *Molecular evolution and the origin of life*. Freeman, San Francisco. 359 pp.

FRAENKEL, G. S. 1959. The raison d'être of secondary plant substances. *Science* 129:1466–1470.

FRANK, P. W. 1968. Life histories and community stability. *Ecology* 49:355–357.

FRAZZETTA, T. H. 1975. *Complex adaptations in evolving populations*. Sinauer, Sunderland, Mass.

FREELAND, W. J. 1974. Vole cycles: another hypothesis. *Amer. Natur.* 108:238–245.

FREELAND, W. J., and D. H. JANZEN. 1974. Strategies in herbivory by mammals: the role of plant secondary compounds. *Amer. Natur.* 108:269–289.

FRETWELL, S. D. 1972. *Populations in a seasonal environment*. Princeton Univ. Press, Princeton, N.J. 217 pp.

FRETWELL, S. D., and H. L. LUCAS, JR. 1969. On territorial behavior and other factors influencing habitat distribution in birds. I. Theoretical development. *Acta Biotheoretica* 19:16–36.

FREY, D. G. 1963. *Limnology in North America*. Univ. of Wisconsin Press, Madison. 734 pp.

FRIED, M., and H. BROESHART. 1967. *The soil-plant system in relation to inorganic nutrition*. Academic Press, New York. 358 pp.

FUTUYMA, D. J. 1973. Community structure and stability in constant environments. *Amer. Natur.* 107:443–446.

FUTUYMA, D. J. 1976. Food plant specialization and environmental predictability in lepidoptera. *Amer. Natur.* 110:285–292.

GADGIL, M., and W. H. BOSSERT. 1970. Life historical consequences of natural selection. *Amer. Natur.* 104:1–24.

GADGIL, M., and O. T. SOLBRIG. 1972. The concept of *r* and *K* selection: evidence from wild flowers and some theoretical considerations. *Amer. Natur.* 106:14–31.

GAFFNEY, P. M. 1975. Roots of the niche concept. *Amer. Natur.* 109:490.

GALLOPIN, G. C. 1972. Structural properties of food webs. Pp. 241–282 in B. Patten (Ed.), *Systems analysis and simulation in ecology*. Vol. II. Academic Press, New York. 592 pp.

GATES, D. M. 1962. *Energy exchange in the biosphere*. Harper & Row, New York. 151 pp.

GATES, D. M. 1965. Energy, plants and ecology. *Ecology* 46:1–13.

GATES, D. M. 1972. *Man and his environment: climate*. Harper & Row, New York. 175 pp.

GATES, D. M., and R. B. SCHMERL. 1975. *Perspectives of biophysical ecology*. Ecological Studies, Vol. 12. Springer-Verlag, New York.

GAUSE, G. F. 1934. *The struggle for existence*. Hafner, New York (reprinted 1964 by Williams & Wilkins, Baltimore, Md.). 163 pp.

GAUSE, G. F. 1935. Experimental demonstration of Volterra's periodic oscillations in the numbers of animals. *J. Exp. Biol.* 12:44–48.

GEIGER, R. 1966. *The climate near the ground*. Harvard Univ. Press, Cambridge, Mass. 611 pp.

GENTRY, A. H. 1969. A comparison of some leaf characteristics of tropical dry forest and tropical wet forest in Costa Rica. *Turrialba* 19:419–428.

GIBB, J. A. 1956. Food, feeding habits, and territory of the Rock Pipit, *Anthus spinoletta*. *Ibis* 98:506–530.

GIBB, J. A. 1960. Populations of tits and goldcrests and their food supply in pine plantations. *Ibis* 102:163–208.

GILBERT, L. E. 1971. Butterfly–plant coevolution: has *Passiflora adenopoda* won the selectional race with Heliconiine butterflies? *Science* 172:585–586.

GILBERT, L. E. 1972. Pollen feeding and reproductive biology of *Heliconius* butterflies. *Proc. Nat. Acad. Sci.* 69:1403–1407.

GILBERT, L. E. 1977. Development of theory in the analysis of insect–plant interactions. Chapter in D. J. Horn, R. Mitchell, and G. R. Stairs (Eds.), *Analysis of ecological systems*. Ohio State Univ. Press.

GILBERT, L. E., and P. H. RAVEN (Eds.) 1975. *Coevolution of animals and plants*. Univ. Texas Press, Austin. 246 pp.

GILBERT, L. E., and M. C. SINGER. 1973. Dispersal and gene flow in a butterfly species. *Amer. Natur.* 107:58–72.

GILBERT, L. E., and M. C. SINGER. 1975. Butterfly ecology. *Ann. Rev. Ecol. Syst.* 6:365–397.

GILL, D. E. 1972. Intrinsic rates of increase, saturation densities, and competitive ability. I. An experiment with *Paramecium*. *Amer. Natur.* 106:461–471.

GILPIN, M. E. 1973. Do hares eat lynx? *Amer. Natur.* 107:727–730.

GILPIN, M. E. 1975a. *Group selection in predator–prey communities*. Princeton Univ. Press, Princeton, New Jersey. 108 pp.

GILPIN, M. E. 1975b. Limit cycles in competition communities. *Amer. Natur.* 109:51–60.

GINDELL, I. 1973. *A new ecophysiological approach to forest–water relationships in arid climates*. Junk, the Hague. 142 pp.

GISBORNE, H. T. 1941. How the wind blows in the forest of northern Idaho. *Northern Rocky Mountain Forest Range Experimental Station*.

GIVNISH, T. J., and G. J. VERMEIJ. 1976. Sizes and shapes of liane leaves. *Amer. Natur.* 110:743–778.

GLEASON, H. A. 1922. On the relation between species and area. *Ecology* 3:158–162.

GLEASON, H. A. 1929. The significance of Raunkiaer's law of frequency. *Ecology* 10:406–408.

GLEASON, H. A., and A. CRONQUIST. 1964. *The natural geography of plants*. Columbia Univ. Press, New York. 420 pp.

GOLLEY, F. B. 1960. Energy dynamics of a food chain of an old-field community. *Ecol. Monogr.* 30:187–206.

GOODMAN, D. 1974. Natural selection and a cost ceiling on reproductive effort. *Amer. Natur.* 108:247–268.

GOODMAN, D. 1975. The theory of diversity–stability relationships in ecology. *Quart. Rev. Biol.* 50:237–266.

GOODMAN, L. A. 1971. On the sensitivity of the intrinsic growth rate to changes in the age-specific birth and death rates. *Theoret. Pop. Biol.* 2:339–354.

GORDON, H. T. 1961. Nutritional factors in insect resistance to chemicals. *Ann. Rev. Entomol.* 6:27–54.

GORDON, M. S. (Ed.) 1972. *Animal physiology: principles and adaptations.* Macmillan, New York. 592 pp.

GRANT, P. R. 1967. Bill length variability in birds of the Tres Marías Islands, Mexico. *Canad. J. Zool.* 45:805–815.

GRANT, P. R. 1971. Variation in the tarsus length of birds in island and mainland regions. *Evolution* 25:599–614.

GRANT, P. R. 1972. Convergent and divergent character displacement. *Biol. J. Linnean Soc.* 4:39–68.

GRASSLE, J. F., and J. P. GRASSLE. 1974. Opportunistic life histories and genetic systems in marine benthic polychaetes. *J. Marine Res.* 32:253–284.

GREEN, R. H. 1969. Population dynamics and environmental variability. *Amer. Zool.* 9:393–398.

GREEN, R. H. 1971. A multivariate statistical approach to the Hutchinsonian niche: bivalve mollusks of central Canada. *Ecology* 52:543–556.

GREIG-SMITH, P. 1964. *Quantitative plant ecology.* (2nd ed.) Butterworth, London. 256 pp.

GRICE, G. D., and A. D. HART. 1962. The abundance, seasonal occurrence and distribution of the epizooplankton between New York and Bermuda. *Ecol. Monogr.* 32:287–307.

GRIFFIN, D. R. 1958. *Listening in the dark.* Yale Univ. Press, New Haven, Conn.

GRINNELL, J. 1917. The niche relationships of the California thrasher. *Auk* 21:364–382.

GRINNELL, J. 1924. Geography and evolution. *Ecology* 5:225–229.

GRINNELL, J. 1928. The presence and absence of animals. *Univ. Calif. Chronicle* 30:429–450. (Reprinted in *Joseph Grinnell's Philosophy of Nature,* Univ. California Press, Berkeley, 1943. pp. 187–208.)

GRODZINSKI, W., and A. GORECKI. 1967. Daily energy budgets of small rodents Pp. 295–314 in K. Petrusewicz (Ed.), *Secondary productivity of terrestrial ecosystems,* Vol. I. Warsaw.

GUNTER, G. 1941. Death of fishes due to cold on the Texas coast, January, 1940. *Ecology* 22:203–208.

GUYTON, A. C., and D. HORROBIN, (Eds.) 1974. *Environmental physiology.* Physiology, Vol. 7, Series One. Butterworth, London. 326 pp.

HAARTMAN, L. V. 1969. Nest-site and evolution of polygamy in European passerine birds. *Ornis Fenn.* 46:1–12.

HADLEY, N. F. (Ed.) 1975. *Environmental physiology of desert organisms.* Dowden, Hutchinson & Ross, Inc. Stroudsburg, Penn. 283 pp.

HAIGH, J., and J. MAYNARD SMITH. 1972. Can there be more predators than prey? *Theoret. Pop. Biol.* 3:290–299.

HAIRSTON, N. G. 1951. Interspecies competition and its probable influence upon the vertical distribution of Appalachian salamanders of the genus *Plethodon. Ecology* 32:266–274.

HAIRSTON, N. G., and G. W. BYERS. 1954. The soil arthropods of a field in southern Michigan: a study in community ecology. *Contrib. Lab. Vert. Biol., Univ. of Michigan* 64:1–37.

HAIRSTON, N. G., F. E. SMITH, and L. B. SLOBODKIN. 1960. Community structure, population control, and competition. *Amer. Natur.* 94:421–425.

HAIRSTON, N. G., J. D. ALLAN, R. K. COLWELL, D. J. FUTUYMA, J. HOWELL, M. D. LUBIN, J. MATHIAS, and J. H. VANDERMEER. 1968. The relationship between species diversity and stability: an experimental approach with protozoa and bacteria. *Ecology* 49:1091–1101.

HALDANE, J. B. S. 1932. *The causes of evolution* (reprinted 1966). Cornell Univ. Press, Ithaca, N.Y. 235 pp.

HALDANE, J. B. S. 1941. *New paths in genetics.* Harper, London. 206 pp.

HAMILTON, T. H. 1961. On the functions and causes of sexual dimorphism in breeding plumage of North American species of warblers and orioles. *Amer. Natur.* 45:121–123.

HAMILTON, T. H., and I. RUBINOFF. 1963. Isolation, endemism, and multiplication of species in the Darwin finches. *Evolution* 17:388–403.

HAMILTON, T. H., and I. RUBINOFF. 1967. On predicting insular variation in endemism and sympatry for the Darwin finches in the Galapagos archipelago. *Amer. Natur.* 101:161–172.

HAMILTON, W. D. 1964. The genetical evolution of social behavior (two parts). *J. Theoret. Biol.* 7:1–52.

HAMILTON, W. D. 1966. The moulding of senescence by natural selection. *J. Theoret. Biol.* 12:12–45.

HAMILTON, W. D. 1967. Extraordinary sex ratios. *Science* 156:477–488.

HAMILTON, W. D. 1970. Selfish and spiteful behaviour in an evolutionary model. *Nature* 228:1218–1220.

HAMILTON, W. D. 1971. Geometry for the selfish herd. *J. Theoret. Biol.* 31:295–311.

HAMILTON, W. D. 1972. Altruism and related phenomena, mainly in insects. *Ann. Rev. Ecol. Syst.* 3:193–232.

HAMILTON, W. J. III. 1973. *Life's color code.* McGraw-Hill, New York. 238 pp.

HARDIN, G. 1960. The competitive exclusion principle. *Science* 131:1292–1297.

HARPER, J. L. 1961a. Approaches to the study of plant competition. *Soc. Exp. Biol. Symp.* 15:1–39.

HARPER, J. L. 1961b. The evolution and ecology of closely related species living in the same area. *Evolution* 15:209–227.

HARPER, J. L. 1967. A Darwinian approach to plant ecology. *J. Ecol.* 55:247–270.

HARPER, J. L. 1969. The role of predation in vegetational diversity. *Brookhaven Symp. Biol.* 22:48–62.

HARPER, J. L., and J. OGDEN. 1970. The reproductive strategy of higher plants. I. The concept of strategy with special reference to *Senecio vulgaris* L. *J. Ecol.* 58:681–698.

HARPER, J. L., and J. WHITE. 1974. The demography of plants. *Ann. Rev. Ecol. Syst.* 5:419–463.

HARPER, J. L., P. H. LOVELL, and K. G. MOORE. 1970. The shapes and sizes of seeds. *Ann. Rev. Ecol. Syst.* 1:327–356.

HASKELL, E. F. 1947. A natural classification of societies. *N.Y. Acad. Sci., Trans.* Series 2, 9:186–196.

HASKELL, E. F. 1949. A clarification of social science. *Main Currents in Modern Thought* 7:45–51.

HAURWITZ, B., and J. M. AUSTIN. 1944. *Climatology.* McGraw-Hill, New York. 410 pp.

HAZEN, W. E. 1964. *Readings in population and community ecology.* (1st ed.) Saunders, Philadelphia. 388 pp.

HAZEN, W. E. 1970. *Readings in population and community ecology.* (2nd ed.) Saunders, Philadelphia. 421 pp.

HEATWOLE, H. 1965. Some aspects of the association of cattle egrets with cattle. *Anim. Behaviour* 13:79–83.

HEER, D. M. 1968. *Society and population.* Prentice-Hall, Englewood Cliffs, N.J.

HEINRICH, B. 1975. Energetics of pollination. *Ann. Rev. Ecol. Syst.* 6:139–170.

HEINRICH, B., and P. H. RAVEN. 1972. Energetics and pollination ecology. *Science* 176:597–602.

HENDERSON, L. J. 1913. *The fitness of the environment.* Macmillan, New York.

HENSLEY, M. M., and J. B. COPE. 1951. Further data on removal and repopulation of the breeding birds in a spruce-fir forest community. *Auk* 68:483–493.

HESPENHIDE, H. 1971. Food preference and the extent of overlap in some insectivorous birds, with special reference to Tyrannidae. *Ibis* 113:59–72.

HESSE, R., W. C. ALLEE, and K. P. SCHMIDT. 1951. *Ecological animal geography.* (2nd ed.) Wiley, New York. 715 pp.

HICKMAN, J. C. 1975. Environmental unpredictability and plastic energy allocation strategies in the annual *Polygonum cascadense* (Polygonaceae) *J. Ecol.* 63:689–701.

HIRSHFIELD, M. F., and D. W. TINKLE. 1975. Natural selection and the evolution of reproductive effort. *Proc. Nat. Acad. Sci.* USA 72:2227–2231.

HOCHACHKA, P. W., and G. N. SOMERO. 1973. Strategies of biochemical adaptation. Saunders, Philadelphia. 358 pp.

HOLDRIDGE, L. R. 1947. Determination of world plant formations from simple climatic data. *Science* 105:367–368.

HOLDRIDGE, L. R. 1959. Simple method for determining potential evapotranspiration from temperature data. *Science* 130:572.

HOLDRIDGE, L. R. 1967. *Life zone ecology.* Tropical Science Center, San Jose, Costa Rica. 124 pp.

HOLLING, C. S. 1959a. The components of predation as revealed by a study of small-mammal predation of the European pine sawfly. *Canad. Entomol.* 91:293–320.

HOLLING, C. S. 1959b. Some characteristics of simple types of predation and parasitism. *Canad. Entomol.* 91:385–398.

HOLLING, C. S. 1961. Principles of insect predation. *Ann. Rev. Entomol.* 6:163–182.

HOLLING, C. S. 1963. An experimental component analysis of population processes. *Mem. Entomol. Soc. Canada* 32:22–32.

HOLLING, C. S. 1964. The analysis of complex population processes. *Canad. Entomol.* 96:335–347.

HOLLING, C. S. 1965. The functional response of predators to prey density and its role in mimicry and population regulation. *Mem. Entomol. Soc. Canada* 45:1–60.

HOLLING, C. S. 1966. The functional response of invertebrate predators to prey density. *Mem. Entomol. Soc. Canada* 48:1–87.

HOLLING, C. S. 1973. Resilience and stability of ecological systems. *Ann. Rev. Ecol. Syst.* 4:1–23.

HOLM, C. H. 1973. Breeding sex ratios, territoriality, and reproductive success in the red-winged blackbird *(Agelaius phoeniceus)*. *Ecology* 54:356–365.

HORN, H. S. 1966. Measurement of overlap in comparative ecological studies. *Amer. Natur.* 100:419–424.

HORN, H. S. 1968a. Regulation of animal numbers: a model counter-example. *Ecology* 49:776–778.

HORN, H. S. 1968b. The adaptive significance of colonial nesting in the Brewer's blackbird *(Euphagus cyanocephalus). Ecology* 49:682–694.

HORN, H. S. 1971. *The adaptive geometry of trees.* Princeton Univ. Press, Princeton, N.J. 144 pp.

HORN, H. S. 1974. The ecology of secondary succession. *Ann. Rev. Ecol. Syst.* 5:25–37.

HORN, H. S. 1975a. Forest succession. *Sci. Amer.* 232(May):90–98.

HORN, H. S. 1975b. Markovian properties of forest succession. Pp. 196–211 in M. L. Cody and J. M. Diamond (Eds.), *Ecology and evolution of communities.* Harvard Univ. Press, Cambridge.

HORN, H. S . 1976. Succession. Chapter 10 (pp. 187–204) in R. M. May (Ed.), *Theoretical ecology: principles and applications.* Blackwell. 317 pp.

HORN, H. S., and R. H. MACARTHUR. 1972. Competition among fugitive species in a harlequin environment. *Ecology* 53:749–752.

HOWARD, H. E. 1920. *Territory in bird life.* Murray, London. 308 pp. (reprinted 1964 by Atheneum, New York. 293 pp.).

HOWARD, R. D. 1974. The influence of sexual selection and interspecific competition on mockingbird song *(Mimus polyglottos). Evolution* 28:428–438.

HOWLAND, H. C. 1962. Structural, hydraulic, and "economic" aspects of leaf venation and shape. In E. E. Bernard and M. R. Kare (Eds.), *Biological prototypes and synthetic systems.* Vol. 1. Cornell Univ. Press, Ithaca, New York, 597 pp.

HUBBELL, S. P. 1971. Of sowbugs and systems: the ecological bioenergetics of a terrestrial isopod. Pp. 269–324 in B. Patten (Ed.), *Systems analysis and simulation in ecology.* Vol I. Academic Press, New York. 607 pp.

HUBBELL, S. P. 1973a. Populations and simple food webs as energy filters. I. One-species systems. *Amer. Natur.* 107:94–121.

HUBBELL, S. P. 1973b. Populations and simple food webs as energy filters. II. Two-species systems. *Amer. Natur.* 107:122–151.

HUEY, R. B., E. R. PIANKA, M. E. EGAN, and L. W. COONS. 1974. Ecological shifts in sympatry: Kalahari fossorial lizards *(Typhlosaurus). Ecology* 55:304–316.

HUEY, R. B., and M. SLATKIN. 1976. Costs and benefits of lizard thermoregulation. *Quart. Rev. Biol.* 51:363–384.

HUFFAKER, C. B. 1958. Experimental studies on predation: dispersion factors and predator–prey oscillations. *Hilgardia* 27:343–383.

HUFFAKER, C. B. 1971. *Biological control.* Plenum, New York. 511 pp.

HURD, L. E., M. V. MELLINGER, L. L. WOLF, and S. J. MCNAUGHTON. 1971. Stability and diversity at three trophic levels in terrestrial successional ecosystems. *Science* 173:1134–1136.

HUTCHINSON, G. E. 1951. Copepodology for the ornithologist. *Ecology* 32:571–577.

HUTCHINSON, G. E. 1953. The concept of pattern in ecology. *Proc. Nat. Acad. Sci.* 105:1–12.

HUTCHINSON, G. E. 1957a. Concluding remarks. *Cold Spring Harbor Symp. Quant. Biol.* 22:415–427.

HUTCHINSON, G. E. 1957b. *A treatise on limnology.* Vol. I. *Geography, physics, and chemistry.* Wiley, New York. 1015 pp.

HUTCHINSON, G. E. 1959. Homage to Santa Rosalia, or why are there so many kinds of animals? *Amer. Natur.* 93:145–159.

HUTCHINSON, G. E. 1961. The paradox of the plankton. *Amer. Natur.* 95:137–145.

HUTCHINSON, G. E. 1965. *The ecological theater and the evolutionary play.* Yale Univ. Press, New Haven, Conn. 139 pp.

HUTCHINSON, G. E. 1967. *A treatise on limnology.* Vol. II. *Introduction to lake biology and the limnoplankton.* Wiley, New York. 1115 pp.

HUTCHINSON, G. E., and R. H. MACARTHUR. 1959. A theoretical ecological model of size distributions among species of animals. *Amer. Natur.* 93:117–125.

INGER, R., and R. K. COLWELL. 1977. Organization of three adjacent tropical communities of amphibians and reptiles in Thailand. *Ecology* 58 (in press).

ISTOCK, C. A. 1967. The evolution of complex life cycle phenomena: an ecological perspective. *Evolution* 21:592–605.

JAEGER, R. G. 1971. Competitive exclusion as a factor influencing the distributions of two species of terrestrial salamanders. *Ecology* 52:632–637.

JANZEN, D. H. 1966. Coevolution of mutualism between ants and acacias in Central America. *Evolution* 20:249–275.

JANZEN, D. H. 1967. Fire, vegetation structure, and the ant-acacia interaction in Central America. *Ecology* 48:26–35.

JANZEN, D. H. 1970. Herbivores and the number of tree species in tropical forests. *Amer. Natur.* 104:501–528.

JANZEN, D. H. 1971a. Euglossine bees as long-distance pollinators of tropical plants. *Science* 171:203–205.

JANZEN, D. H. 1971b. Seed predation by animals. *Ann. Rev. Ecol. Syst.* 2:465–492.

JANZEN, D. H. 1976. *Ecology of plants in the tropics.* Edward Arnold, London. 66 pp.

JELGERSMA, S. 1966. Sea-level changes during the last 10,000 years. Pp. 54–71 in J. S. Sawyer (Ed.), *World climate from 8,000 to 0 B.C. Proc. Int. Symp. on World Climate 8,000 to 0 B.C.* Imperial College, London 1966. Royal Meterological Society, London.

JENNY, H. 1941. *Factors of soil formation.* McGraw-Hill, New York. 281 pp.

JOFFE, J. S. 1949. *Pedology.* Pedology, New Brunswick, N.J. 662 pp.

JOHNSON, M. P., and S. A. COOK. 1968. "Clutch size" in buttercups. *Amer. Natur.* 102:405–411.

JOHNSON, M. P., L. G. MASON, and P. H. RAVEN. 1968. Ecological parameters and plant species diversity. *Amer. Natur.* 102:297–306.

JOHNSTON, R. F. 1954. Variation in breeding season and clutch size in song sparrows of the Pacific coast. *Condor* 56:268–273.

JONES, D. A. 1962. Selective eating of the acyanogenic form of the plant *Lotus corniculatus* L. by various animals. *Nature* 193:1109–1110.

JONES, D. A. 1966. On the polymorphism of cyanogenesis in *Lotus corniculatus*. Selection by animals. *Canad. J. Genet. Cytol.* 8:556–567.

JONES, E. W. 1956. Ecological studies on the rain forest of southern Nigeria. *J. Ecol.* 44:83–117.

KEITH, L. B. 1963. *Wildlife's ten-year cycle.* Univ. of Wisconsin Press, Madison. 201 pp.

KEITH, L. B. 1974. Some features of population dynamics in mammals. *Proc. Int. Congr. Game Biol.* Stockholm 11:17–58.

KENDEIGH, S. C. 1961. *Animal Ecology*. Prentice-Hall, Englewood Cliffs, N.J. 468 pp.

KERSHAW, K. A. 1964. *Quantitative and dynamic ecology*. Arnold, London. 183 pp.

KETTLEWELL, H. B. D. 1956. Further selection experiments on industrial melanism in the Lepidoptera. *Heredity* 10:287–301.

KETTLEWELL, H. B. D. 1958. Industrial melanism in the Lepidoptera and its contribution to our knowledge of evolution. *Proc. 10th Int. Congr. Entomol.* 2:831–841.

KEYFITZ, N. 1968. *Introduction to the mathematics of population*. Addison-Wesley, Reading, Mass. 450 pp.

KEYFITZ, N., and W. FLIEGER. 1971. *Populations: facts and methods of demography*. Freeman, San Francisco.

KING, C. E. 1971. Resource specialization and equilibrium population size in patchy environments. *Proc. Nat. Acad. Sci.* 68:2634–2637.

KING, C. E., and W. W. ANDERSON. 1971. Age-specific selection. II. The interaction between r and K during population growth. *Amer. Natur.* 105:137–156.

KIRCHER, H. W., and W. B. HEED. 1970. Phytochemistry and host plant specificity in *Drosophila*. Pp. 191–209 in C. Steelink and V. C. Runeckles (Eds.), *Recent advances in phytochemistry* Vol. 3. Appleton, New York.

KIRCHER, H. W., W. B. HEED, J. S. RUSSELL, and J. GROVE. 1967. Senita cactus alkaloids: their significance to Sonoran desert *Drosophila* ecology. *J. Insect Physiol.* 13:1869–1874.

KLOMP, H. 1970. The determination of clutch-size in birds. *Ardea* 58:1–124.

KLOPFER, P. H. 1962. *Behavioral aspects of ecology*. Prentice-Hall, Englewood Cliffs, N.J. 171 pp.

KLOPFER, P. H., and R. H. MACARTHUR. 1960. Niche size and faunal diversity. *Amer. Natur.* 94:293–300.

KLOPFER, P. H., and R. H. MACARTHUR. 1961. On the causes of tropical species diversity: niche overlap. *Amer. Natur.* 95:223–226.

KNIGHT, C. B. 1965. *Basic concepts of ecology*. Macmillan, New York. 468 pp.

KOHN, A. J. 1959. The ecology of *Conus* in Hawaii. *Ecol. Monogr.* 29:47–90.

KOHN, A. J. 1968. Microhabitats, abundance and food of *Conus* on atoll reefs in the Maldive and Chagos Islands. *Ecology* 49:1046–1062.

KOLMAN, W. A. 1960. The mechanism of natural selection for the sex ratio. *Amer. Natur.* 94:373–377.

KOZLOVSKY, D. G. 1968. A critical evaluation of the trophic level concept. I. Ecological efficiencies. *Ecology* 49:48–60.

KREBS, C. J. 1964. The lemming cycle at Baker Lake, Northwest Territories, during 1959–62. *Arctic Inst. of North America Tech. Paper No. 15*. 104 pp.

KREBS, C. J. 1966. Demographic changes in fluctuating population *of Microtus californicus*. *Ecol. Monogr.* 36:239–273.

KREBS, C. J. 1970. *Microtus* population biology: behavioral changes associated with the population cycle in *M. ochrogaster* and *M. pennsylvanicus*. *Ecology* 51:34–52.

KREBS, C. J. 1972. *Ecology: the experimental analysis of distribution and abundance*. Harper & Row, New York. 694 pp.

KREBS, C. J., and K. T. DELONG. 1965. A *Microtus* population with supplemental food. *J. Mammal.* 46:566–573.

KREBS, C. J., B. L. KELLER, and J. H. MYERS. 1971. *Microtus* population densities and soil nutrients in southern Indiana grasslands. *Ecology* 52:660–663.

KREBS, C. J., B. L. KELLER, and R. H. TAMARIN. 1969. *Microtus* population biology: I. Demographic changes in fluctuating populations of *M. ochrogaster* and *M. pennsylvanicus* in southern Indiana, 1965–1967. *Ecology* 50:587–607.

KURTÉN, B. 1969. Continental drift and evolution. *Sci. Amer.* 220 (March) 54–64.

LACK, D. 1945. The ecology of closely related species with special reference to cormorant (*Phalacrocorax carbo*) and shag (*P. aristotelis*). *J. Anim. Ecol.* 14:12–16.

LACK, D. 1947. *Darwin's finches.* Cambridge Univ. Press, Cambridge, England. 204 pp. (Reprinted 1961 by Harper & Row, New York. 204 pp.)

LACK, D. 1954. *The natural regulation of animal numbers.* Oxford Univ. Press, New York. 343 pp.

LACK, D. 1966. *Population studies of birds.* Oxford Univ. Press, New York. 341 pp.

LACK, D. 1968. *Ecological adaptations for breeding in birds.* Methuen, London. 409 pp.

LACK, D. 1971. *Ecological isolation in birds.* Blackwell, Oxford. 404 pp.

LAWLOR, L. R. 1977. A comment on randomly constructed model ecosystems. *Amer. Natur.* 111 (in press).

LAWLOR, L. R., and J. MAYNARD SMITH. 1976. The coevolution and stability of competing species. *Amer. Natur.* 110:79–99.

LEIGH, E. G., JR. 1965. On the relation between the productivity, biomass, diversity, and stability of a community. *Proc. Nat. Acad. Sci.* 53:777–783.

LERNER, I. M., and F. K. HO. 1961. Genotype and competitive ability of *Tribolium* species. *Amer. Natur.* 95:329–343.

LESLIE, P. H. 1945. On the use of matrices in certain population mathematics. *Biometrika* 33:183–212.

LESLIE, P. H. 1948. Some further notes on the use of matrices in population mathematics. *Biometrika.* 35:213–245.

LESLIE, P. H., and T. PARK. 1949. The intrinsic rate of natural increase of *Tribolium castaneum* Herbst. *Ecology* 30:469–477.

LEVIN, S. 1974. Dispersion and population interactions. *Amer. Natur.* 108:207–228.

LEVINS, R. 1964. The theory of fitness in a heterogeneous environment. IV. The adaptive significance of gene flow. *Evolution* 18:635–638.

LEVINS, R. 1966. The strategy of model building in population biology. *Amer. Sci.* 54:421–431.

LEVINS, R. 1968. *Evolution in changing environments.* Princeton Univ. Press, Princeton, N.J. 120 pp.

LEVINS, R. 1970. Extinction. Pp. 75–108 in M. Gerstenhaber (Ed.), *Some mathematical questions in biology.* Amer. Math. Soc.

LEVINS, R. 1975. Evolution in communities near equilibrium. Pp. 16–50 in M. L. Cody and J. M. Diamond (Eds.), *Ecology and evolution of communities.* Harvard Univ. Press, Cambridge.

LEVITT, J. 1972. *Responses of plants to environmental stresses.* Academic Press, New York.

LEWONTIN, R. C. 1965. Selection for colonizing ability. Pp. 77–94 in H. G. Baker and G. L. Stebbins (Eds.), *The genetics of colonizing species.* Academic Press, New York. 588 pp.

LEWONTIN, R. C. 1969. The meaning of stability. *Brookhaven Symp. Biol.* 22:13–24.

LEWONTIN, R. C. 1970. The units of selection. *Ann. Rev. Ecol. Syst.* 1:1–18.

LEWONTIN, R. C. 1974. *The genetic basis of evolutionary change.* Columbia Univ. Press, New York. 346 pp.

LIEBIG, J. 1840. *Chemistry in its application to agriculture and physiology.* Taylor and Walton, London.

LINDEMANN, R. I. 1942. The trophic-dynamic aspect of ecology. *Ecology* 23:399–418.

LOTKA, A. J. 1922. The stability of the normal age distribution. *Proc. Nat. Acad. Sci.* 8:339–345.

LOTKA, A. J. 1925. *Elements of physical biology.* Williams and Wilkins, Baltimore. (Reprinted as *Elements of mathematical biology* in 1956 by Dover, New York.) 460 pp.

LOTKA, A. J. 1956. *Elements of mathematical biology.* Dover, New York. 465 pp.

LOUCKS, O. L. 1970. Evolution of diversity, efficiency, and community stability. *Amer. Zool.* 10:17–25.

LOWRY, W. P. 1969. *Weather and life: an introduction to biometeorology.* Academic Press, New York. 305 pp.

LUCKINBILL, L. S. 1973. Coexistence in laboratory populations of *Paramecium aurelia* and its predator *Didinium nasutum. Ecology* 54:1320–1327.

LUCKINBILL, L. S. 1974. The effects of space and enrichment on a predator–prey system. *Ecology* 55:1142–1147.

MACARTHUR, R. H. 1955. Fluctuations of animal populations, and a measure of community stability. *Ecology* 36:533–536.

MACARTHUR, R. H. 1957. On the relative abundance of bird species. *Proc. Nat. Acad. Sci.* 43:293–295.

MACARTHUR, R. H. 1958. Population ecology of some warblers of northeastern coniferous forests. *Ecology* 39:599–619.

MACARTHUR, R. H. 1959. On the breeding distribution pattern of North American migrant birds. *Auk* 76:318–325.

MACARTHUR, R. H. 1960a. On the relative abundance of species. *Amer. Natur.* 94:25–36.

MACARTHUR, R. H. 1960b. On the relation between reproductive value and optimal predation. *Proc. Nat. Acad. Sci.* 46:143–145.

MACARTHUR, R. H. 1961. Population effects of natural selection. *Amer. Natur.* 95:195–199.

MACARTHUR, R. H. 1962. Some generalized theorems of natural selection. *Proc. Nat. Acad. Sci.* 48:1893–1897.

MACARTHUR, R. H. 1964. Environmental factors affecting bird species diversity. *Amer. Natur.* 98:387–397.

MACARTHUR, R. H. 1965. Patterns of species diversity. *Biol. Rev.* 40:510–533.

MACARTHUR, R. H. 1968. The theory of the niche. Pp. 159–176 in R. C. Lewontin (Ed.), *Population biology and evolution.* Syracuse Univ. Press, Syracuse N.Y. 205 pp.

MACARTHUR, R. H. 1970. Species packing and competitive equilibrium for many species. *Theoret. Pop. Biol.* 1:1–11.

MACARTHUR, R. H. 1971. Patterns of terrestrial bird communities. Chapter 5 (pp. 189–221) in D. S. Farner and J. R. King (Eds.), *Avian Biology.* Vol. I. Academic Press, New York. 586 pp.

MACARTHUR, R. H. 1972. *Geographical ecology: patterns in the distribution of species.* Harper & Row, New York. 269 pp.

MACARTHUR, R. H., and J. H. CONNELL. 1966. *The biology of populations.* Wiley, New York. 200 pp.

MACARTHUR, R. H., and R. LEVINS. 1964. Competition, habitat selection, and character displacement in a patchy environment. *Proc. Nat. Acad. Sci.* 51:1207–1210.

MACARTHUR, R. H., and R. LEVINS. 1967. The limiting similarity, convergence, and divergence of coexisting species. *Amer. Natur.* 101:377–385.

MACARTHUR, R. H., and J. W. MACARTHUR. 1961. On bird species diversity. *Ecology* 42:594–598.

MACARTHUR, R. H., and E. R. PIANKA. 1966. On optimal use of a patchy environment. *Amer. Natur.* 100:603–609.

MACARTHUR, R. H., and E. O. WILSON. 1963. An equilibrium theory of insular zoogeography. *Evolution* 17:373–387.

MACARTHUR, R. H., and E. O. WILSON. 1967. *The theory of island biogeography.* Princeton Univ. Press, Princeton, N.J. 203 pp.

MACARTHUR, R. H., J. M. DIAMOND, and J. R. KARR. 1972. Density compensation in island faunas. *Ecology* 53:330–342.

MACARTHUR, R. H., J. W. MACARTHUR, and J. PREER. 1962. On bird species diversity. II. Prediction of bird census from habitat measurements. *Amer. Natur.* 96:167–174.

MACARTHUR, R. H., H. RECHER, and M. CODY. 1966. On the relation between habitat selection and species diversity. *Amer. Natur.* 100:319–332.

MACFADYEN, A. 1963. *Animal ecology.* Pitman, London. 344 pp.

MACMAHON, J. A. 1976. Species and guild similarity of North American desert mammal faunas: a functional analysis of communities. Pp. 133–148 in D. W. Goodall (Ed.) *Evolution of desert biota.* Univ. Texas Press, Austin. 250 pp.

MACHIN, K. E., and H. W. LISSMAN. 1960. The mode of operation of the electric receptors in *Gymnarchus niloticus. J. Exp. Biol.* 37:801–811.

MCKEY, D. 1974. Adaptive patterns in alkaloid physiology. *Amer. Natur.* 108:305–320.

MCINTOSH, R. P. 1967. The continuum concept of vegetation. *Bot. Rev.* 33:130–187.

MCLAREN, I. A. 1971. *Natural regulation of animal populations.* Atherton, New York. 195 pp.

MCNAB, B. K. 1963. Bioenergetics and the determination of home range size. *Amer. Natur.* 97:133–140.

MCNAUGHTON, S. J., and L. L. WOLF. 1970. Dominance and the niche in ecological systems. *Science* 167:131–139.

MAGUIRE, B. 1963. The passive dispersal of small aquatic organisms and their colonization of isolated bodies of water. *Ecol. Monog.* 33:161–185.

MAGUIRE, B. 1967. A partial analysis of the niche. *Amer. Natur.* 101:515–523.

MAGUIRE, B. 1971. Phytotelmata: Biota and community structure determination in plant-held waters. *Ann. Rev. Ecol. Syst.* 2:439–464.

MAGUIRE, B. 1973. Niche response structure and the analytic potentials of its relationship to the habitat. *Amer. Natur.* 107:213–246.

MAIN, A. R. 1976. Adaptation of Australian vertebrates to desert conditions. Chapter 5 (pp. 101–131) in D. W. Goodall (Ed.), *Evolution of Desert Biota.* Univ. Texas Press, Austin. 250 pp.

MALY, E. J. 1969. A laboratory study of the interaction between the predatory rotifer *Asplanchna* and *Paramecium. Ecology* 50:59–73.

MANN, K. H. 1969. The dynamics of aquatic ecosystems. *Adv. Ecol. Res.* 6:1–81.

MARGALEF, R. 1958a. Information theory in ecology. *Gen. Syst.* 3:36–71.

MARGALEF, R. 1958b. Temporal succession and spatial heterogeneity in phyto-plankton. In Buzzati-Traverso (Ed), *Perspectives in marine biology.* Univ. of California Press, Berkeley. 621 pp.

MARGALEF, R. 1963. On certain unifying principles in ecology. *Amer. Natur.* 97:357–374.

MARGALEF, R. 1968. *Perspectives in ecological theory.* Univ. of Chicago Press, Chicago. 111 pp.

MARGALEF, R. 1969. Diversity and stability: a practical proposal and a model of interdependence. *Brookhaven Symp. Biol.* 22:25–37.

MARTIN, P., and P. J. MEHRINGER, JR. 1965. Pleistocene pollen analysis and biogeography of the southwest. Pp. 433–451 in Wright and Fry (Eds.), *The quaternary of the U.S.* Princeton Univ. Press, Princeton, N.J.

MAY, R. M. 1971. Stability in multi-species community models. *Math. Biosci.* 12:59–79.

MAY, R. M. 1973. *Stability and complexity in model ecosystems.* Princeton Univ. Press, Princeton, N.J.

MAY, R. M. 1974. On the theory of niche overlap. *Theoret. Pop. Biol.* 5:297–332.

MAY, R. M. 1975a. Patterns of species abundance and diversity. Chapter 4 (pp. 81–120) in M. L. Cody and J. M. Diamond (Eds.), *Ecology and evolution of communities.* Harvard Univ. Press, Cambridge.

MAY, R. M. 1975b. Some notes on estimating the competition matrix, α. *Ecology* 56:737–741.

MAY, R. M. 1975c. Stability in ecosystems: some comments. Pp. 161–168 in W. H. van Dobben and R. H. Lowe-McConnell (Eds.) *Unifying concepts in ecology.* D. W. Junk, The Hague.

MAY, R. M (Ed.) 1976a. *Theoretical ecology: principles and applications.* Blackwell, Oxford. 317 pp.

MAY, R. M. 1976b. Estimating r: a pedagogical mote. *Amer. Natur.* 110:496–499.

MAY, R. M., and R. H. MACARTHUR. 1972. Niche overlap as a function of environmental variability. *Proc. Nat. Acad. Sci.* 69:1109–1113.

MAYNARD SMITH, J. 1956. Fertility, mating behavior, and sexual selection in *Drosophila subobscura. J. Genet.* 54:261–279.

MAYNARD SMITH, J. 1958. *The theory of evolution.* Penguin, Baltimore. 320 pp.

MAYNARD SMITH, J. 1964. Group selection and kin selection: a rejoinder. *Nature* 201:1145–1147.

MAYNARD SMITH, J. 1968. *Mathematical ideas in biology.* Cambridge Univ. Press, Cambridge. 152 pp.

MAYNARD SMITH, J. 1971. The origin and maintenance of sex. Pp. 163–175 in G. C. Williams (Ed.), *Group selection.* Aldine, Chicago. 210 pp.

MAYNARD SMITH, J. 1974. *Models in ecology.* Cambridge Univ. Press. 146 pp.

MAYNARD SMITH, J. 1976. A comment on the Red Queen. *Amer. Natur.* 110:325–330.

MAYR, E. 1959. Where are we? *Cold Spring Harbor Symp. Quant. Biol.* 24:1–14.

MAYR, E. 1961. Cause and effect in biology. *Science* 134:1501–1506.

MEDAWAR, P. B. 1957. *The uniqueness of the individual.* Methuen, London. 191 pp.

MENDEL, G. 1865. Versuche über Pflanzenhybriden. *Verh. naturforsch. Verein Brunn* 4:3–17. (Translated and reprinted in W. Bateson. 1909. *Mendel's principles of heredity.* Cambridge Univ. Press, Cambridge.)

MENGE, B. A. 1972a. Foraging strategy of a starfish in relation to actual prey availability and environmental predictability. *Ecol. Monogr.* 42:25–50.

MENGE, B. A. 1972b. Competition for food between two intertidal starfish species and its effect on body size and feeding. *Ecology* 53:635–644.

MENGE, B. A. 1974. Effect of wave action and competition on brooding and reproductive effort in the seastar, *Leptasterias hexactis, Ecology* 55:84–93.

MENGE, B. A., and J. P. SUTHERLAND. 1976. Species diversity gradients: synthesis of the roles of predation, competition, and temporal heterogeneity. *Amer. Natur.* 110:351–369.

MENGE, J. L., and B. A. MENGE. 1974. Role of resource allocation, aggression and spatial heterogeity in coexistence of two competing intertidal starfish. *Ecol. Monog.* 44:189–209.

MERRIAM, C. H. 1890. Results of a biological survey of the San Francisco mountain region and the desert of the Little Colorado, Arizona. *North American Fauna* 3:1–113.

MERTZ, D. B. 1970. Notes on methods used in life-history studies. Pp. 4–17 in J. H. Connell, D. B. Mertz, and W. W. Murdoch (Eds.), *Readings in ecology and ecological genetics.* Harper & Row, New York. 397 pp.

MERTZ, D. B. 1971a. Life history phenomena in increasing and decreasing populations. Pp. 361–399 in E. C. Pielou and W. E. Waters (Eds.), *Statistical ecology.* Vol. II. *Sampling and modeling biological populations and population dynamics.* Pennsylvania State Univ. Press, University Park.

MERTZ, D. B. 1971b. The mathematical demography of the California condor population. *Amer. Natur.* 105:437–453.

MERTZ, D. B. 1975. Senescent decline in flour beetle strains selected for early adult fitness. *Physiological Zoology* 48:1–23.

METTLER, L. E., and T. G. GREGG. 1969. *Population genetics and evolution.* Prentice-Hall, Englewood Cliffs, N.J. 212 pp.

MEYER, B. S., D. B. ANDERSON, and R. H. BOHNING. 1960. *Introduction to plant physiology.* Van Nostrand, New York. 784 pp.

MILLAR, J. S. 1973. Evolution of litter size in the pika, *Ochotona princeps. Evolution* 27:134–143.

MILLER, P. L. 1977. Quantitative plant ecology. In D. J. Horn, R. Mitchell, and G. R. Stairs (Eds.), *Analysis of ecological systems.* Ohio State Univ. Press.

MILLER, R. S. 1964. Ecology and distribution of pocket gophers *(Geomyidae)* in Colorado. *Ecology* 45:256–272.

MILLER, R. S. 1967. Pattern and process in competition. *Adv. Ecol. Res.* 4:1–74.

MILNE, A. 1961. Definition of competition among animals. Pp. 40–61 in F. L. Milthorpe (Ed.), *Mechanisms in biological competition.* Symp. Soc. Exp. Biol. No. 15. Cambridge Univ. Press, London.

MILSUM, J. H. 1973. A short note on "stability in multi-species community models." *Math. Biosci.* 17:189–190.

MILTHORPE, F. L. (Ed.) 1961. *Mechanisms in biological competition.* Symp. Soc. Exp. Biol. No. 15. Cambridge Univ. Press, London. 365 pp.

MOHR, C. O. 1940. Comparative populations of game, fur and other mammals. *Amer. Midl. Natur.* 24:581–584.

MOHR, C. O. 1943. Cattle droppings as ecological units. *Ecol. Monogr.* 13:275–298.

MOONEY, H. A., O. BJÖRKMAN, and J. BERRY. 1975. Photosynthetic adaptations to high temperature. Pp. 138–151 in N. Hadley (Ed.), *Environmental physiology of desert organisms.* Dowden, Hutchinson & Ross, Inc. Stroudsburg, Penn.

MORSE, D. H. 1971. The insectivorous bird as an adaptive strategy. *Ann. Rev. Ecol. Syst.* 2:177–200.

MOTOMURA, I. 1932. A statistical treatment of associations (in Japanese). *Japan. J. Zool.* 44:379–383.

MURDOCH, W. W. 1966a. Community structure, population control, and competition—a critique. *Amer. Natur.* 100:219–226.

MURDOCH, W. W. 1966b. Population stability and life history phenomena. *Amer. Natur.* 100:5–11.

MURDOCH, W W. 1969. Switching in general predators: experiments on predator specificity and stability of prey populations. *Ecol. Monogr.* 39:335–354.

MURDOCH, W. W. 1970. Population regulation and population inertia. *Ecology* 51:497–502.

MURDOCH, W. W., F. C. EVANS, and C. H. PETERSON. 1972. Diversity and pattern in plants and insects. *Ecology* 53:819–829.

MURPHY, G. I. 1968. Pattern in life history and the environment. *Amer. Natur.* 102:391–403.

NATIONAL ACADEMY OF SCIENCE. 1969. *Eutrophication: causes, consequences and correctives.* Int. Symp. Eutrophication, Washington, D.C. 661 pp.

NEILL, W. E. 1972. Effects of size-selective predation on community structure in laboratory aquatic microcosms. Ph.D. Dissertation, Univ. of Texas, Austin. 177 pp.

NEILL, W. E. 1974. The community matrix and interdependence of the competition coefficients. *Amer. Natur.* 108:399–408.

NEILL, W. E. 1975. Experimental studies of microcrustacean competition, community composition and efficiency of resource utilization. *Ecology* 56:809–826.

NEWBIGIN, M. I. 1936. *Plant and animal geography.* Methuen, London. 298 pp.

NEWELL, N. D. 1949. Phyletic size increase: an important trend illustrated by fossil invertebrates. *Evolution* 3:103–124.

NEYMAN, J., T. PARK, and E. L. SCOTT. 1956. Struggle for existence. The *Tribolium* model: biological and statistical aspects. Pp. 41–79 in *Proc. 3rd Berkeley symp. on mathematical statistics and probability.* Vol. IV. Univ. California Press, Berkeley.

NICHOLSON, A. J. 1933. The balance of animal populations. *J. Anim. Ecol.* 2:132–178.

NICHOLSON, A. J. 1954. An outline of the dynamics of animal populations. *Aust. J. Zool.* 2:9–65.

NICHOLSON, A. J. 1957. The self-adjustment of populations to change. *Cold Spring Harbor Symp. Quant. Biol.* 22:153–173.

ODUM, E. P. 1959. *Fundamentals of ecology.* (2nd ed.) Saunders, Philadelphia. 564 pp.

ODUM, E. P. 1963. *Ecology.* Holt, Rinehart and Winston, New York. 152 pp.

ODUM, E. P. 1968. Energy flow in ecosystems: a historical review. *Amer. Zool.* 8:11–18.

ODUM, E. P. 1969. The strategy of ecosystem development. *Science* 164:262–270.

ODUM, E. P. 1971. *Fundamentals of ecology.* (3rd ed.) Saunders, Philadelphia. 574 pp.

ODUM, H. T. 1971. *Environment, power, and society.* Wiley, New York. 331 pp.

OPARIN, A. I. 1957. *The origin of life on the earth.* (3rd ed.) Oliver and Boyd, London. 495 pp.

OOSTING, H. J. 1958. *The study of plant communities.* (2nd ed.) Freeman, San Francisco. 440 pp.

ORIANS, G. H. 1962. Natural selection and ecological theory. *Amer. Natur.* 96:257–263.

ORIANS, G. H. 1969a. The number of bird species in some tropical forests. *Ecology* 50:783–797.

ORIANS, G. H. 1969b. On the evolution of mating systems in birds and mammals. *Amer. Natur.* 103:589–603.

ORIANS, G. H. 1971. Ecological aspects of behavior. Chapter 11 (Pp. 513–546) in D. S. Farner and J. R. King (Eds.) *Avian biology,* vol. I. Academic Press, New York. 586 pp.

ORIANS, G. H. 1972. The adaptive significance of mating systems in the Icteridae. *Proc. XV Int. Ornith. Congr.* 389–398.

ORIANS, G. H. 1974. An evolutionary approach to the study of ecosystems. Pp. 198–200 in *Structure, functioning and management of ecosystems.* Proc. First Int. Congr. Ecol., The Hague, Netherlands.

ORIANS, G. H. 1975. Diversity, stability and maturity in natural ecosystems. Pp. 139–150 in W. H. VanDobben and R. H. Lowe-McConnell (Eds.), *Unifying concepts in ecology.* W. Junk, The Hague.

ORIANS, G. H., and H. S. HORN. 1969. Overlap in foods and foraging of four species of blackbirds in the potholes of central Washington. *Ecology* 50:930–938.

ORIANS, G. H., and N. PEARSON. 1977. On the theory of Central place foraging. In D. J. Horn, R. Mitchell, and G. R. Stairs, (Eds.), *Analysis of ecological systems.* Ohio State Univ. Press, Columbus.

ORIANS, G. H., and O. T. SOLBRIG. 1977. A cost-income model of leaves and roots with special reference to arid and semi-arid areas. *Amer. Natur.* 111:677–690.

ORIANS, G. H., and M. F. WILLSON. 1964. Interspecific territories of birds. *Ecology* 45:736–745.

OTTE, D. 1975. Plant preference and plant succession. A consideration of evolution of plant preference in *Schistocerca. Oecologia* 18:129–144.

OTTE, D., and K. WILLIAMS. 1972. Environmentally induced color dimorphisms in grasshoppers, *Syrbula admirablis, Dichromorpha viridis,* and *Chortophaga viridifasciata. Ann. Entomol. Soc. Amer.* 65:1154–1161.

PAINE, R. T. 1966. Food web complexity and species diversity. *Amer. Natur.* 100:65–76.

PAINE, R. T. 1971. The measurement and application of the calorie to ecological problems. *Ann. Rev. Ecol. Syst.* 2:145–164.

PARK, T. 1948. Experimental studies of interspecific competition. I. Competition between populations of flour beetles *Tribolium confusum* Duval and *T. castaneum* Herbst. *Physiol. Zool.* 18:265–308.

PARK, T. 1954. Experimental studies of interspecific competition. II. Temperature, humidity, and competition in two species of *Tribolium. Physiol. Zool.* 27:177–238.

PARK, T. 1962. Beetles, competition, and populations. *Science* 138:1369–1375.

PARK, T., P. H. LESLIE, and D. B. MERTZ. 1964. Genetic strains and competition in populations of *Tribolium. Physiol. Zool.* 37:97–162.

PARKER, B. C., and B. L. TURNER. 1961. "Operational niche" and "community-interaction values" as determined from *in vitro* studies of some soil algae. *Evolution* 15:228–238.

PARKHURST, D. F., and O. L. LOUCKS. 1971. Optimal leaf size in relation to environment. *J. Ecol.* 60:505–537.

PATTEN, B. C. 1959. An introduction to the cybernetics of the ecosystem: the trophic-dynamic aspect. *Ecology* 40:221–231.

PATTEN, B. C. 1961. Competitive exclusion. *Science* 134:1599–1601.

PATTEN, B. C. 1962. Species diversity in net phytoplankton of Raritan Bay. *J. Marine Res.* 20:57–75.

PATTEN, B. C. (Ed.) 1971. *Systems analysis and simulation in ecology.* Vol. I. Academic Press, New York. 607 pp.

PATTEN, B. C. (Ed.) 1972. *Systems analysis and simulation in ecology.* Vol. II. Academic Press, New York. 592 pp.

PATTEN, B. C. (Ed.) 1975. *Systems analysis and simulation in ecology.* Vol. III. Academic Press, New York.

PATTEN, B. C. (Ed.) 1976. *Systems analysis and simulation in ecology.* Volume IV. Academic Press, New York. 608 pp.

PATTEN, D. T. and E. M. SMITH. 1975. Heat flux and the thermal regime of desert plants. Pp. 1–19 in H. F. Hadley (Ed.) *Environmental physiology of desert organisms.* Dowden, Hutchinson & Ross, Stroudsburg Penn. 283 pp.

PAULING, L. 1970. *Vitamin C and the common cold.* Freeman.

PEARL, R. 1922. *The biology of death.* Lippincott, Philadelphia. 275 pp.

PEARL, R. 1927. The growth of populations. *Quart. Rev. Biol.* 2:532–548.

PEARL, R. 1928. *The rate of living.* Knopf, New York.

PEARL, R. 1930. *The biology of population growth.* Knopf, New York. 260 pp.

PEARSON, O. P. 1948. Metabolism and energetics. *Sci. Monthly* 66:131–134.

PERKINS, E. J. 1974. *The biology of estuaries and coastal waters.* Academic Press, New York. 678 pp.

PERRINS, C. M. 1964. Survival of young swifts in relation to brood-size. *Nature* 201:1147–1149.

PERRINS, C. M. 1965. Population fluctuations and clutch size in the great tit, *Parus major* L. *J. Anim. Ecol.* 34:601–647.

PETERSON, C. H. 1975. Stability of species and of community for the benthos of two lagoons. *Ecology* 56:958–965.

PHILLIPSON, J. 1966. *Ecological energetics.* Edward Arnold, London. 57 pp.

PIANKA, E. R. 1966a. Latitudinal gradients in species diversity: a review of concepts. *Amer. Natur.* 100:33–46.

PIANKA, E. R. 1966b. Convexity, desert lizards, and spatial heterogeneity. *Ecology* 47:1055–1059.

PIANKA, E. R. 1969. Sympatry of desert lizards (*Ctenotus*) in western Australia. *Ecology* 50:1012–1030.

PIANKA, E. R. 1970. On r and K selection. *Amer. Natur.* 104:592–597.

PIANKA, E. R. 1971a. Species diversity. Pp. 401–406 in *Topics in the study of life: the bio source book.* Harper & Row, New York. 482 pp.

PIANKA, E. R. 1971b. Ecology of the agamid lizard *Amphibolurus isolepis* in Western Australia. *Copeia* 1971:527–536.

PIANKA, E. R. 1972. r and K selection or b and d selection? *Amer. Natur.* 106:581–588.

PIANKA, E. R. 1973. The structure of lizard communities. *Ann. Rev. Ecol. Syst.* 4:53–74.

PIANKA, E. R. 1974. Niche overlap and diffuse competition. *Proc. Nat. Acad. Sci.* USA 71:2141–2145.

PIANKA, E. R. 1975. Niche relations of desert lizards. Chapter 12 (pp. 292–314) in M. Cody and J. M. Diamond (Eds.), *Ecology and evolution of communities.* Harvard Univ. Press, Cambridge.

PIANKA, E. R. 1976a. Competition and niche theory. Chapter 7 (pp. 114–141) in R. M. May (Ed.), *Theoretical ecology: principles and applications.* Blackwell.

PIANKA, E. R. 1976b. Natural selection of optimal reproductive tactics. *Amer. Zool.* 16:775–784.

PIANKA, E. R., and W. S. PARKER. 1975a. Ecology of horned lizards: a review with special reference to *Phrynosoma platyrhinos. Copeia* 1975:141–162.

PIANKA, E. R., and W. S. PARKER. 1975b. Age-specific reproductive tactics. *Amer. Natur.* 109:453–464.

PIANKA, E. R., R. B. HUEY, and L. R. LAWLOR. 1977. Niche segregation in desert lizards. In D. J. Horn, R. Mitchell, and G. R. Stairs (Eds.), *Analysis of ecological systems.* Ohio State Univ. Press, Columbus.

PICKETT, S. T. A. 1976. Succession: an evolutionary interpretation. *Amer. Natur.* 110:107–119.

PIELOU, E. C. 1969. *An introduction to mathematical ecology.* Wiley-Interscience, New York. 286 pp.

PIELOU, E. C. 1972. Niche width and niche overlap: a method for measuring them. *Ecology* 53:687–692.

PIELOU, E. C. 1975. *Ecological diversity.* Wiley, New York. 162 pp.

PIELOU, E. C. 1974. *Population and community ecology: principles and methods.* Gordon and Breach, New York. 424 pp.

PIMENTEL, D. 1968. Population regulation and genetic feedback. *Science* 159:1432–1437.

PITELKA, F. A. 1964. The nutrient-recovery hypothesis for arctic microtine cycles. I. Introduction Pp. 55–56 in D. J. Crisp (Ed.), *Grazing in terrestrial and marine environments.* Brit. Ecol. Soc. Symposium.

PITTENDRIGH, C. S. 1961. Temporal organization in living systems. *Harvey Lecture Series* 56:93–125. Academic Press, New York.

PLATT, J. R. 1964. Strong inference. *Science* 146:347–353.

PONNAMPERUMA, C. 1972. *The origins of life.* Dutton, New York. 215 pp.

POOLE, R. W. 1974. *An introduction to quantitative ecology.* McGraw-Hill, New York. 532 pp.

PORTER, W. P., and D. M. GATES. 1969. Thermodynamic equilibria of animals with environment. *Ecol. Mong.* 39:227–244.

PORTER, W. P., J. W. MITCHELL, W. A. BECKMAN, and C. B. DEWITT. 1973. Behavioral implications of mechanistic ecology—thermal and behavioral modeling of desert ectotherms and their microenvironment. *Oecologia* 13:1–54.

POULSON, T. L., and D. D. CULVER. 1969. Diversity in terrestrial cave communities. *Ecology* 50:153–158.

PRESTON, F. W. 1948. The commonness and rarity of species. *Ecology* 29:254–283.

PRESTON, F. W. 1960. Time and space and the variation of species. *Ecology* 41:611–627.

PRESTON, F. W. 1962a. The canonical distribution of commonness and rarity. I. *Ecology* 43:185–215.

PRESTON, F. W. 1962b. The canonical distribution of commonness and rarity. II. *Ecology* 43:410–432.

PRICE, G., and J. MAYNARD SMITH. 1973. The logic of animal conflict. *Nature* 246:15–18.

PRICE, P. W. 1975. *Insect ecology.* Wiley, New York. 514 pp.

PROSSER, C. L. (Ed.) 1973. *Comparative animal physiology.* Saunders, Philadelphia. 428 pp.

PULLIAM, H. R. 1974. On the theory of optimal diets. *Amer. Natur.* 108:50–65.

RAND, A. S. 1967. Predator–prey interactions and the evolution of aspect diversity. *Atas do Simposio sobre a Biota Amazonica* 5:73–83.

RANDOLPH, P. A., J. C. RANDOLPH, and C. A. BARLOW. 1975. Age-specific energetics of the pea aphid, *Acyrothosiphon pisum. Ecology* 56:359–369.

RAPPORT, D. J. 1971. An optimization model of food selection. *Amer. Natur.* 105:575–587.

RAUNKAIER, C. 1934. *The life form of plants and statistical plant geography.* Clarendon, Oxford. 632 pp.

RECHER, H. F. 1969. Bird species diversity and habitat diversity in Australia and North America. *Amer. Natur.* 103:75–80.

REICHLE, D. (Ed.) 1970. *Analysis of temperate forest ecosystems.* Springer-Verlag, Heidelberg, Berlin. 304 pp.

RHOADES, D. F., and R. G. CATES. 1976. Toward a general theory of plant antiherbivore chemistry. In J. Wallace and R. Mansell (Eds.), *Biochemical interactions between plants and insects.* Recent Advances in Phytochemistry, Vol. 10.

RICHARDS, B. N. 1974. *Introduction to the soil ecosystem.* Longman, New York. 266 pp.

RICHARDS, P. W. 1952. *The tropical rain forest.* Cambridge Univ. Press, New York. 450 pp.

RICKLEFS, R. E. 1966. The temporal component of diversity among species of birds. *Evolution* 20:235–242.

RICKLEFS, R. E. 1973. *Ecology.* Chiron Press, Portland, Oregon. 861 pp.

RICKLEFS, R. E. 1977. Environmental heterogeneity and plant species diversity: a hypothesis. *Amer. Natur.* 111:376–381.

RICKLEFS, R. E., and G. W. COX. 1972. The taxon cycle in the land bird fauna of the West Indies. *Amer. Natur.* 106:195–219.

RICKLEFS, R. E., and K. O'ROURKE. 1975. Aspect diversity in moths: a temperate-tropical comparison. *Evolution* 29:313–324.

ROOT, R. B. 1967. The niche exploitation pattern of the blue-gray gnatcatcher. *Ecol. Monog.* 37:317–350.

ROSEN, R. 1967. *Optimality principles in biology.* Plenum, New York. 198 pp.

ROSENZWEIG, M. L. 1968. Net primary productivity of terrestrial communities: prediction from climatological data. *Amer. Natur.* 102:67–74.

ROSENZWEIG, M. L. 1971. The paradox of enrichment: destabilization of exploitation ecosystems in ecological time. *Science* 171:385–387.

ROSENZWEIG, M. L. 1973a. Exploitation in three trophic levels. *Amer. Natur.* 107:275–294.

ROSENZWEIG, M. L. 1973b. Evolution of the predator isocline. *Evolution* 27:84–94.

ROSENZWEIG, M. L., and R. H. MACARTHUR. 1963. Graphical representation and stability conditions of predator–prey interactions. *Amer. Natur.* 97:209–223.

ROSS, H. H. 1957. Principles of natural coexistence indicated by leafhopper populations. *Evolution* 11:113–129.

ROSS, H. H. 1958. Further comments on niches and natural coexistence. *Evolution* 12:112–113.

ROTHSTEIN, S. I. 1973. The niche-variation model—is it valid? *Amer. Natur.* 107:598–620.

ROUGHGARDEN, J. 1971. Density-dependent natural selection. *Ecology* 52:453–468.

ROUGHGARDEN, J. 1972. Evolution of niche width. *Amer. Natur.* 106:683–718.

ROUGHGARDEN, J. 1974a. Species packing and the competition function with illustrations from coral reef fish. *Theoret. Pop. Biol.* 5:163–186.

ROUGHGARDEN, J. 1974b. Niche width: biogeographic patterns among *Anolis* lizard populations. *Amer. Natur.* 108:429–442.

ROUGHGARDEN, J. 1974c. The fundamental and realized niche of a solitary population. *Amer. Natur.* 108:232–235.

ROUGHGARDEN, J. 1976. Resource partitioning among competing species: a coevolutionary approach. *Theoret. Pop. Biol.* 9:388–424.

ROUGHGARDEN, J., and M. FELDMAN. 1975. Species packing and predation pressure. *Ecology* 56:489–492.

ROYAMA, T. 1969. A model for the global variation of clutch size in birds. *Oikos* 20:562–567.

ROYAMA, T. 1970. Factors governing the hunting behaviour and selection of food by the great tit *(Parus major* L.*). J. Anim. Ecol.* 39:619–668.

RUIBAL, R. 1961. Thermal relations of five species of tropical lizards. *Evolution* 15:98–111.

RUIBAL, R., and R. PHILIBOSIAN. 1970. Eurythermy and niche expansion in lizards. *Copeia* 1970:645–653.

RUSSELL-HUNTER, W. D. 1970. *Aquatic productivity: an introduction to some basic aspects of biological oceanography and limnology.* Macmillan, New York. 306 pp.

RUTTNER, F. 1953. *Fundamentals of limnology.* Univ. of Toronto Press, Toronto Canada. 242 pp.

RYDER, V. 1954. On the morphology of leaves. *Bot. Rev.* 20:263–276.

SALE, P. 1974. Overlap in resource use and interspecific competition. *Oecologia* 17:245–256.

SALISBURY, E. J. 1942. *The reproductive capacity of plants; studies in quantitative biology.* Bell and Sons, London. 244 pp.

SADLIER, R. M. 1973. *Reproduction of vertebrates.* Academic Press, New York.

SALT, G. W. 1967. Predation in an experimental protozoan population *(Woodruffia-Paramecium). Ecol. Monogr.* 37:113–144.

SALTHE, S. N. 1972. *Evolutionary biology.* Holt, Rinehart and Winston, New York. 437 pp.

SAVAGE, J. M. 1958. The concept of ecologic niche with reference to the theory of natural coexistence. *Evolution* 12:111–121.

SAWYER, J. S. (Ed.) 1966. *World climate from 8,000 to 0 B.C. Proc. Int. Symp. on World Climate 8,000 to 0 B.C.,* Imperial College, London. Royal Meteorological Society, London. 229 pp.

SCHAFFER, W. M. 1974. Selection for optimal life histories: effects of age structure. *Ecology* 55:291–303.

SCHAFFER, W. M., and R. H. TAMARIN. 1973. Changing reproductive rates and population cycles in lemmings and voles. *Evolution* 27:111–124.

SCHALLER, F. 1968. *Soil animals* Univ. of Michigan Press, Ann Arbor. 144 pp.

SCHMIDT-NIELSEN, K. 1964. *Desert animals: physiological problems of heat and water.* Oxford Univ. Press, London. 277 pp.

SCHMIDT-NIELSEN, K. 1972. Locomotion: energy cost of swimming, flying, and running. *Science* 177:222–228.

SCHMIDT-NIELSEN, K. 1975. *Animal physiology: adaptation and environment.* Cambridge Univ. Press, London. 699 pp.

SCHMIDT-NIELSEN, K., and W. R. DAWSON. 1964. Terrestrial animals in dry heat: desert reptiles. Pp. 467–480 in D. B. Dill (Ed.), *Handbook of physiology,* Section 4: *Adaptation to the environment.* Amer. Physiol. Soc., Washington, D.C.

SCHOENER, A. 1974. Experimental zoogeography: colonization of marine mini-islands. *Amer Natur.* 108:715–738.

SCHOENER, T. W. 1965. The evolution of bill size differences among sympatric congeneric species of birds. *Evolution* 19:189–213.

SCHOENER, T. W. 1967. The ecological significance of sexual dimorphism in size in the lizard *Anolis conspersus. Science* 155:474–477.

SCHOENER, T. W. 1968a. The *Anolis* lizards of Bimini: resource partitioning in a complex fauna. *Ecology* 49:704–726.

SCHOENER, T. W. 1968b. Sizes of feeding territories among birds. *Ecology* 49:123–141.

SCHOENER, T. W. 1969a. Models of optimal size for solitary predators. *Amer. Natur.* 103:277–313.

SCHOENER, T. W. 1969b. Optimal size and specialization in constant and fluctuating environments: an energy-time approach. *Brookhaven Symp. Biol.* 22:103–114.

SCHOENER, T. W. 1970. Nonsynchronous spatial overlap of lizards in patchy habitats. *Ecology* 51:408–418.

SCHOENER, T. W. 1971. Theory of feeding strategies. *Ann. Rev. Ecol. Syst.* 2:369–404.

SCHOENER, T. W. 1973. Population growth regulated by intraspecific competition for energy or time: some simple representations. *Theoret. Pop. Biol.* 4:56–84.

SCHOENER, T. W. 1974a. Resource partitioning in ecological communities. *Science* 185:27–39.

SCHOENER, T. W. 1974b. The compression hypothesis and temporal resource partitioning. *Proc. Nat. Acad. Sci. USA* 71:4169–4172.

SCHOENER, T. W. 1975a. Competition and the form of habitat shift. *Theoret. Pop. Biol.* 5:265–307.

SCHOENER, T. W. 1975b. Presence and absence of habitat shift in some widespread lizard species. *Ecol. Monog.* 45:232–258.

SCHOENER, T. W. 1976a. The species–area relation within archipelagos: models and evidence from island land birds. *Proc. 16th Int. Ornith. Congr.*

SCHOENER, T. W. 1976b. Alternatives to Lotka-Volterra competition: models of intermediate complexity. *Theoret. Pop. Biol.* 10:309–333.

SCHOENER, T. W. 1977. Competition and the niche. In D. W. Tinkle and C. Gans (Eds.), *Biology of the Reptilia.* Academic Press, New York.

SCHOENER, T. W., and G. C. GORMAN. 1968. Some niche differences in three Lesser Antillean lizards of the genus *Anolis. Ecology* 49:819–830.

SCHOENER, T. W., and D. JANZEN. 1968. Notes on environmental determinants of tropical versus temperate insect size patterns. *Amer. Natur.* 102:207–224.

SCHULTZ, A. M. 1964. The nutrient-recovery hypothesis for arctic microtine cycles. Pp. 57–68 in D. J. Crisp (Ed.), *Grazing in terrestrial and marine environments.* Birt. Ecol. Soc. Symposium.

SCHULTZ, A. M. 1969. A study of an ecosystem: the arctic tundra. Pp. 77–93 in G. Van Dyne (Ed.), *The ecosystem concept in natural resource management.* Academic Press, New York.

SEIFERT, R. P., and F. H. SEIFERT. 1976. A community matrix analysis of *Heliconia* insect communities. *Amer. Natur.* 110:461–483.

SELANDER, R. K. 1965. On mating systems and sexual selection. *Amer. Natur.* 99:129–141.

SELANDER, R. K. 1966. Sexual dimorphism and differential niche utilization in birds. *Condor* 68:113–151.

SELANDER, R. K. 1972. Sexual selection and dimorphism in birds. Pp. 180–230 in B. G. Campbell (Ed.), *Sexual selection and the descent of man (1871–1971).* Aldine-Atherton, Chicago.

SELANDER, R. K., and W. E. JOHNSON. 1972. Genetic variation among vertebrate species. *Proc. XVII Int. Congr. Zool.*, 1972. (Abridged version in *Ann. Rev. Ecol. Syst.* 4:75–91.)

SHANNON, C. E. 1948. The mathematical theory of communication. Pp. 3–91 in Shannon and Weaver (Eds.), *The mathematical theory of communication.* Univ. Illinois Press, Urbana. 117 pp.

SHELDON, A. L. 1972. Comparative ecology of *Arcynopteryx* and *Diura* (Plecoptera) in a California stream. *Arch. Hydrobiol.* 69:521–546.

SHELFORD, V. E. 1913a. *Animal communities in temperate America.* Univ. of Chicago Press, Chicago. 368 pp.

SHELFORD, V. E. 1913b. The reactions of certain animals to gradients of evaporating power and air. A study in experimental ecology. *Biol. Bull.* 25:79–120.

SHELFORD, V. E. 1963. *The ecology of North America.* Univ. of Illinois Press, Urbana. 610 pp.

SHEPPARD, P. M. 1951. Fluctuations in the selective value of certain phenotypes in the polymorphic land snail *Cepaea nemoralis. Heredity* 5:125–134.

SHEPPARD, P. M. 1959. *Natural selection and heredity.* Hutchinson Univ. Library, London. 212 pp.

SHIMWELL, D. W. 1971. *Description and classification of vegetation.* Univ. Washington Press, Seattle. 264 pp.

SHUGART, H. H., and B. C. PATTEN. 1972. Niche quantification and the concept of niche pattern. Pp. 284–327 in B. Patten (Ed.), *Systems analysis and simulation in ecology.* Vol. II. Academic Press, New York. 592 pp.

SIMBERLOFF, D. S. 1974. Equilibrium theory of island biogeography and ecology. *Ann. Rev. Ecol. Syst.* 5:161–182.

SIMBERLOFF, D. S., and E. O. WILSON. 1970. Experimental zoogeography of islands. A two-year record of colonization. *Ecology* 51:934–937.

SIMPSON, E. H. 1949. Measurement of diversity. *Nature* 163:688.

SIMPSON, G. G. 1969. Species density of North American recent mammals. *Syst. Zool.* 13:57–73.

SKUTCH, A. F. 1949. Do tropical birds rear as many young as they can nourish? *Ibis* 91:430–455.

SKUTCH, A. F. 1967. Adaptive limitation of the reproductive rate of birds. *Ibis* 109:579–599.

SLATKIN, M. 1974. Competition and regional coexistence. *Ecology* 55:128–134.

SLOBODKIN, L. B. 1960. Ecological energy relationships at the population level. *Amer. Natur.* 94:213–236.

SLOBODKIN, L. B. 1962. Energy in animal ecology. *Adv. Ecol. Res.* 1:69–101.

SLOBODKIN, L. B. 1962. *Growth and regulation of animal populations.* Holt, Rinehart and Winston, New York. 184 pp.

SLOBODKIN, L. B. 1968. How to be a predator. *Amer. Zool.* 8:43–51.

SMITH, A. D. 1940. A discussion of the application of a climatological diagram, the hythergraph, to the distribution of natural vegetation types. *Ecology* 21:184–191.

SMITH, C. C. 1968. The adaptive nature of social organization in the genus of tree squirrels *Tamiasciurus. Ecol. Monogr.* 38:31–63.

SMITH, C. C. 1970. The coevolution of pine squirrels *(Tamiasciurus)* and conifers. *Ecol. Monogr.* 40:349–371.

SMITH, F. E. 1952. Experimental methods in population dynamics. A critique. *Ecology* 33:441–450.

SMITH, F. E. 1954. Quantitative aspects of population growth. Pp. 277–294 in E. Boell (Ed.), *Dynamics of growth processes.* Princeton Univ. Press, Princeton, N.J. 307 pp.

SMITH, F. E. 1961. Density dependence in the Australian thrips. *Ecology* 42:403–407.

SMITH, F. E. 1963a. Population dynamics in *Daphnia magna* and a new model for population growth. *Ecology* 44:651–663.

SMITH, F. E. 1963b. Density dependence. *Ecology* 44:220.

SMITH, F. E. 1970a. Analysis of ecosystems. Pp. 7–18 in D. Reichle (Ed.), *Analysis of temperate forest ecosystems.* Springer, Berlin.

SMITH, F. E. 1970b. Effects of enrichment in mathematical models. Pp. 631–645 in *Eutrophication: causes, consequences, correctives.* National Acad. Sciences, Washington, D.C.

SMITH, F. E. 1972. Spatial heterogeneity, stability, and diversity in ecosystems. Pp. 309–335 in E. S. Deevey (Ed.), *Growth by intussusception: Ecological essays in honor of G. Evelyn Hutchinson. Trans. Conn. Acad. Arts. Sci.* 44:1–443.

SMITH, N. 1968. The advantage of being parasitized. *Nature* 219:690–694.

SMITH, R. 1966. *Ecology and field biology.* Harper & Row, New York. 686 pp.

SMOUSE, P. E. 1971. The evolutionary advantages of sexual dimorphism. *Theoret. Pop. Biol.* 2:469–481.

SNELL, T. W., and D. G. BURCH. 1975. The effects of density on resource partitioning in *Chamaesyce hirta* (Euphorbiacae). *Ecology* 56:742–746.

SNELL, T. W., and C. E. KING. 1977. Lifespan and fecundity patterns in rotifers: the paradox of reproduction. *Evolution* 30 (in press).

SOKAL, R. R. 1970. Senescence and genetic load: evidence from *Tribolium. Science* 167:1733–1734.

SOLOMON, M. E. 1949. The natural control of animal populations. *J. Anim. Ecol.* 18:1–32.

SOLOMON, M. E. 1972. *Population dynamics.* Edward Arnold, London.

SOMERO, G. N. 1969. Enzymic mechanisms of temperature compensation. *Amer. Natur.* 103:517–530.

SOULÉ, M. 1971. The variation problem: the gene flow–variation hypothesis. *Taxon* 20:37–50.

SOULÉ, M., and B. R. STEWART. 1970. The "niche-variation" hypothesis: a test and alternatives. *Amer. Natur.* 104:85–97.

SOUTHWOOD, T. R. E. 1966. *Ecological methods with particular reference to the study of insect populations.* Methuen, London. 391 pp.

SPINAGE, C. A. 1972. African ungulate life tables. *Ecology* 53:645–652.

ST. AMANT, J. L. S. 1970. The detection of regulation in animal populations. *Ecology* 51:823–828.

STAHL, E. 1888. Pflanzen und Schnecken. Biolosche Studie über die Schutzmittel der Pflanzen gegen Schneckenfrass. *Jena Z. Med. Naturw.* 22:557–684.

STEARNS, S. C. 1976. Life-history tactics: a review of the ideas. *Quart. Rev. Biol.* 51:3–47.

STEWART, R. E., and J. W. ALDRICH. 1951. Removal and repopulation of breeding birds in a spruce-fir community. *Auk* 68:471–482.

STROBECK, C. 1973. n-species competition. *Ecology* 54:650–654.

SUTHERLAND, J. P. 1974. Multiple stable points in natural communities. *Amer. Natur.* 108:859–873.

SVERDRUP, H. U., M. W. JOHNSON, and R. H. FLEMING. 1942. *The oceans: their physics, chemistry, and general biology.* Prentice-Hall, Englewood Cliffs, N.J. 1087 pp.

TAMARIN, R. H., and C. J. KREBS. 1969. *Microtus* population biology. II. Genetic changes at the transferrin locus in fluctuating populations of two vole species. *Evolution* 23:183–211.

TAYLOR, G. 1920. *Australian meteorology.* Clarendon Press, Oxford. 312 pp.

TAYLOR, H. M., R. S. GOURLEY, C. E. LAWRENCE, and R. S. KAPLAN. 1974. Natural selection of life history attributes: an analytical approach. *Theoret. Pop. Biol.* 5:104–122.

TEAL, J. M. 1962. Energy flow in the salt marsh ecosystem of Georgia. *Ecology* 43:614–624.

TERBORGH, J. 1971. Distribution on environmental gradients: theory and a preliminary interpretation of distributional patterns in the avifauna of the Cordillera Vilcabamba, Peru. *Ecology* 52:23–40.

TERBORGH, J. 1974a. Faunal equilibria and the design of wildlife preserves. In F. Golley and E. Medina (Eds.), *Tropical ecological systems: trends in terrestrial and aquatic research.* Springer-Verlag, New York.

TERBORGH, J. 1974b. Preservation of natural diversity: the problem of extinction prone species. *BioScience* 24:715–722.

TERBORGH, J., and J. M. DIAMOND. 1970. Niche overlap in feeding assemblages of New Guinea birds. *Wilson Bull.* 82:29–52.

TERBORGH, J., and J. S. WESKE. 1969. Colonization of secondary habitats by Peruvian birds. *Ecology* 50:765–782.

THORNTHWAITE, C. W. 1948. An approach toward a rational classification of climate. *Geogr. Rev.* 38:55–94.

TINBERGEN, N. 1957. The functions of territory. *Bird Study* 4:14–27.

TINKLE, D. W. 1967. The life and demography of the side-blotched lizard, *Uta stansburiana. Misc. Publ. Mus. Zool., Univ. Mich.* No. 132. 182 pp.

TINKLE, D. W. 1969. The concept of reproductive effort and its relation to the evolution of life histories of lizards. *Amer. Natur.* 103:501–516.

TINKLE, D. W., H. M. WILBUR, and S. G. TILLEY. 1970. Evolutionary strategies in lizard reproduction. *Evolution* 24:55–74.

TOSI, J. A. 1964. Climatic control of terrestrial ecosystems: a report on the Holdridge model. *Econ. Geogr.* 40:173–181.

TRAMER, E. J. 1969. Bird species diversity: components of Shannon's formula. *Ecology* 50:927–929.

TREWARTHA, G. T. 1943. *An introduction to weather and climate.* McGraw-Hill, New York. 545 pp.

TRIVERS, R. L. 1971. The evolution of reciprocal altruism. *Quart. Rev. Biol.* 46:35–57.

TRIVERS, R. L. 1972. Parental investment and sexual selection. Pp. 136–179 in B. G. Campbell (Ed.), *Sexual selection and the descent of man (1871–1971)*. Aldine-Atherton, Chicago.

TRIVERS, R. L. 1974. Parent-offspring conflict. *Amer. Zool.* 14:249– 264.

TRIVERS, R. L., and D. E. WILLARD. 1973. Natural selection of parental ability to vary the sex ratio of offspring. *Science* 179:90–92.

TUCKER, V. A. 1975. The energetic cost of moving about. *Amer. Sci.* 63:413–419.

TULLOCK, G. 1970. The coal tit as a careful shopper. *Amer. Natur.* 104:77–80.

TURK, A., J. TURK, and J. T. WITTES. 1972. *Ecology pollution environment.* Saunders, Philadelphia. 217 pp.

TURNER, F. B., G. A. HODDENBACH, P. A. MEDICA, and J. R. LANNOM. 1970. The demography of the lizard *Uta stansburiana* (Baird and Girard), in southern Nevada. *J. Anim. Ecol.* 39:505–519.

TURNER, F. B., R. I. JENNRICH, and J. D. WEINTRAUB. 1969. Home ranges and body size of lizards. *Ecology* 50:1076–1081.

UDVARDY, M. D. F. 1959. Notes on the ecological concepts of habitat, biotope, and niche. *Ecology* 40:725–728.

UDVARDY, M. D. F. 1969. *Dynamic zoogeography with special reference to land animals.* Van Nostrand Reinhold, New York. 445 pp.

ULFSTRAND, S. 1977. Foraging niche dynamics and overlap in a guild of passerine birds in a south Swedish coniferous woodland. *Oecologia* 27:23–45.

UNITED NATIONS. 1968. *Demographic Year Book.* U.N., New York.

UNITED STATES DEPARTMENT OF AGRICULTURE. 1941. *Climate and man.* Washington, D.C. 1248 pp.

USHER, M. B., and M. H. WILLIAMSON (Eds.) 1974. *Ecological stability.* Halstead Press, New York.

UTIDA, S. 1957. Population fluctuation, an experimental and theoretical approach. *Cold Spring Harbor Symp. Quant. Biol.* 22:139–151.

VAN DYNE, G. M. (Ed.) 1966. *The ecosystem concept in natural resource management.* Academic Press, New York. 383 pp.

VAN VALEN, L. 1965. Morphological variation and width of the ecological niche. *Amer. Natur.* 94:377–390.

VAN VALEN, L. 1971. Group selection and the evolution of dispersal. *Evolution* 25:591–598.

VAN VALEN, L., and P. R. GRANT. 1970. Variation and niche width reexamined. *Amer. Natur.* 104:589–590.

VANDERMEER, J. H. 1968. Reproductive value in a population of arbitrary age distribution. *Amer. Natur.* 102:586–589.

VANDERMEER, J. H. 1969. The competitive structure of communities: an experimental approach with protozoa. *Ecology* 50:362–371.

VANDERMEER, J. H. 1970. The community matrix and the number of species in a community. *Amer. Natur.* 104:73–83.

VANDERMEER, J. H. 1972a. On the covariance of the community matrix. *Ecology* 53:187–189.

VANDERMEER, J. H. 1972b. Niche theory. *Ann. Rev. Ecol. Syst.* 3:107–132.

VANDERMEER, J. H. 1973. Generalized models of two species interactions: a graphical analysis. *Ecology* 54:809–818.

VANDERMEER, J. H. 1975. Interspecific competition: a new approach to the classical theory. *Science* 188:253–255.

VAURIE, C. 1951. Adaptive differences between two sympatric species of nuthatches *(Sitta). Proc. Int. Ornithol. Congr.* 19:163–166.

VERNBERG, F. J. 1975. *Physiological adaptation to the environment.* Intext Educational Publishers, New York. 576 pp.

VERNBERG, F. J., and W. B. VERNBERG. (Eds.) 1974. *Pollution and the physiological ecology of estuarine and coastal water organisms.* Academic Press, New York. 426 pp.

VERNER, J. 1964. Evolution of polygamy in the long-billed marsh wren. *Evolution* 18:252–261.

VERNER, J. 1965. Breeding biology of the long-billed marsh wren. *Condor* 67:6–30.

VERNER, J., and G. H. ENGELSEN. 1970. Territories, multiple nest building, and polygyny in the long-billed marsh wren. *Auk* 87:557–567.

VERNER, J., and M. F. WILLSON. 1966. The influence of habitats on mating systems of North American passerine birds. *Ecology* 47:143–147.

VOGEL, S. 1970. Convective cooling at low air speeds and the shape of broad leaves. *J. Exp. Bot.* 21:91–101.

VOLTERRA, V. 1926a. Fluctuations in the abundance of a species considered mathematically. *Nature* 188:558–560.

VOLTERRA, V. 1926b. Variazioni e fluttuazioni del numero d'individui in specie animali conviventi. *Mem. Acad. Lincei* 2:31–113.

VOLTERRA, V. 1931. Variation and fluctuations of the number of individuals in animal species living together. Appendix (pp. 409–448) in R. N. Chapman (1939), *Animal ecology.* McGraw-Hill, New York.

WADE, M. J. 1976. Group selection among laboratory populations of *Tribolium. Proc. Nat. Acad. Sci.* 73:4604–4607.

WADE, M. J. 1977. An experimental study of group selection. *Evolution* 31:134–153.

WAKSMAN, S. A. 1952. *Soil microbiology.* Wiley, New York. 356 pp.

WALD, G. 1964. The origins of life. *Proc. Nat. Acad. Sci.* 52:595–611.

WALLACE, A. R. 1876. *The geographical distribution of animals* (2 volumes). Hafner, New York. 503 pp. and 607 pp. (Reprinted in 1962.)

WALLACE, B. 1973. Misinformation, fitness, and kin selection. *Amer. Natur.* 107:1–7.

WALTER, H. 1939. Grassland, Savanne und Busch der Arideren teile Afrikas in ihrer ökologischen Bedingtheit. *Jahrbucher für wissenschaftliche Botanik* 87:750–860.

WANGERSKY, P. J., and W. J. CUNNINGHAM. 1956. On time lags in equations of growth. *Proc. Nat. Acad. Sci.* 42:699–702.

WARBURG, M. 1965. The evolutionary significance of the ecological niche. *Oikos* 16:205–213.

WATERMAN, T. H. 1968. Systems theory and biology—view of a biologist. In M. D. Mesarovic (Ed.), *Systems theory and biology.* Proc. 3rd Syst. Symp. Case Inst. Tech. Springer-Verlag, New York. 408 pp.

WATT, K. E. F. 1964. Comments on fluctuations of animal populations and measures of community stability. *Canad. Entomol.* 96:1434–1442.

WATT, K. E. F. 1965. Community stability and the strategy of biological control. *Canad. Entomol.* 97:887–895.

WATT, K. E. F. (Ed.) 1966. *Systems analysis in ecology*. Academic Press, New York. 276 pp.

WATT, K. E. F. 1968. *Ecology and resource management*. McGraw-Hill, New York. 450 pp.

WATT, K. E. F. 1973. *Principles of environmental science*. McGraw-Hill, New York. 319 pp.

WATTS, D. 1971. *Principles of biogeography*. McGraw-Hill, New York. 402 pp.

WEATHERLEY, A. H. 1963. Notions of niche and competition among animals, with special reference to freshwater fish. *Nature* 197:14–17.

WEAVER, J. E., and F. E. CLEMENTS. 1938. *Plant ecology*. (2nd ed.) McGraw-Hill, New York. 601 pp.

WEEDON, J. S., and J. B. FALLS. 1959. Differential responses of male ovenbirds to recorded songs of neighboring and more distant individuals. *Auk* 76:343–351.

WELCH, P. S. 1952. *Limnology*. (2nd ed.) McGraw-Hill, New York. 538 pp.

WELLINGTON, W. G. 1957. Individual differences as a factor in population dynamics: the development of a problem. *Canad. J. Zool.* 35:293–323.

WELLINGTON, W. G. 1960. Qualitative changes in natural populations during changes in abundance. *Canad. J. Zool.* 38:289–314.

WERNER, E. E., and D. J. HALL. 1974. Optimal foraging and size selection of prey by the bluegill sunfish. *Ecology* 55:1042–1052.

WETZEL, R. G. 1975. *Limnology*. Saunders, Philadelphia. 743 pp.

WEYL, P. K. 1970. *Oceanography: an introduction to the marine environment*. Wiley, New York. 535 pp.

WHITESIDE, M. C., and R. B. HAINSWORTH. 1967. Species diversity in chydorid *(Cladocera)* communities. *Ecology* 48:664–667.

WHITTAKER, R. H. 1953. A consideration of climax theory: the climax as a population and pattern. *Ecol. Monogr.* 23:41–78.

WHITTAKER, R. H. 1962. Classification of natural communities. *Bot. Rev.* 28:1–239.

WHITTAKER, R. H. 1965. Dominance and diversity in land plant communities. *Science* 147:250–260.

WHITTAKER, R. H. 1967. Gradient analysis of vegetation. *Biol. Rev.* 42:207–264.

WHITTAKER, R. H. 1969. Evolution of diversity in plant communities. *Brookhaven Symp. Biol.* 22:178–196.

WHITTAKER, R. H. 1970. *Communities and ecosystems*. Macmillan, New York. 162 pp.

WHITTAKER, R. H. 1972. Evolution and measurement of species diversity. *Taxon* 21:213–251.

WHITTAKER, R. H., and P. P. FEENY. 1971. Allelochemics: chemical interactions between species. *Science* 171:757–770.

WHITTAKER, R. H., and S. A. LEVIN. (Eds.) 1975. *Niche: theory and application*. Dowden, Hutchinson & Ross, New York.

WHITTAKER, R. H., and G. M. WOODWELL. 1971. Evolution of a natural communities. Pp. 137–159 in J. A. Wiens (Ed.), *Ecosystem structure and function*. Proc. 31st Ann. Biol. Coll., Oregon State Univ. Press.

WHITTAKER, R. H., S. A. LEVIN, and R. B. ROOT. 1973. Niche, habitat, and ecotope. *Amer. Natur.* 107:321–338.

WHITTAKER, R. H., R. B. WALKER, and A. R. KRUCKEBERG. 1954. The ecology of serpentine soils. *Ecology* 35:258–288.

WHITTOW, G. C. (Ed.) 1970. *Comparative physiology of thermoregulation*. Academic Press, New York.

WIEGERT, R. C. 1968. Thermodynamic considerations in animal nutrition. *Amer. Zool.* 8:71–81.

WIENS, J. A. 1966. Group selection and Wynne-Edward's hypothesis. *Amer. Sci.* 54:273–287.

WIESER, W. (Ed.) 1973. *Effects of temperature on ectothermic organisms.* Springer-Verlag, Berlin. 298 pp.

WILBUR, H. M. 1972. Competition, predation, and the structure of the *Ambystoma-Rana sylvatica* community. *Ecology* 53:3–21.

WILBUR, H. M. 1977. Propagule size, number, and dispersion pattern in *Ambystoma* and *Asclepias. Amer. Natur.* 111 43–68.

WILBUR, H. M., D. W. TINKLE, and J. P. COLLINS. 1974. Environmental certainty, trophic level and resource availability in life history evolution. *Amer. Natur.* 108:805–817.

WILEY, R. H. 1974. Evolution of social organization and life-history patterns among grouse. *Quart. Rev. Biol.* 49:201–227.

WILLIAMS, C. B. 1944. Some applications of the logarithmic series and the index of diversity to ecological problems. *J. Ecol.* 32:1–44.

WILLIAMS, C. B. 1953. The relative abundance of different species in a wild animal population. *J. Anim. Ecol.* 22:14–31.

WILLIAMS, C. B. 1964. *Patterns in the balance of nature.* Academic Press, New York. 324 pp.

WILLIAMS, G. C. 1957. Pleiotropy, natural selection, and the evolution of senescence. *Evolution* 11:398–411.

WILLIAMS, G. C. 1966a. *Adaptation and natural selection.* Princeton Univ. Press, Princeton, N.J. 307 pp.

WILLIAMS, G. C. 1966b. Natural selection, the costs of reproduction, and a refinement of Lack's principle. *Amer. Natur.* 100:687–690.

WILLIAMS, G. C. 1971. *Group selection.* Aldine-Atherton, Chicago. 210 pp.

WILLIAMS, G. C. 1975. *Sex and evolution.* Princeton Univ. Press, Princeton, New Jersey. 200 pp.

WILLIAMSON, M. 1967. Introducing students to the concepts of population dynamics. Pp. 169–176 in J. M. Lambert (Ed.), *The teaching of ecology.* Symp. Brit. Ecol. Soc. No. 7.

WILLIAMSON, M. 1971. *The analysis of biological populations.* Edward Arnold, London. 180 pp.

WILLIAMSON, P. 1971. Feeding ecology of the red-eyed vireo (*Vireo olivaceous*) and associated foliage gleaning birds. *Ecol. Monog.* 41:129–152.

WILLSON, M. F. 1969. Avian niche size and morphological variation. *Amer. Natur.* 103:531–542.

WILLSON, M. F. 1971. Life history consequences of death rates. *The Biologist* 53:49–56.

WILLSON, M. F. 1972a. Evolutionary ecology of plants: a review. I. Introduction and energy budgets. *The Biologist* 54:140–147.

WILLSON, M. F. 1972b. Evolutionary ecology of plants: a review. II. Ecological life histories. *The Biologist* 54:148–162.

WILLSON, M. F. 1973a. Evolutionary ecology of plants: a review. III. Ecological genetics and life history. *The Biologist* 55:1–12.

WILLSON, M. F. 1973b. Evolutionary ecology of plants: a review. IV. Niches and competition. *The Biologist* 55:74–82.

WILLSON, M. F. 1973c. Evolutionary ecology of plants: a review. V. Plant/animal interactions. *The Biologist* 55:89–105.

WILLSON, M. F., and E. R. PIANKA. 1963. Sexual selection, sex ratio, and mating system. *Amer. Natur.* 97:405–407.

WILSON, D. S. 1975. A theory of group selection. *Proc. Nat. Acad. Sci. USA* 72:143–146.

WILSON, E. O. 1961. The nature of the taxon cycle in the Melanesian ant fauna. *Amer. Natur.* 95:169–193.

WILSON, E. O. 1969. The species equilibrium. *Brookhaven Symp. Biol.* 22:38–47.

WILSON, E. O. 1971. *The insect societies.* Belknap Press, Cambridge, Mass. 548 pp.

WILSON, E. O. 1973. Group selection and its significance for ecology. *BioScience* 23:631–638.

WILSON, E. O. 1975. *Sociobiology: the new synthesis.* Harvard Univ. Press, Cambridge. 697 pp.

WILSON, E. O. 1976. The central problems of sociobiology. Pp. 205–217 in R. M. May (Ed.) *Theoretical ecology: principles and applications.* Blackwell.

WILSON, E. O., and W. H. BOSSERT. 1971. *A primer of population biology.* Sinauer, Stamford, Conn. 192 pp.

WILSON, E. O., and E. O. WILLIS. 1975. Applied biogeography. Chapter 18 (pp. 522–534) in M. L. Cody and J. M. Diamond (Eds.), *Ecology and evolution of communities.* Harvard Univ. Press, Cambridge.

WILSON, J. T. 1971. Continental drift. Pp. 88–92 in *Topics in animal behavior, ecology, and evolution.* Harper & Row, New York. 184 pp.

WILSON, J. T. (Ed.) 1973. *Continents adrift. A collection of articles from Scientific American.* Freeman, San Francisco. 172 pp.

WISEMAN, J. D. H. 1966. Evidence for recent climatic changes in cores from the ocean bed. In J. S. Sawyer (Ed.), *World climate from 8,000 to 0* B.C. Imperial College, London, 1966. Royal Meteorological Society, London.

WITTENBERGER, J. F. 1976. The ecological factors selecting for polygyny in altricial birds. *Amer. Natur.* 110:779–799.

WITTOW, G. C. (Ed.) 1970. *Comparative physiology of thermoregulation.* Academic Press, New York.

WOLF, L. L., and F. R. HAINSWORTH. 1971. Time and energy budgets of territorial hummingbirds. *Ecology* 52:980–988.

WOLF, L. L., F. R. HAINSWORTH, and F. G. STILES. 1972. Energetics of foraging: rate and efficiency of nectar extraction by hummingbirds. *Science* 176:1351–1352.

WOODWELL, G. M. and H. SMITH (Eds.) 1969. *Diversity and stability in ecological systems. Brookhaven Symp. Biol.* No. 22. Upton, N.Y. 264 pp.

WOODWELL, G. M., and R. H. WHITTAKER. 1968. Primary production in terrestrial communities. *Amer. Zool.* 8:19–30.

WRIGHT, H. E., and D. FREY (Eds) 1965. *The quaternary of the United States.* Princeton Univ. Press, Princeton, N.J.

WRIGHT, S. 1931. Evolution in Mendelian populations. *Genetics* 16:97–159.

WYNNE-EDWARDS, V. C. 1955. Low reproductive rates in birds, especially seabirds. *Acta XI Int. Orn. Congr., Basel* 1954:540–547.

WYNNE-EDWARDS, V. C. 1962. *Animal dispersion in relation to social behaviour.* Oliver and Boyd, Edinburgh. 653 pp.

WYNNE-EDWARDS, V. C. 1964. Group selection and kin selection. *Nature* 201:1145–1147.

WYNNE-EDWARDS, V. C. 1965a. Self-regulating systems in populations of animals. *Science* 147:1543–1548.

WYNNE-EDWARDS, V. C. 1965b. Social organization as a population regulator. *Symp. Zool. Soc. London* 14:173–178.

YOUSEF, M. K., S. M. HORVATH, and R. W. BULLARD (Eds.) 1972. *Physiological adaptations. Desert and mountain.* Academic Press, New York. 258 pp.

ZEUTHEN, E. 1953. Oxygen uptake as related to body size in organisms. *Quart. Rev. Biol.* 28:1–12.

ZWEIFEL, R. G., and C. H. LOWE. 1966. The ecology of a population of *Xantusia vigilis,* the desert night lizard. *Amer. Mus. Novitates* 2247:1–57.

Index

Abundance
  along ecoclines, 60–65
  factors determining, 305
  and niche breadth, 253
  of prey of different sizes, 262–263
  and relative importance, 286–287
  and species diversity, 286–287
Acacia-ant interaction, 224, 227
Acclimation, 75, 85
Acclimatization, 75
Activity, temporal patterns of, 83,
      190–193, 247, 268, 289–290
Actual evapotranspiration, 47–52
Actual rate of increase, 113–117,
      177–178, 182
Adaptation, 83–86. *See also*
      Counteradaptations
  to local conditions, 145
  of a predator-prey relationship, 201
Adaptive radiations, 267, 320–322
Adaptive suites, 92–94
Adiabatic cooling, 23–25, 31–32
Adiabatic warming, 23–25
Aerobic processes in a deep lake,
      70–71
AET. *See* Actual evapotranspiration
African ungulates, survivorship
      curves of, 102
Age class zero, 103, 108
Age distributions, 108–110, 139–140
Age of first reproduction, α, 104
Age of last reproduction, ω, 104
Age-specific expectation of future
      offspring. *See* Reproductive
      value
Age-specific fecundity and mortality,
      98–104
Age specific selection, 125–141
Age specificity, 98–110, 125–141
Age structure, 98–110, 113

Aging, 139–141
Agriculture, 299
Air. *See also* Wind
  adiabatic cooling of, 23–25, 31–32
  cooling, 20–25
  descending, 20, 23–25
  effects of currents on heat and water
      exchange, 46
  equatorial zone of rising, 20, 23,
      27–28
  movements, 20–28, 46
  and movements of gametes, 158
  pollution, 216
  surface currents, 20–21, 27
  temperature changes of, 20–25, 89
  temperature and water vapor, 23–25
Alarm calls, 169, 201, 218–220
Alkaloids, 222
Allele, 6–8, 140–141
  modifiers, 140–141
  time of expression, 140–141
  wild type, 7
Allocation
  principle of, 82, 138, 149
  of time and energy, 9–10, 82,
      148–150, 152–155, 174,
      257–266
Allopatry, 194
Alonella globulosa, reproductive
      value of, 107
Alpha. *See also* Age of first
      reproduction; Competition,
      coefficients
  matrix, 274
  values, 175–182, 189, 199, 274–276
Altricial, 161
Altruism, 167–169
  pseudo-, 167–169
  reciprocal, 169
  true, 167

79 80 81 9 8 7 6 5 4 3